OPTICAL SPECTROSCOPY OF GLASSES

PHYSIC AND CHEMISTRY OF MATERIALS WITH LOW-DIMENSIONAL STRUCTURES

Series C: Molecular Structures

Managing Editor

I. ZSCHOKKE, *Institute of Physics, The University of Basel, Basel, Switzerland*

Advisory Editorial Board

D. HAARER, *University of Bayreuth*

W. SIEBRAND, *National Research Council of Canada, Division of Chemistry, Ottawa*

J. H. VAN DER WAALS, *Huygens Laboratorium, Leiden*

GENERAL EDITOR: E. MOOSER

OPTICAL SPECTROSCOPY OF GLASSES

Edited by

I. ZSCHOKKE

*University of Basel Institute of Physics,
Department of Condensed Matter Physics*

D. REIDEL PUBLISHING COMPANY

A MEMBER OF THE KLUWER  ACADEMIC PUBLISHERS GROUP

DORDRECHT / BOSTON / LANCASTER / TOKYO

Library of Congress Cataloging in Publication Data

Optical spectroscopy of glasses.

    (Physics and chemistry of materials with low-dimensional structures. Series C,
Molecular structures)
    Includes index.
    1. Glass—Spectra. 2. Glass—Optical properties.
I. Zschokke, I., 1933–   II. Series.
QC464.G55068   1986   530.4′1   86–17712
ISBN 90–277–2231–5

Published by D. Reidel Publishing Company.
P.O. Box 17, 3300 AA Dordrecht, Holland.

Sold and distributed in the U.S.A. and Canada
by Kluwer Academic Publishers,
101 Philip Drive, Assinippi Park, Norwell, MA 02061, U.S.A.

In all other countries, sold and distributed
by Kluwer Academic Publishers Group,
P.O. Box 322, 3300 AH Dordrecht, Holland.

All Rights Reserved
© 1986 by D. Reidel Publishing Company, Dordrecht, Holland
No part of the material protected by this copyright notice may be reproduced or
utilized in any form or by any means, electronic or mechanical,
including photocopying, recording or by any information storage and
retrieval system, without written permission from the copyright owner.

Printed in The Netherlands

EDITORIAL PREFACE

During the last fifteen years the field of the investigation of glasses has experienced a period of extremely rapid growth, both in the development of new theoretical approaches and in the application of new experimental techniques. After these years of intensive experimental and theoretical work our understanding of the structure of glasses and their intrinsic properties has greatly improved. In glasses we are confronted with the full complexity of a disordered medium. The glassy state is characterised not only by the absence of any long-range order; in addition, a glass is in a non-equilibrium state and relaxation processes occur on widely different time scales even at low temperatures. Therefore it is not surprising that these complex and novel physical properties have provided a strong stimulus for work on glasses and amorphous systems.

The strikingly different properties of glasses and of crystalline solids, e.g. the low-temperature behaviour of the heat capacity and the thermal conductivity, are based on characteristic degrees of freedom described by the so-called two-level systems. The random potential of an amorphous solid can be represented by an ensemble of asymmetric double minimum potentials. This ensemble gives rise to a new class of low-lying excitations unique to glasses. These low-energy modes arise from tunneling through a potential barrier of an atom or molecule between the two minima of a double-well. Although the microscopic nature of this two-state tunneling system is not yet clear, this concept has proven to be a very promising one for a better understanding of the fundamental processes and many anomalous properties. For experimental investigations advanced high-resolution laser techniques have proven to be a very powerful tool for studying these new characteristical phenomena.

The purpose of this volume is to provide a review of the current knowledge that has resulted from the great activity in the experimental and theoretical investigation of the static and dynamic spectroscopic properties of transparent glasses. We tried to find a balance between theoretical and experimental contributions. The majority of the articles are concerned with the shape and linewidth of optical transitions of guest impurities in a glass. The correlation between the coupling of a guest to the ensemble of two level systems of the host medium and the temperature and time dependence of the observed homogeneous optical linewidth of the guest, is currently a central question. Besides the scientific interest, the understanding of the general properties of optically active glasses has immediate technical consequences e.g. for laser technology, ceramics etc. To impart a feeling on this area, one of the contributions gives a state-of-the-art review of recent advances on the spectroscopic properties of a great variety of insulating inorganic glasses.

In the final article of the book the role of disorder, including fractal concepts, and the influence of random walks on the dynamics and the relaxation phenomena in an amorphous system are analysed. This difficult problem is closely related to the transport properties in glassy materials, and therefore is of fundamental importance.

In a book like this, the scope is necessarily limited. Much of the extensive work which provided the foundations of this field and the names of the numerous scientists who have made key contributions to the development of our present understanding and the material presented in this book, are only found in the extensive lists of references. This book is intended to be of immediate interest to material scientists, physicists, chemists and scientists concerned with modern problems of condensed matter physics.

I wish to express my sincere gratitude to the authors for the careful preparation of their manuscript and their valuable collaboration in assembling this book.

Basel, July 1986   MRS I. ZSCHOKKE-GRÄNACHER

TABLE OF CONTENTS

EDITORIAL PREFACE ... v

S. K. LYO / Dynamical Theory of Optical Linewidths in Glasses ... 1
1. Introduction ... 1
2. Model Hamiltonian ... 2
3. TLS Line-Broadening Mechanism ... 5
   3.1. Diagonal Modulation ... 5
   3.2. Off-Diagonal Modulation ... 7
4. Homogeneous Linewidth ... 9
   4.1. Spectral Function ... 9
   4.2. Short-Time Behavior ... 9
   4.3. Long-Time Behavior ... 12
5. Microscopic Theory ... 14
   5.1. Green's Functions ... 14
   5.2. Diagonal Modulation ... 16
   5.3. Off-Diagonal Modulation ... 18
   5.4. General Treatment ... 19
6. Conclusions ... 19

WILLIAM M. YEN / Optical Spectroscopy of Ions in Inorganic Glasses ... 23
1. Introduction ... 23
2. Inorganic Glass Structure and Composition ... 24
   2.1. Definition of the Glassy State ... 24
   2.2. Terminology and Structure ... 25
   2.3. Coloring and Activation of Glasses ... 29
   2.4. Composition and Phase Separation ... 30
3. Optical Properties of Impurity Centers in Inorganic Glass ... 31
   3.1. Spectra of Ions in Solids ... 33
   3.2. Inhomogeneous Contributions in the Spectra of Solids ... 35
   3.3. Conventional Spectroscopy of Ions in Glasses ... 36
4. Laser Spectroscopy of Ions in Glasses ... 48
   4.1. Static Spectroscopic Studies and Structure ... 48
   4.2. Radiative and Non-Radiative Transitions ... 54
   4.3. Thermalization and Homogeneous Linewidths ... 54
   4.4. Energy Transfer of Optical Excitation in Glasses ... 58
5. Concluding Remarks ... 59

P. REINEKER and K. KASSNER / Model Calculation of Optical Dephasing in Glasses ... 65

1. Introduction ... 65
   1.1. Anomalous Low-Temperature Properties of Amorphous Solids — Experiments and Interpretations ... 65
   1.2. Optical Low-Temperature Properties of Glasses ... 66
   1.3. Organization of the Paper ... 68
2. The Model and its Hamiltonian ... 69
   2.1. Theoretical Description of the Dynamics of Glasses ... 69
   2.2. Interaction between a Guest Molecule and the TLSs ... 77
   2.3. The Total Model ... 78
3. Optical Line Shape Calculated with Mori's Formalism ... 87
   3.1. Correlation Functions for the Optical Line Shape ... 87
   3.2. Calculation of Correlated Functions with Mori's Formalism ... 88
4. Guest Molecule Coupled to a Single TLS ... 93
   4.1. Dipole Moment Operator ... 93
   4.2. Equations for Correlation Functions ... 95
   4.3. Evaluation of the Coefficients (Debye Model) ... 97
   4.4. Solution to the Eigenvalue Problem ... 101
   4.5. Dependence of the Eigenvalues on System Parameters ... 104
   4.6. Analytical Approximations ... 107
5. Line Shape ... 111
6. Averaging over Two-Level Systems ... 113
   6.1. Averaging Procedures ... 113
   6.2. Numerical Averaging of the Linewidth ... 114
   6.3. Analytical Approximations for the Averaged Linewidth ... 118
   6.4. Averaging of the Line Shape ... 125
7. Coupling of the Impurity to Several Two-Level Systems ... 128
   7.1. The System without Phonons: Eigenvalues and Eigenstates ... 128
   7.2. Representation of the Dipole Moment Operator ... 129
   7.3. Calculation of the Correlation Function Evolution Matrix ... 130
   7.4. Approximate Calculation of the Eigenvalues ... 132
   7.5. Line Shape Formula ... 133
   7.6. Linewidth Calculation: Comparison with Experiment ... 137
   7.7. Numerical Line Shapes ... 141
8. Concluding Remarks ... 142

J. FRIEDRICH and D. HAARER / Structural Relaxation Processes in Polymers and Glasses as Studied by High Resolution Optical Spectroscopy ... 149
1. Introduction ... 149
2. The 'Site-Memory' Function ... 152
3. The Non-Equilibrium Nature of Glasses and its Relation to Optical Properties ... 155
4. Dynamic and Adiabatic Optical Relaxation Processes ... 159
5. Reversibility and Irreversibility ... 160

6.	The Residual Linewidth	163
7.	Spectral Diffusion and Structural Relaxation: Model Description	163
	7.1. The Decay Law of Persistent Spectral Holes in Amorphous Solids	164
	7.2. The Time Evolution of the Optical Width	166
8.	The Logarithmic Decay Law and its Relation to Other Dispersive Time Dependencies	171
9.	Experimental Investigation of Spontaneous Structural Relaxation Processes	174
	9.1. The Photochemical Systems	174
	9.2. Experimental Results	175
	9.3. Investigation of the Microscopic Rate Parameters of the Logarithmic Law: the Deuteration Effect	179
	9.4. Polarization Diffusion	184
10.	Field Effects and Spectral Diffusion Phenomena	188
	10.1. Electric Field Effects for Molecules with Inversion Symmetry	189
	10.2. Electric Field Effects for Molecules without Inversion Symmetry	192
	10.3. Hole-Burning Experiments under External Pressure	193

A. BLUMEN, J. KLAFTER, and G. ZUMOFEN / Models for Reaction Dynamics in Glasses — 199

1.	Introduction	199
2.	Relaxation Viewed as Chemical Reaction: the Kinetic Approach	201
3.	A Parallel Relaxation Scheme: the Direct Transfer	206
	3.1. Regular Lattices with Impurities	206
	3.2. Restricted Geometries	209
	3.3. Fractals	213
4.	Parallel-Sequential Schemes: Random Walks	223
	4.1. Random Walks on Regular Lattices	223
	4.2. Random Walks on Fractals	230
5.	Continuous-Time Random Walks (CTRW)	235
6.	Ultrametric Spaces (UMS)	246
7.	The Bimolecular Reactions $A + A \to 0$, $A + B \to 0$ ($A_0 = B_0$)	250
	7.1. Bimolecular Reactions on Regular Lattices and on Fractals	252
	7.2. Bimolecular Reactions: CTRW and Ultrametric Spaces	256
8.	Conclusions	260

INDEX — 267

DYNAMICAL THEORY OF OPTICAL LINEWIDTHS IN GLASSES

S. K. LYO

Sandia National Laboratories, Albuquerque, N.M. 87185, U.S.A.

1. Introduction

Many physical properties of amorphous solids are strikingly different from those of crystalline solids at low temperatures [1]. One of these anomalous properties, which is the subject of this chapter, is the low-temperature homogeneous optical linewidth of an optically active ion imbedded in a glass matrix. Recently there has been a considerable amount of experimental work [2—10] in this area due to advances in high-resolution spectroscopic techniques. The linewidths have been determined by fluorescence line-narrowing in inorganic glasses [2, 3] and by photochemical [5—9] and non-photochemical [10] hole-burning in organic glasses.

The linewidths are typically of the order of 30—500 MHz at 1 K and are orders of magnitude larger than those found in the crystalline solids at low temperatures. In some organic glasses, holewidths as large as a few cm^{-1} (10—100 GHz) were measured at 2 K [10b]. In amorphous materials the linewidth follows an unusual power-law temperature (T) dependence:

$$\hbar\Delta\omega \propto T^m \qquad (1)$$

with the exponents in the range $1 \leq m \leq 2.2$ for various combinations of ions and amorphous hosts. The typical exponents observed so far are $m = 1$ [4, 7b, 10], $m = 1.33$ [7a], $m = 1.5$ [8], and $m \approx 2.0$ [2, 3, 6] at low temperatures (roughly below 20 K). In inorganic glasses [2, 3], the same power-law (i.e. $m \approx 2.0$) extends to room temperature. This behavior is in clear contrast with the standard T^7 dependence arising from two-phonon Raman processes at low temperatures or the activated exponential temperature dependence arising from Orbach processes in ordered solids [11]. The quadratic temperature dependence observed in inorganic glasses can be explained by a two-phonon Raman mechanism at high temperatures [2]. More detailed descriptions of the data are given in other chapters of this book.

After years of theoretical attempts to explain the anomaly [10, 12—21], it seems to be clear that the explanation is not attributable to conventional excitations known in crystalline solids alone. Naturally, we look into the possible role of the new class of low-energy excitations unique to glasses, namely, the so-called two-level systems (here designated as TLS in both singular and plural) which are known to give rise to other anomalies [1]. The low-energy modes arise from tunneling of an atom or a group of atoms between two energy minima of an asymmetric double-well through a potential barrier. A distribution of the energy asymmetry of the well depths and of the barrier heights gives a broad distribution of energies and a nearly constant density of states for TLS [22, 23]. Although the original TLS model was proposed for explaining very low temperature phenomena (e.g. below 1 K) [22, 23], we are extending it to higher temperatures. The advantage of using TLS rather than phonons for interactions with optical ions lies in the fact that the ion-TLS coupling as well as the density of states of TLS does not decrease with decreasing temperature as rapidly as those of phonons. While phonons are not coupled directly to optical ions in our model, they play an important role in inducing tunneling-transitions in TLS.

The homogeneous linewidths are much smaller than the background inhomogeneous linewidths, which are of the order of a few hundred cm^{-1} in glasses [7, 24]. We will be concerned with the dynamical broadening of a line out of the quasi-uniform background (i.e. away from the wings), arising from modulations of the optical levels by TLS. The present treatment is based on earlier work by the author [16] for the short-time behavior of the spectral function and on recent work [25] for the long-time behavior. Emphasis is given to a simple picture for the line-broadening mechanism.

We begin with a brief description of the model Hamiltonian by defining the optical levels of an ion, the TLS Hamiltonian, ion—TLS interaction, and TLS—phonon coupling. This is followed by studies of two types of basic TLS—line modulation (i.e. dephasing) mechanisms using analogies with nuclear spin relaxation and second-order perturbation theory. These results are then used to describe the short-time and the long-time behaviors of homogeneous linewidths. This simple approach is then justified by a full Green's function theory. The paper is concluded with a brief summary.

2. Model Hamiltonian

The optical line at the energy $\hbar\omega_0$ is chosen by a laser line and corresponds to optical transitions between the ground ($n = 0$) and excited ($n = 1$) levels of impurity ions. The optical excitation energy $\varepsilon_{01} = \varepsilon_1 - \varepsilon_0$ (Figure 1) is much larger than other energy parameters such as the Debye acoustic phonon energy and the energy (E) of TLS. In a crystalline solid, the energy ε_{01} is modulated by lattice vibrations which cause phase interruptions and thereby linewidths. However, this contribution is too small to explain the linewidths in glasses as mentioned in Section 1. Therefore, we propose an impurity dephasing mechanism by TLS in glasses.

The basic idea is that a TLS atom or a group of atoms makes a rapid phonon-assisted tunneling motion between two local wells as illustrated in Figure 1. As a result, its coupling to the optical level n ($= 0, 1$) undulates by a small amount C_n, yielding a net modulation of $C_1 - C_0$ of the resonant optical transition energy. This mechanism is displayed in Figure 1 by solid lines connecting the optical levels and the atom in the wells. The motion associated with a TLS may also be rotational. The coupling between the ion and TLS is achieved by electric multipolar or phonon-exchange interactions [13, 16, 21, 26].

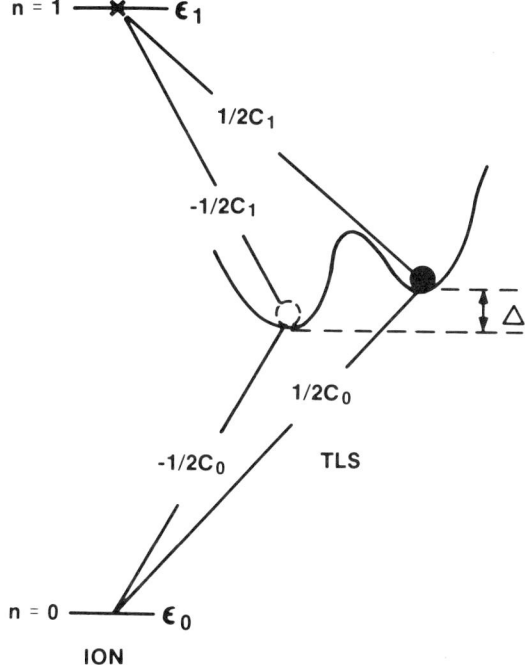

Fig. 1. Interaction of the ground ($n = 0$) and excited ($n = 1$) levels of an ion with an atom in TLS shown by an asymmetric double-well.

The dynamical part of the coupling of TLS to the nth level of the ion is then written conveniently by a 2×2 matrix [14, 16]:

$$\frac{1}{2} C_n \begin{bmatrix} 1 & 0 \\ 0 & -1 \end{bmatrix}_{\text{well}}, \tag{2}$$

where C_n is the difference of the coupling strengths at the two wells. The subscript 'well' means that we are in a local site representation; the wave function (ϕ_L) in the left well is orthogonal to that (ϕ_R) in the right well. The diagonal elements represent the coupling strengths at the right well (+1) and left well (−1) apart from a constant static part. The off-diagonal elements are proportional to the overlap of the wave functions and are ignored.

For TLS we use the Hamiltonian [22, 23]

$$\frac{1}{2} \begin{bmatrix} \Delta & t' \\ t' & -\Delta \end{bmatrix}_{\text{well}}, \tag{3}$$

again given in the site representation. Here Δ and $t'/2$ are the energy asymmetry of the wells and the tunneling integral, respectively. The energy needed for hopping between the wells is provided by the deformation potential [22]

$$\frac{1}{2} B\varepsilon \begin{bmatrix} 1 & 0 \\ 0 & -1 \end{bmatrix}_{\text{well}}, \tag{4}$$

where ε is the strain and B the difference in the deformation potential constants for the two unperturbed wells.

The TLS Hamiltonian in (3) is diagonalized by the eigenfunctions $\Phi_\pm$ related to the site representation wave functions $\phi_{R,L}$ by

$$\Phi_s = t'[2E(E - s\Delta)]^{-1/2} \phi_R - s \left(\frac{E - s\Delta}{2E} \right)^{1/2} \phi_L, \quad (s = \pm 1), \tag{5}$$

with the eigenvalues $\frac{1}{2}sE$ and $E = [\Delta^2 + t'^2]^{1/2}$. The quantities in (2) and (4) can be transformed into the new diagonal representation by (5). The total Hamiltonian then reads in the new diagonal $\Phi_\pm$-representation [16]:

$$H = \sum_n \varepsilon_n \psi_n^\dagger \psi_n + \frac{1}{2} E\sigma^z + \frac{1}{2} \sum_n \psi_n^\dagger \psi_n \sum_\alpha V_n^\alpha \sigma^\alpha +$$

$$+ \sum_q \left(n_q + \frac{1}{2} \right) \hbar\omega_q + \frac{1}{2} \sum_\alpha f^\alpha \varepsilon \sigma^\alpha. \tag{6}$$

In (6), $\psi_n^\dagger (\psi_n)$ is a Fermion creation (destruction) operator and $\psi_n^\dagger \psi_n = P_n$ ($= 0, 1$) is the probability for the occupation of the optical level n. A pseudo spin-$\frac{1}{2}$ representation is used for TLS, which will be referred to as 'spin' at times hereafter. The quantity σ denotes the Pauli spin matrices with $\sigma^\pm = \sigma^x \pm i\sigma^y$. The superscript α is summed over $\alpha = z, +$, and $-$. The first three terms in (6) then represent an unperturbed optical ion, a single spin (TLS) with energy E, and interaction between them, respectively. The fourth term describes the phonon bath, with ω_q and n_q standing for the angular frequency and the occupation number of phonons with crystal momentum q. Finally, the last term in (6) represents the TLS–phonon coupling. The strain ε is assumed to be small. The spin–ion and spin–phonon coupling constants V and f are given by

$$V_n^z = \frac{C_n \Delta}{E}, \quad V_n^\pm = \frac{\frac{1}{2} C_n t'}{E} \tag{7}$$

and

$$f^z = \frac{B\Delta}{E}, \quad f^\pm = \frac{\frac{1}{2} B t'}{E}. \tag{8}$$

Although a single spin is shown in (6) for simplicity, summation over all spins is implicitly assumed. The model Hamiltonian (6) forms the basis of our analysis.

The TLS of interest are those for which the energy barrier is sufficiently large (i.e. $t' < \Delta$) so that resonant tunneling between the two wells does not occur. For these TLS (i.e. $t' < \Delta$) the correction to the eigenvalues due to tunneling is negligible so that the density of states of TLS is determined essentially by the distribution of Δ [22]. It is seen from (7) that the diagonal coupling is always larger than the off-diagonal coupling (i.e. $V_n^z > 2V_n^\pm$). Furthermore, for a small t' and a large Δ, the diagonal coupling is large and the off-diagonal coupling is small. This is easily understood from the fact that a large energy asymmetry, Δ, spatially separates the two eigenfunctions into the unperturbed wells (i.e. $\Phi_\pm \sim \phi_\pm$ in (5)), thereby enhancing the amount of the modulation of the coupling energy during phonon-assisted transitions.

3. TLS Line-Broadening Mechanism

The line-broadening mechanism is different for diagonal (V^z) and off-diagonal ($V^\pm$) spin–ion couplings. Although one can affect the other in a certain parameter regime through renormalization of the optical and TLS structures and by an interference effect, we attempt to contrast the differences by first considering modulation by the diagonal coupling in the absence of the off-diagonal coupling and vice versa for conceptual clarity. Fortunately this kind of description seems to be relevant to a wide class of systems at low temperatures. A discussion of the full combined effect of the diagonal and off-diagonal couplings will be postponed to Section 5.4.

3.1. Diagonal modulation

The diagonal part of spin–ion coupling (i.e. the third term of (6)) reads:

$$\frac{1}{2} \sum_n P_n V_n^z \sigma^z. \tag{9}$$

As a result of this coupling, the resonant optical transition will be shifted to $\varepsilon_{01} - \frac{1}{2}V^z$ and $\varepsilon_{01} + \frac{1}{2}V^z$ when the TLS is in the spin-down (i.e. $\sigma^z = -1$) state and spin-up (i.e. $\sigma^z = +1$) state, respectively. Here V^z is defined by $V^z = V_1^z - V_0^z$. As the atom rattles between the two wells, the resonant optical energy of the ion switches back and forth by an amount $\pm V^z$. The problem is then similar to the relaxation of a nuclear spin (corresponding to the ion) coupled to fluctuating electron spin σ^z, yielding a linewidth [17, 27]

$$\hbar \Delta \omega' = \hbar^{-1} \left(\frac{V^z}{2}\right)^2 \mathrm{Re} \int_0^\infty \langle \sigma^z(t)\sigma^z(0) - \langle \sigma^z \rangle^2 \rangle \mathrm{e}^{-i\omega t} \, \mathrm{d}t$$

$$= \frac{(V^z)^2}{4\hbar} \mathrm{sech}^2\left(\frac{\beta E}{2}\right) \frac{\tau}{1 + (\omega \tau)^2}. \tag{10}$$

Here Re means the real part, τ is the spin-lattice relaxation time of TLS,

$\beta = (k_B T)^{-1}$, and k_B is Boltzmann's constant. The quantity τ is a random parameter determined by the spin energy E and the tunneling integral t'. The 'nuclear spin' energy equals the optical shift $\hbar\omega = \frac{1}{2} V^z$, yielding

$$\hbar \Delta\omega' = (V^z)^2 \operatorname{sech}^2\left(\frac{\beta E}{2}\right) \frac{\hbar\tau}{4\hbar^2 + (\tau V^z)^2}. \tag{11}$$

When the damping of the TLS ($\hbar\tau^{-1}$) is much larger than V^z, namely, the spectral shift, the expression in (11) reduces to [2a, 16]

$$\hbar \Delta\omega'_a = (\tfrac{1}{2} V^z)^2 \hbar^{-1} \operatorname{sech}^2\left(\frac{\beta E}{2}\right) \tau. \tag{12a}$$

According to this result, the line becomes narrower for a faster phonon-assisted tunneling between the wells (i.e. for a smaller τ). This result of fast modulation is similar to motional narrowing in the spin relaxation problem.

On the other hand, if the damping of the TLS is much smaller than the spectral shift, we find from (11) [16]

$$\hbar \Delta\omega'_b = \operatorname{sech}^2\left(\frac{\beta E}{2}\right) \hbar\tau^{-1}. \tag{12b}$$

Therefore, in the slow modulation regime, the line broadens as the modulation rate (τ^{-1}) becomes faster (e.g. with increasing temperature). Also, the linewidth in (12b) is independent of the ion–TLS coupling strength! Of course, this does not mean that a spin at an infinite distance away from the ion gives a finite contribution to the linewidth. It means, however, that as long as the spin–ion coupling strength V^z is larger than the damping of the spin, it yields the linewidth in (12b) independent of the coupling strength (i.e. ion–spin distance). While a full microscopic proof of this claim with be given for an arbitrarily large V^z later in Section 5, this interesting result deserves an immediate and simple explanation, which is given below.

In order to understand how the linewidth is independent of ion–spin coupling strength for a large coupling, we consider second-order perturbation processes (Figure 2). The circled numbers there represent the order in the perturbation chain. In Figure 2(a) the ion makes a virtual interaction of strength $\frac{1}{2} V^z$ with the spin in the upper level in step 1. In step 2 the spin flips (by the last term in (6): $\frac{1}{2} f^- \varepsilon \sigma^-$) to the lower level by emitting a phonon q. In Figure 2(b) the steps 1 and 2 are reversed. The t-matrices for these processes are given by

$$t_{a,b} = \frac{\frac{1}{2} V^z}{\frac{1}{2} V^z} f^+ \langle n_q + 1 | \varepsilon | n_q \rangle, \tag{13}$$

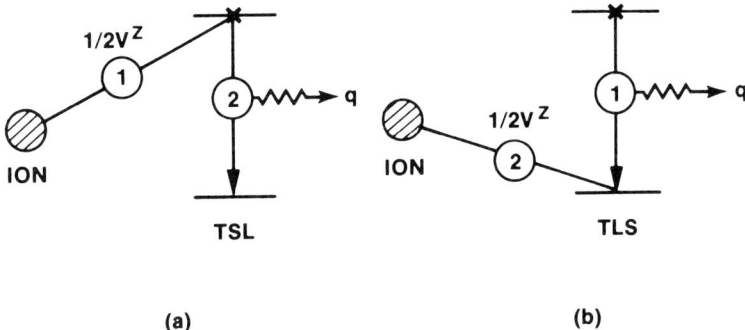

Fig. 2. Two-step processes for diagonal modulation. Directed wiggly lines and circled numbers indicate phonon-emission and the sequence in the perturbation chain, respectively.

where the quantity in the denominator denotes the intermediate energy which includes the optical shift. It is immediately seen from (13) that the quantity V^z cancels out. There are two more processes similar to those in Figure 2, but with the spin initially in the lower level. In these processes one phonon is absorbed to conserve the energy. The dephasing rate from these processes yields $\Delta\omega'_b$; using

$$\Gamma^{\pm} = 2\pi \sum_q (f^+)^2 |\langle n_q \pm 1 | \varepsilon | n_q \rangle|^2 \delta(E - \hbar\omega_q). \tag{14}$$

and

$$\hbar\tau^{-1} = \Gamma^- + \Gamma^+ = \Gamma^-[1 + e^{\beta E}], \tag{15}$$

we obtain the relationship in (12b). The quantities $\Gamma^{\pm}$ indicate the linewidths of the upper (+) and the lower (−) spin levels. The second equality in (15) is due to detailed balance.

3.2. OFF-DIAGONAL MODULATION

The contribution to the linewidth by the off-diagonal coupling $V_n^{\pm}$ (i.e. the third term of (6)) is evaluated in a similar way by employing the two-step processes [14] illustrated in Figure 3 for the case where the spin is initially in the upper level. In Figure 3(a) a virtual phonon of momentum q is emitted in step 1 (by the last term in (6): $\frac{1}{2} f^z \varepsilon \sigma^z$) and then the spin is flipped to the lower level by the ion–spin interaction $V_n^{\pm}$ in step 2. In Figure 3(b) the steps 1 and 2 are reversed so that one phonon is emitted in the lower level. The t-matrices for these processes are given by

$$t'_a = t'_b = \frac{V_n^-}{2E} f^z \langle n_q + 1 | \varepsilon | n_q \rangle. \tag{16}$$

The quantity E in the denominator represents the intermediate energy. Again, there are two more one-phonon-absorption processes similar to those in Figure 3

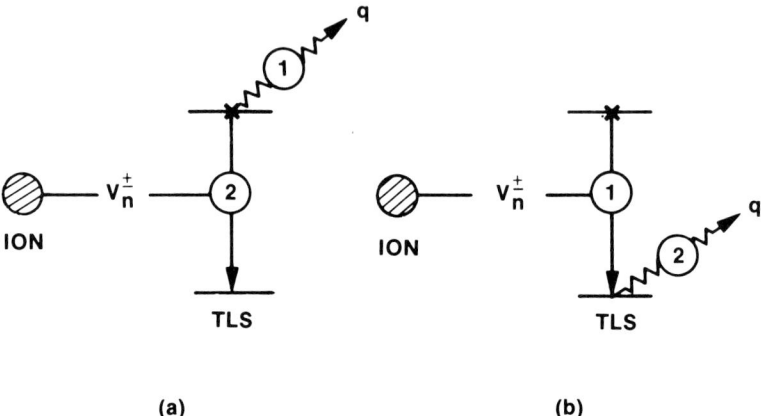

Fig. 3. Two-step processes for off-diagonal modulation. Directed wiggly lines and circled numbers indicate phonon-emission and the sequence in the perturbation chain, respectively.

but with the spin initially in the lower level. Combining these processes, and using (7), (8), (14), and (15), we find

$$\hbar \Delta\omega'' = \frac{1}{2} \hbar\tau^{-1} \sum_n \left(\frac{V_n^z}{E}\right)^2 \operatorname{sech}^2\left(\frac{\beta E}{2}\right). \tag{17}$$

This result is qualitatively different from that of diagonal modulation (i.e. Equation (11)) in that (1) the line always broadens with increasing rate (τ^{-1}) of modulation and that (2) only strongly coupled TLS adjacent to the ion make dominant contributions to $\Delta\omega''$ because of the strong quadratic dependence on V_n^z.

The result in (17) was originally proposed by Lyo and Orbach [14] (and studied in more detail by later authors [15, 20]) to explain the T^2-dependence of the linewidth data [2] of inorganic glasses in a wide temperature range $7 < T < 300$ K. The quadratic temperature dependence is obtained from (17) by using $\tau^{-1} \propto E^3$ (see (28)) and summing over E for a constant density of states of TLS. When the thermal energy $k_B T$ is much larger than the maximum TLS energy, a linear temperature dependence is obtained [14]. However, at low temperatures relevant to most of the recent data of organic glasses measured by hole-burning below 20 K down to 1 K or below, the above quadratic temperature will be modified. This follows from the fact that at low temperatures the coupling energy V_n^z in (17) is not small compared to the thermal TLS energy E. The latter should then be replaced by the renormalized energy

$$E_n = [(E + V_n^z)^2 + 4(V_n^+)^2]^{1/2}, \tag{18}$$

which is obtained from the first three terms in (6). Also, note that $\Delta\omega''$ is smaller than $\Delta\omega'_b$ in (12b). Since the damping $\hbar\tau^{-1}$ ($\propto E^3 \sim T^3$) becomes very small at low temperatures, the expression in (12b) is valid even for weakly coupled

TLS with $V^z \sim \hbar\tau^{-1}$. In this regime, $\Delta\omega''$ is smaller than $\Delta\omega'_b$ by a factor $\sim (V_n^z/E)^2 \ll 1$ and will not be considered until Section 5.3.

4. Homogeneous Linewidth

4.1. SPECTRAL FUNCTION

The lineshape is obtained by Fourier-transform of the spectral function:

$$F(t) = \exp\left(-i\omega_0 t - \frac{t}{2}\sum_{\text{TLS}} \Delta\omega'_i\right), \qquad (19)$$

where $\Delta w'_i$ is the dephasing rate given by (11) and the subscript i denotes TLS. The quantity ω_0 represents the frequency of the probe laser line. A microscopic study of the spectral function $F(t)$ will be presented later in Section 5.

The laser line excites the optical ions with perturbed energies resonant with $\hbar\omega_0$ out of the broad inhomogeneous background. The quantity $F(t)$ represents the spectral function of ions in the packet $\hbar\omega_0$. The perturbed energy of the ion equals the unperturbed resonance energy ε_{01} plus the sum of the spectral shifts $\frac{1}{2}V^z\sigma^z$ (cf. (9)) from all TLS interacting with it. A spin-flip from $\sigma^z = -1$ (+1) to $\sigma^z = +1$ (−1) will cause a spectral shift, $+V^z$ ($-V^z$). Therefore as a spin flips up (down), a line will be pulled up (down) out of the packet. Namely, only a down (up)-spin can pull up (down) a line, and these two processes should be weighted properly by Boltzmann factors. The quantity $\Delta\omega'$ in (19) contains this effect.

The quantity of interest is

$$\Phi(t) = \langle F(t) \rangle. \qquad (20)$$

Here the angular brackets denote the configuration average and the average over the random TLS parameters. By carrying out the average [28], we find

$$\Phi(t) = \exp[-\tfrac{2}{3}\pi n \psi(t) - i\omega_0 t], \qquad (21)$$

where n is the density of TLS and

$$\psi(t) = 6\int_0^\infty r^2 (1 - \langle e^{-t\Delta\omega'/2}\rangle)\,dr. \qquad (22)$$

In (22) $\Delta\omega'$ indicates the dephasing rate (defined in Equation (11)) by a single TLS of energy E at a distance r from the ion, and the angular brackets now denote averaging over the TLS parameters. The function $\psi(t)$ behaves differently over different time scales, which are determined by the dephasing time. In the following sections we study two extreme time limits analytically.

4.2. SHORT-TIME BEHAVIOR

Within a time shorter than the shortest spin-lattice relaxation time τ, the function

$\psi(t)$ in (22) reduces to

$$\psi(t) = 3t \int_0^\infty r^2 \langle \Delta\omega' \rangle \, dr, \tag{23}$$

yielding, in view of (11),

$$\psi(t) = 3t \int dE \, \rho(E) \int_0^\infty dr \, r^2 V(r)^2 \, \text{sech}^2 \left(\frac{\beta E}{2} \right) \left\langle \frac{\tau}{4\hbar^2 + \tau^2 V(r)^2} \right\rangle_E. \tag{24}$$

The subscript E of the angular brackets means that the average is for TLS with energy E. In (24), $\rho(E)$ is the normalized density of states of TLS and $V(r) = V^z$. This approximation is suitable for the situation where the dephasing rate is larger than the spin-lattice relaxation rate appropriate to high density (n) systems and to a very low temperature regime. The latter arises from the fact that the spin relaxation rate decreases faster (i.e. as $\propto T^3$) than the dephasing rate. The actual domain of validity of this approximation for a given system depends on its maximum spin-lattice relaxation rate, which is not well known in general.

For ion–TLS interaction we consider the multipolar form [16]:

$$V(r) = \frac{b}{r^s}. \tag{25}$$

Inserting (25) in (24), we obtain

$$\psi(t) = t a_s \left(\frac{b}{2\hbar} \right)^{3/s} \int dE \, \rho(E) \, \text{sech}^2 \left(\frac{\beta E}{2} \right) \langle \tau^{3/s - 1} \rangle_E. \tag{26}$$

where

$$a_s = \frac{3}{s} \int_0^\infty x^{3/s - 1} (1 + x^2)^{-1} \, dx.$$

Using (15) and

$$\Gamma^\pm = \pi \left(\frac{E}{\hbar \omega_D} \right)^3 D' \left[n(E) + \frac{1 \pm 1}{2} \right] \tag{27}$$

obtained from (14) and (8), we find [22]

$$\tau^{-1} = \pi \hbar^{-1} \left(\frac{E}{\hbar \omega_D} \right)^3 D' \coth \left(\frac{\beta E}{2} \right). \tag{28}$$

In (27), ω_D is the Debye frequency, $n(E)$ the Boson function, and

$$D' = \frac{3B^2}{Mc^2}\left(\frac{t'}{E}\right)^2. \tag{29}$$

Here M and c are the mass of a unit cell and the sound velocity, respectively. Inserting (28) in (26), defining $\rho(E) = \rho_0 E^\mu$ and $D = \langle D' \rangle_E$, we find

$$\psi(t) = \frac{3\Delta\omega t}{4\pi n}. \tag{30}$$

The lineshape is Lorentzian and the homogeneous linewidth is given by

$$\hbar\Delta\omega = \frac{16}{3}\pi^2 n a_s D\hbar\omega_D I_{s,\mu}\left(\frac{b}{2\pi D}\right)^{3/s}\rho_0(k_B T)^\mu\left(\frac{T}{\theta_D}\right)^{4-(9/s)} \tag{31a}$$

In (31a) θ_D is the Debye temperature and

$$I_{s,\mu} = \int_0^\infty \frac{x^{3+\mu-(9/s)}e^x\,\mathrm{d}x}{(e^x+1)^{1+(3/s)}(e^x-1)^{1-(3/s)}}.$$

The constants a_s and $I_{s,\mu}$ are of the order of unity (e.g. $a_3 = \pi/2$, $I_{3,0} = 0.5$, $I_{4,0} = 0.684$). The quantity D in (31) is independent of E and can be evaluated by following the method given in Reference 22.

For a dipolar form of interaction (i.e. $s = 3$) (31a) reduces to

$$\hbar\Delta\omega = \tfrac{4}{3}\pi^2 n I_{3,\mu} b\rho_0(k_B T)^{1+\mu}. \tag{31b}$$

Note that the linewidth is independent of the mechanism of the spin-lattice relaxation and TLS parameters. The origin of this interesting result will be discussed later For $n\rho_0 = 2 \times 10^{20}/(\text{eV cm}^3)$, $b = 5.8 \times 10^{-36}$ erg cm^3 [21] and $\mu = 0$, we estimate $\Delta\omega = 100$ MHz at 1 K.

The temperature dependence of the homogeneous linewidth in (31) can be understood in the following way: From (11) and (25) it is clear that only those TLS lying within the radius

$$r_c = \left(\frac{\tau b}{2\hbar}\right)^{1/s} \tag{32}$$

from the ion make a significant contribution, which is given by $\hbar\Delta\omega'_b$ in (12b). The total contribution then arises equally from all TLS within r_c and equals

$$\frac{4}{3}\pi r_c^3 n \operatorname{sech}^2\left(\frac{\beta E}{2}\right)\hbar\tau^{-1}. \tag{33}$$

It should be noted at this point that the dominant part of (33), and therefore the linewith, comes from TLS just inside the radius r_c rather than from TLS near the origin (i.e. optical ion), because the volume grows as $\propto r^3$. This is also true for the long-time behavior to be discussed later. We now assume that

$$\tau^{-1} \propto E^{1+d}, \qquad (34)$$

where E^d comes from the density of states of Bosons that relax the spin ($d = 2$ for acoustic phonons). The total linewidth is then obtained by inserting (32) and (34) in (33), replacing $E \sim k_B T$, and multiplying by the number of active TLS $\sim T^{1+\mu}$, yielding

$$\hbar \Delta\omega \propto T^{(1+d)(1-(3/s))+1+\mu}. \qquad (35)$$

The temperature dependence in (35) is identical to that in (31) for phonon-assisted (i.e. $d = 2$) spin relaxation. For the special case of dipolar interaction (i.e. $s = 3$) the quantity in (33) is rewritten in view of (32) as

$$\frac{2}{3} \pi bn \operatorname{sech}^2\left(\frac{\beta E}{2}\right). \qquad (33')$$

This quantity is independent of τ^{-1}. As a result, the linewidth is not affected by the details of the spin-lattice relaxation mechanism.

We caution at this point that there is a certain assumption in extending the lower limit of the radial integration to zero in (23). If the range r is so small that $V(r)$ is comparable to E, namely, $r < r_1 = (b/E)^{1/s}$, renormalizing TLS energies according to (18) is necessary. However, at low temperatures the ratio

$$\left(\frac{r_1}{r_c}\right)^3 = \left(\frac{2\hbar}{\tau E}\right)^{3/s}$$

is very small, because the damping is always much smaller than the energy for any well-defined excitations. Therefore, TLS can be treated as being unperturbed even within the radius r_1 with a negligible error $\sim (r_1/r_c)^3$.

As mentioned, the linewidth is quasi-linear in temperature for a dipolar form of interaction irrespective of d: $\hbar \Delta\omega \propto T^{1+\mu}$. For a quadrupolar type of interaction (i.e. $s = 4$), we find a quasi-quadratic temperature dependence $\hbar \Delta\omega \propto T^{1.75+\mu}$ for phonon-assisted (i.e. $d = 2$) spin transitions. On the other hand, if we use Alexander–Orbach [29] spectral dimensionality for fracton-assisted spin transitions (i.e. $d = 1/3$), which may be relevant to organic glasses, we find $\hbar \Delta\omega \propto T^{1.33+\mu}$ [19]. These temperature dependences are similar to those exhibited by the data discussed in Section 1.

4.3. LONG-TIME BEHAVIOR

We now consider the situation where the time scale is much larger than the shortest spin-lattice relaxation time τ (τ_{min}) but smaller than the longest τ (τ_{max}). The expression in (22) is rewritten as

$$\psi(t) = \left\langle \left(\frac{2\hbar}{b\tau}\right)^{-3/s} f(u) \right\rangle, \qquad (36)$$

where

$$u = \frac{t}{2\tau} \operatorname{sech}^2\left(\frac{\beta E}{2}\right), \qquad (37)$$

and

$$f(u) = \int_0^\infty \left[1 - \exp\left(-\frac{u}{1 + x^{2s/3}}\right)\right] dx. \qquad (38)$$

The function $f(u)$ increases monotonically with u and behaves as $f(u) \propto u$ for $u \ll 1$, and as $f(u) \propto u^{3/(2s)}$ for $u \gg 1$. The result in (36) is essentially the same as that obtained by Huber et al. [21] by an alternative method. Following the averaging procedures of these authors [21], we find

$$\psi(t) \sim \left(\frac{bt}{4\hbar}\right)^{3/s} g_s h_s \rho_0 (k_B T)^{1+\mu}, \qquad (39)$$

where ($g_3 = 0.5$)

$$g_s = \int_0^\infty \frac{e^{3x/s} dx}{(e^x + 1)^{6/s}},$$

and

$$h_s = \int_0^\infty \frac{f(u) du}{u^{1+(3/s)}}.$$

The upper (lower) limit of the integration for h_s is essentially the ratio t/τ_{min} (t/τ_{max}) which is approximated as ∞ (0). The spectral function is then obtained from (39) and (21) [21]:

$$\Phi(t) = \exp[-c'_s T^{1+\mu} t^{3/s}], \qquad (40a)$$

with

$$c'_s = \frac{2}{3}\pi n \left(\frac{b}{4\hbar}\right)^{3/s} g_s h_s \rho_0 k_B^{1+\mu}. \qquad (40b)$$

The lineshape is Lorentzian only for $s = 3$. In this case the linewidth is given by the same expression as in (31b) except for a factor of the order of unity and is quasi-linear in temperature $\hbar \Delta \omega \sim T^{1+\mu}$, while $\hbar \Delta \omega \sim T^{4(1+\mu)/3}$ for $s = 4$.

5. Microscopic Theory

5.1. GREEN'S FUNCTIONS

The spectral function is defined by the retarded Green's function [30]

$$F(t) = \theta(t)\langle [(\psi_1 \psi_0^\dagger)_t, \psi_0 \psi_1^\dagger] \rangle = \langle\langle (\psi_1 \psi_0^\dagger)_t \rangle\rangle, \quad (41a)$$

where $\theta(t)$ is the unit step function, and $(\cdots)_t$ stands for the Heisenberg representation. The square and angular brackets in (41a) denote the commutator and the thermodynamic average, respectively. The expression following the second equality in (41a) is introduced for the simplicity of notation for a later purpose. The function $F(t)$ is the probability that an optical ion excited at time $t = 0$ will be found in the same excited state at time t. The lineshape is found by the Fourier transform

$$F_\omega = \frac{1}{i\hbar} \int F(t)\, e^{i\omega t}\, dt. \quad (41b)$$

The subscript ω will be omitted for Fourier-transformed quantities hereafter.

The equation of motion for (41a) is found using the Hamiltonian in (6). When Fourier-transformed, it yields

$$\Omega F = 1 + \tfrac{1}{2} V^+ \mathcal{F}_+ + \tfrac{1}{2} V^\circ F^\circ, \quad (42)$$

where $\Omega = \hbar\omega - \varepsilon_{01}$, $\mathcal{F}_\pm = F^+ \pm F^-$, and F^a is the Fourier transform of

$$F^a(t) = \langle\langle (\psi_1 \psi_0^\dagger \sigma^a)_t \rangle\rangle. \quad (43)$$

The superscript z is denoted by '°' in (42) and (43) (i.e. $V^z = V^\circ$, $\sigma^z = \sigma^\circ$). The new functions F^a are evaluated again through their equations of motion, which in turn generate higher order Green's functions. These procedures are continued until all Green's functions generated form a closed set, which is achieved by decoupling approximations exact in the limit of zero strain.

For this purpose, it is necessary to write down the strain in terms of phonon creation $(b_q^\dagger)$ and destruction (b_q) operators:

$$\varepsilon = \sum_q (\varepsilon_q^+ + \varepsilon_q^-), \quad (44)$$

where $\varepsilon_q^+ = \varepsilon_q b_q^\dagger$ and $\varepsilon_q^- = \varepsilon_q b_q$. The q-component of the strain ε_q contains the usual momentum dependence of the strain (i.e., $\varepsilon_q \propto \sqrt{q}$).

Some of the Green's functions generated by equations of motion are given by

$$\Omega F^\circ = \langle \sigma^\circ \rangle + u^+ \mathcal{F}_- + f^+ \mathcal{G}_- + \tfrac{1}{2} V^\circ F \tag{45a}$$

$$\Omega \mathcal{F}_+ + (E + u^\circ) \mathcal{F}_- = \langle \sigma^+ \rangle + \langle \sigma^- \rangle + 2V^+ F - f^\circ \mathcal{G}_- \tag{45b}$$

$$(E + u^\circ) \mathcal{F}_+ + \Omega \mathcal{F}_- = \langle \sigma^+ \rangle - \langle \sigma^- \rangle + 4u^+ F^\circ + 4f^+ G^\circ - f^\circ \mathcal{G}_+, \tag{45c}$$

where $u^a = \tfrac{1}{2}(V_1^a + V_0^a)$, $\mathcal{G}_\pm = G^+ \pm G^-$, and G^a are the Fourier transforms of

$$G^a(t) = \langle\!\langle (\psi_1 \psi_0^\dagger \varepsilon \sigma^a)_t \rangle\!\rangle. \tag{46a}$$

It is convenient to introduce ($p = \pm$)

$$G(t) = \langle\!\langle (\psi_1 \psi_0^\dagger \varepsilon)_t \rangle\!\rangle \tag{46b}$$

$$G_{p,q}(t) = \langle\!\langle (\psi_1 \psi_0^\dagger \varepsilon_q^p)_t \rangle\!\rangle \tag{46c}$$

$$G_{p,q}^a(t) = \langle\!\langle (\psi_1 \psi_0^\dagger \varepsilon_q^p \sigma^a)_t \rangle\!\rangle. \tag{46d}$$

The function $G(G^a)$ equals the sum of $G_{p,q}(G_{p,q}^a)$ over p and q. The G-functions satisfy the following set of equations:

$$\Omega_{p,q} G_{p,q} = A + \tfrac{1}{2} V^+ \mathcal{G}_{+,p,q} + \tfrac{1}{2} V^\circ G_{p,q}^\circ \tag{47a}$$

$$\Omega_{p,q} G_{p,q}^\circ = A_\circ + u^+ \mathcal{G}_{-,p,q} + \tfrac{1}{2} V^\circ G_{p,q} \tag{47b}$$

$$\Omega_{p,q} \mathcal{G}_{+,p,q} + (E + u^\circ) \mathcal{G}_{-,p,q} = A_- + 2V^+ G_{p,q} \tag{47c}$$

$$(E + u^\circ) \mathcal{G}_{+,p,q} + \Omega_{p,q} \mathcal{G}_{-,p,q} = A_+ + 4u^+ G_{p,q}^\circ, \tag{47d}$$

where $\Omega_{p,q} = \Omega + p\hbar\omega_q$, $\mathcal{G}_{\pm,p,q} = G_{p,q}^+ \pm G_{p,q}^-$. The quantities A_n are defined by

$$A_+ = \lambda_q(-f^\circ \mathcal{F}_+ + 4f^+ F^\circ) \tag{48a}$$

$$A_- = -f^\circ \lambda_q \mathcal{F}_- - 2p\varepsilon_q^2 f^+ F \tag{48b}$$

$$A_\circ = \lambda_q f^+ \mathcal{F}_- - \tfrac{1}{2} p\varepsilon_q^2 f^\circ F \tag{48c}$$

$$A = -\tfrac{1}{2} p\varepsilon_q^2(f^\circ F^\circ + f^+ \mathcal{F}_+), \tag{48d}$$

with

$$\lambda_q = \varepsilon_q^2(n_q + \tfrac{1}{2}). \tag{49}$$

Here $p = \pm 1$ when p is written as a factor.

Our goal is to find the spectral function F using the four sets of equations (42), (45), (47), and (48). For this purpose, we solve (47) for $G_{p,q}$, $G_{p,q}^\circ$, and $\mathcal{G}_{\pm,p,q}$ in terms of the A_n's. Inserting these results together with (48) into (45), we now solve (42) and (45) for F, F°, and $\mathcal{F}_\pm$. In the following two sections we analyze the diagonal and off-diagonal modulations separately, assuming that either V^z or $V^\pm$ is non-vanishing each time. Their combined effect will be considered in Section 5.4.

5.2. DIAGONAL MODULATION

As is expected, solving for four sets of Green's functions is not a simple task. When $V^\pm = 0$, the problem is somewhat simplified, yielding [16]

$$F = \frac{\Omega + \frac{1}{2}V^z(p_+ - p_-) - i\hbar\tau^{-1}}{\Omega^2 - (\frac{1}{2}V^z)^2 - i\Omega\hbar\tau^{-1} - \frac{1}{4}iV^z(\Gamma_+ - \Gamma_-)}, \qquad (50)$$

where $\hbar\tau^{-1} = \frac{1}{2}(\Gamma_+ + \Gamma_-)$, $\Gamma_\pm = \Gamma_0^\pm + \Gamma_1^\pm$ and

$$\Gamma_n^\pm = 2\pi \sum_q (f^+\varepsilon_q)^2 \left(n_q + \frac{(1 \pm 1)}{2} \right) \delta(\hbar\omega_q - E_n). \qquad (51)$$

The quantities $\Gamma_n^\pm$ represent the levelwidths arising from one-phonon emission (upper sign) and absorption (lower sign) processes within the spin doublet $\varepsilon_n \pm \frac{1}{2}E_n$ of the nth optical level shown in Figure 4. The quantity τ^{-1} in (50) reduces to the spin-lattice relaxation rate defined earlier in (15) in the limit of small spin–ion coupling. The quantities $p_\pm = \frac{1}{2}[1 \pm \langle\sigma^z\rangle]$ equal the Boltzmann factors for the ground state spin-doublet.

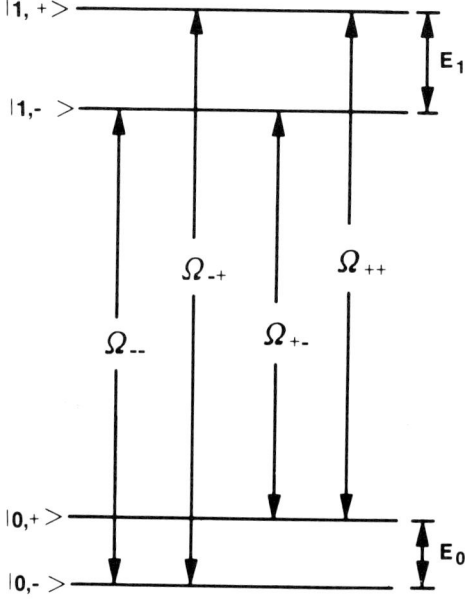

Fig. 4. Ion–TLS coupled states and four possible optical transitions.

For a very rapid modulation, so that $\hbar\tau^{-1} \gg \frac{1}{2}V^z$, the spectral function in (50) reduces to [16],

$$F = [\Omega - \tfrac{1}{2}V^z(p_+ - p_-) - \tfrac{1}{2}i\hbar\Delta\omega'_a]^{-1}. \tag{52}$$

The dephasing rate $\Delta\omega'_a$ is defined in (12a). In this case we find only one line at $\frac{1}{2}V^z(p_+ - p_-)$: This spectral shift is just the thermal average of those by up-spins and down-spins. The line narrows as the spin transitions become faster, as discussed in Section 3.1 (Figure 5(a)).

On the other hand, for a very slow modulation, so that $\hbar\tau^{-1} \ll \frac{1}{2}V^z$, the spectral function in (50) reduces to [16]

$$F = \sum_{\pm} p_{\pm}[\Omega - (\pm\tfrac{1}{2}V^z) - \tfrac{1}{2}i\Gamma_{\pm}]^{-1}. \tag{53}$$

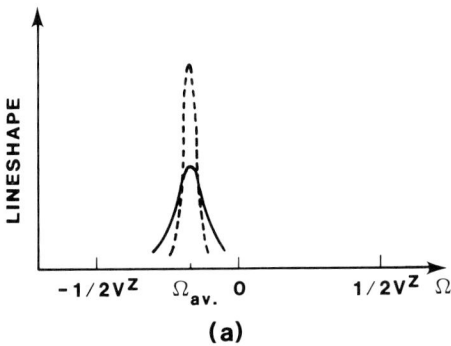

(a)

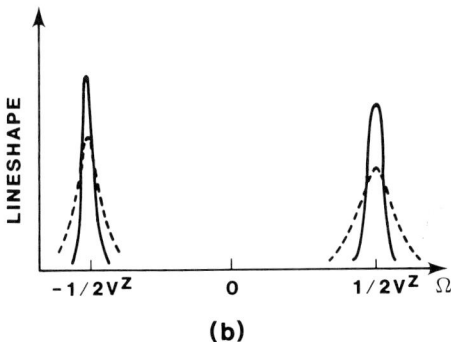

(b)

Fig. 5. Schematic illustration of (a) fast and (b) slow modulations. There is only one resonance line at the average shift $\Omega_{av} = \frac{1}{2}V^z(p_+ - p_-)$ for a fast modulation: the line narrows as the modulation becomes faster (indicated by the dashed curve). There are two resonance lines for a slow modulation: the lines broaden as the modulation becomes faster (indicated by the dashed curves).

In this case, we find two discrete lines of widths $\Gamma_\pm$ each shifted by $\pm\frac{1}{2}V^z$ with probabilities $p_\pm$. Each line broadens as the spin transition becomes faster (Figure 5(b)). The poles have contributions from TLS in up-spin and down-spin states. As a spin in the packet of the reasonance line flips down (up) by one-phonon emission (absorption), a line is pulled down (up) out of the resonance (i.e. laser) line, thereby contributing $p_+\Gamma_+$ ($p_-\Gamma_-$) to the linewidth:

$$\hbar \Delta \omega'_a = p_+\Gamma_+ + p_-\Gamma_- \tag{54}$$

summed over all TLS. The expression on the right-hand side of (54) coincides with that in (12b).

5.3. OFF-DIAGONAL MODULATION

When the off-diagonal coupling $V_n^\pm$ is non-vanishing, the spectral function F has four poles corresponding to the four possible optical transitions between the ground and excited spin-doublets of the ion shown in Figure 4 [16]. The real parts of the poles equal

$$\Omega_{aa'} = \tfrac{1}{2}(-\alpha E_0 + \alpha' E_1), \tag{55}$$

where $\alpha = \pm (\alpha' = \pm)$ denotes the ground (excited) state doublet of the ion. The spin energies E_n are defined in (18). After a tedious and lengthy algebra we find

$$F = \sum_{aa'} \frac{A_{aa'}}{\Omega - \Omega_{aa'} - \tfrac{1}{2}i(\tilde{\Gamma}_0^a + \tilde{\Gamma}_1^{a'})}, \tag{56}$$

where

$$\tilde{\Gamma}_n^a = 2\pi \sum_q \left\{\frac{f^z V_n^+ - f^+(E + V_n^z)}{E_n}\right\}^2 \varepsilon_q^2 \left[n_q + \frac{(1 \pm 1)}{2}\right] \delta(\hbar\omega_q - E_n) \tag{57}$$

and

$$A_{aa'} = \frac{\tilde{p}_a}{2E_0 E_1} \{E_0 E_1 + \alpha\alpha'[(V_0^z + E)(V_1^z + E) + 4V_0^+ V_1^+]\}. \tag{58}$$

Here $\tilde{p}_a$ is the Boltzmann factor for the spin doublet of the optical ground state:

$$\tilde{p}_a = \frac{1}{2} \exp\left(-\frac{\alpha\beta E_0}{2}\right) \operatorname{sech}\left(\frac{\beta E_0}{2}\right). \tag{59}$$

The diagonal coupling V_n^z which is zero in the present analysis of off-diagonal modulation is inserted in the above results for a later purpose. In (56) the quantities $A_{aa'}$, $\Omega_{aa'}$, and $\tilde{\Gamma}_0^a + \tilde{\Gamma}_1^{a'}$ are then to be interpreted as transition probability, the spectral shift, and the linewidth for the $\alpha \to \alpha'$ transition. These interpretations are proved by a direct method in the next section. The net linewidth is then given by the weighted average of the linewidths from the four transitions:

$$\hbar \Delta \omega = \sum_{aa'} A_{aa'} (\tilde{\Gamma}_0^a + \tilde{\Gamma}_1^{a'}). \tag{60}$$

5.4. GENERAL TREATMENT

We now consider the combined effect of diagonal and off-diagonal ion–TLS interactions using a more intuitive approach. Denoting the first three terms of the Hamiltonian in (6) as H_n for the nth level occupation, we diagonalize H_n as [16]

$$H_n |n, \pm\rangle = \{\varepsilon_n \pm \tfrac{1}{2} E_n\} |n, \pm\rangle. \tag{61}$$

Here the eigenvectors $|n, \pm\rangle$ stand for the wave functions for the spin-doublet of the nth optical level (Figure 4). They are obtained by a transformation similar to those transformations in (5).

The spin-phonon Hamiltonian ($H_{\text{TLS-ph}}$), namely, the last term in (6) is rewritten as

$$\langle n, \pm | H_{\text{TLS-ph}} | n, \mp \rangle = \left\{ \frac{f^z V_n^+ - f^+(E + V_n^z)}{E_n} \right\} \varepsilon = \frac{f^+ E \varepsilon}{E_n}. \tag{62}$$

The levelwidths of the spin-doublet arising from one-phonon emission (+) and absorption (−) transitions through the matrix elements in (62) are then given by $\tilde{\Gamma}_n^\pm$ (defined in Equation (57)). Furthermore, the occurrence probability of the optical transition between $|0, \alpha\rangle$ and $|1, \alpha'\rangle$ equals $\tilde{p}_a |\langle 0, \alpha | 1, \alpha' \rangle|^2$ which coincides with $A_{aa'}$ (defined in Equation (58)). The lineshape resulting from modulation by a *single* spin is Lorentzian. Therefore, the net linewidth of a given optical transition equals the sum of the levelwidths of the initial and final states. The expression in (60) then represents the total contribution from the four optical transitions properly averaged. It reduces to the earlier specific results obtained for diagonal and off-diagonal modulations. For diagonal modulation, however, it contains only the line-broadening part relevant to strong coupling $V^z \gg 2\hbar\tau^{-1}$. This means that the coupled ion–TLS state can be treated as a coherent state (as in (61)) only when the coupling is sufficiently strong (i.e. $V^z \gg 2\hbar\tau^{-1}$) for diagonal modulation.

6. Conclusions

We have examined impurity dephasing mechanism by TLS in glasses at low temperatures. The linewidth arises from modulations of the optical ground and excited levels via phonon-assisted tunneling motion in random two-level systems.

The dephasing mechanism was studied from three different points of view, namely, (1) by using analogies with nuclear spin relaxation by electron spins, (2) by a simple perturbation approach, and (3) by a Green's function method, all of them leading to the same result.

For diagonal modulation (which was argued to be important at low temperatures), short-time and long-time behaviors have been studied analytically. The short-time approximation assumes that the dephasing rate is faster than the TLS spin-lattice relaxation rate. This assumption is relevant to samples with large density of states of TLS per volume at low temperatures. On the other hand, the long-time approximation is valid in the opposite regime where the dephasing rate is smaller than the maximum spin-lattice relaxation rate. The results predict the power-law temperature dependences with the exponents similar to those observed. For studies of the combined effect of diagonal and off-diagonal modulations, a numerical analysis is necessary.

Acknowledgment

This work was supported by the U.S. Department of Energy under Contract Number DE-ACO4-76-DPOO789.

References

1. For a review see W. A. Phillips (ed.), *Amorphous Solids. Low Temperature Properties*, Vol. 24 of *Topics in Current Physics*, Springer, Berlin (1981).
2. (a) P. M. Selzer, D. L. Huber, D. S. Hamilton, W. M. Yen, and M. J. Weber, *Phys. Rev. Lett.* **36**, 813 (1976).
   (b) J. Hegarty and W. M. Yen, *Phys. Rev. Lett.* **43**, 1126 (1979).
3. P. Avouris, A. Campion, and M. A. El-Sayed, *J. Chem. Phys.* **67**, 3397 (1977); J. R. Morgan and M. A. El-Sayed, *Chem. Phys. Lett.* **84**, 213 (1981).
4. R. M. Macfarlane and R. M. Shelby, *Opt. Commun.* **45**, 46 (1983).
5. B. M. Kharlamov, R. I. Personov, and L. A. Bykovskaya, *Opt. Commun.* **12**, 191 (1974); B. M. Kharlamov, L. A. Bykovskaya, and R. I. Personov, *Chem. Phys. Lett.* **50**, 407 (1977).
6. J. Friedrich and D. Haarer, *Chem. Phys. Lett.* **74**, 503 (1980); J. Friedrich, H. Scheerer, B. Zickendraht-Wendelstadt, and D. Haarer, *J. Chem. Phys.* **74**, 2260 (1981) and *J. Am. Chem. Soc.* **103**, 1030 (1981); J. Friedrich, H. Wolfrum, and D. Haarer, *J. Chem. Phys.* **77**, 2309 (1982); J. Friedrich and D. Haarer, *J. Chem. Phys.* **76**, 61 (1982) and *Angew. Chem. Int. Engl.* **23**, 113 (1984); W. Breinel, J. Friedrich, and D. Haarer, *J. Chem. Phys.* **80**, 3496 (1984) and *Chem. Phys. Lett.* **106**, 487 (1984).
7. (a) H. P. H. Thijssen, A. I. M. Dicker, and S. Völker, *Chem. Phys. Lett.* **92**, 7 (1982); H. P. H. Thijssen, S. Völker, M. Schmidt, and H. Port, *Chem. Phys. Lett.* **94**, 537 (1983); H. P. H. Thijssen, R. van den Berg, and S. Völker, *Chem. Phys. Lett.* **97**, 295 (1983).
   (b) H. P. H. Thijssen, R. E. van den Berg, and S. Völker, *Chem. Phys. Lett.* **103**, 23 (1983).
8. A. A. Gorokhovskii, Ya V. Kikas, V. V. Pal'm, and L. A. Rebane, *Sov. Phys. Sol. St.* **23**, 602 (1981) [*Fiz. Tverd. Tela (Leningrad)* **23**, 1040 (1981)].
9. R. Jankowiak and H. Bassler, *Chem. Phys. Lett.* **95**, 124 (1983); ibid. **95**, 310 (1983).
10 (a) J. M. Hays and G. J. Small, *Chem. Phys. Lett.* **54**, 435 (1978), and *Chem. Phys.* **27**, 151 (1978).

(b) J. M. Hays, R. P. Stout, and G. J. Small, *J. Chem. Phys.* **73**, 4129 (1980) and **74**, 4266 (1981).
11. G. F. Imbush, and R. Kopelman, in *Laser Spectroscopy of Solids*, Vol. 49 of *Topics in Applied Physics*, Springer, Berlin (1981), pp. 1—36.
12. G. J. Small, in *Spectroscopy and Excitation Dynamics of Condensed Molecular Systems* (eds V. M. Agranovich and R. M. Hochstrasser), North-Holland, Amsterdam (1983).
13. T. L. Reinecke, *Solid State Commun.* **32**, 1103 (1979).
14. S. K. Lyo and R. Orbach, *Phys. Rev.* **B22**, 4223 (1980).
15. P. Reineker and H. Morawitz, *Chem. Phys. Lett.* **86**, 359 (1982); H. Morawitz and P. Reineker, *Solid State Commun.* **42**, 609 (1982).
16. S. K. Lyo, *Phys. Rev. Lett.* **48**, 688 (1982).
17. S. K. Lyo, in *Electronic Excitations and Interaction Processes in Organic Molecular Aggregates* (eds P. Reineker, H. Haken, and H. C. Wolf), Vol. 49 of Springer Series in Solid State Sciences, Springer, Berlin (1983), pp. 215—226.
18. B. Jackson and R. Silbey, *Chem. Phys. Lett.* **99**, 381 (1983).
19. S. K. Lyo and R. Orbach, *Phys. Rev.* **B29**, 2300 (1984).
20. P. Reineker, H. Morawitz, and K. Kassner, *Phys. Rev.* **B29**, 4546 (1984).
21. D. L. Huber, M. M. Broer, and B. Golding, *Phys. Rev. Lett.* **52**, 2281 (1984).
22. P. W. Anderson, B. I. Halperin, and C. M. Varma, *Philos. Mag.* **25**, 1 (1972).
23. W. A. Phillips, *J. Low Temp. Phys.* **7**, 351 (1972).
24. L. A. Riseberg, *Phys. Rev. Lett.* **28**, 789 (1972).
25. S. K. Lyo, Unpublished (1985).
26. J. Joffrin and A. Levlut, *J. Physique* **36**, 811 (1975).
27. See also L. M. Molenkamp and D. A. Wiersma, *J. Chem. Phys.* **83**, 1 (1985).
28. S. Chandrasekhar, *Rev. Mod. Phys.* **15**, 1 (1943).
29. S. Alexander and R. Orbach, *J. Phys. (Paris) Lett.* **43**, L625 (1982).
30. D. E. McCumber and M. D. Sturge, *J. Appl. Phys.* **34**, 1682 (1963).

OPTICAL SPECTROSCOPY OF IONS IN INORGANIC GLASSES

WILLIAM M. YEN

*Department of Physics, University of Wisconsin,
Madison, Wisconsin 53706, U.S.A.*

1. Introduction

This review concerns itself with a limited subset of the large variety of disordered systems which are the subject of this volume. Specifically, attention is given here to insulating glasses and concentration is focused on the optical properties of such glasses which have been activated by the introduction of centers or impurities. Insulating glasses belonging to this class are encountered in a host of every day common and technical usages, anywhere in fact from window panes to beverage containers to laser glasses.

Glass has a very long and interesting history [1]. Archaeological sites have yielded fragments of glass vessels traceable to 1500 BC. Recipes for glass compositions can be found in Assyrian cuniform tablets dating to approximately the sixth century BC. The content of glasses as revealed by these ancient tablets do not differ dramatically from some glasses which are in common usage today. Of course, until the advent of chemistry and physics, glass-making was considered a proprietary craft replete with trade secrets which were guarded jealously. The craft flourished in locations determined by the availability of resources, i.e. sand, soda or potash and fuels, and the state of the art pretty much paralleled the ascendancy and decline of this or that civilization. Beginning in the Near East, the craft propagated East and West following trade routes [2]. With the dawning of the industrial revolution, demand for glasses for housing and commercial purposes increased enormously. Driven by this along with an expanding number of technical and scientific applications, glass technology gradually evolved into the industry which we now take for granted. True scientific understanding of the nature of the vitreous state, nevertheless, had to await the development of quantum physics and chemistry [3, 4].

Because materials used in glass melts were not purified, unintentional contamination with traces of metal ions in the end product was inevitable, thus colored or activated glasses probably also date to the beginning of the glass craft. Glass-makers made a connection on the relation between coloration and composition totally on an alchemical basis and it was precisely these secrets which were most closely guarded. A Roman cup (Lycurgus), dated 400 BC, which shows a dramatic color dichroism was recently analyzed. Along with elements found in standard glass compositions, i.e. silicon, alkalis, boron, magnesium, aluminum and oxygen, trace amounts of iron, titanium, lead, antimony, tin, silver and gold were detected. There is no way of ascertaining which elements entered accidentally and which were introduced intentionally, though it can be surmised that Ag and Au which account for the dichroism in this case belong to the latter. The Middle Ages saw an increase of production of colored and tinted glasses fueled by the installation of massive stained glass windows in cathedrals and palaces throughout Europe. Compounds such as cobalt oxide, copper sulfate and ferric oxide were employed to produce a wide array of colors, many of these procedures and processes replicated those developed in earlier times. Once again, a fundamental understanding of the processes responsible for these sometimes spectacular optical properties did not evolve till earlier in this century. That this is the case should come as no surprise, the origin of and processes resulting in the well-known R-line fluorescence in crystalline ruby remained controversial until 1954 [5], for example.

The insulating glasses of concern in this review are prototypical of materials in the glassy or vitreous state. This state is characterized by the absence of any long-range order, i.e. no positional order exists over a macroscopic distance. The ground state of these materials consists of a filled valence band with a forbidden gap intervening before additional allowed and empty electronic states are encountered (conduction band). For the purpose of this review, the glasses of interest will invariably possess a large forbidden gap, $E_g \approx 6$ eV. In their inactivated state, these glasses will appear clear to the eye. In other words, the glass matrix is generally optically inert in the visible and it is used as a host to those various centers and ions which provide those spectral properties of interest (see Figure 5).

This review then will be organized as follows: in Section 2, we will discuss briefly the structure and terminology applicable to inorganic glasses; this will be followed by a description of the types of centers which are normally used to activate inorganic glass. In Section 3, we will consider the nature of the optical transitions and the spectra of activated glasses. The recent advances attained in our understanding static and dynamic spectroscopic properties of doped glasses are presented in Section 4. The latter were derived mostly through laser spectroscopic means. The review is concluded with remarks on some future directions for these studies.

In a review such as this, the scope is by necessity limited and much of the extensive work which provided the foundation of this field of study will be more or less taken for granted. A comprehensive review of the status of spectroscopic studies in glasses appeared about a decade ago [6], and interested readers are referred to this and others [7, 8] for details on some of the background we will be assuming. The review's main focus will be concentrated on developments in the past ten years or so, even then space limitations prevent us from doing justice to all the workers who have contributed to the evolution of our understanding. Again reviews exist of these developments [9].

2. Inorganic Glass Structure and Composition

2.1. Definition of the Glassy Composition

The glassy or vitreous state is a manifestation of the more general and inclusive amorphous state of matter [6]. Certain requirements are placed on the glassy state which restrict the number of materials which are known as glasses; some of these requirements are vague so that the boundary line between glasses and certain supercooled liquids and rapidly quenched materials is at times not distinct.

A glass is required to possess those mechanical properties which characterize a solid. Further, the glassy state, though intrinsically a non-equilibrium configuration, is also required to be stable against crystallization over some lengthy period of time. The state is also to be reproducible when external parameters such as the temperature are recycled.

A somewhat more quantitative definition of the glassy state may be given in terms of a glass transition temperature, T_g [6]. An amorphous solid is said to undergo a glass transition if the solid amorphous phase shows a sudden change in the derivative thermodynamic properties from crystal to liquid-like values as the temperature is changed. The point at which the sudden change occurs is defined as T_g. The difficulty with this definition, it has been pointed out, is two-fold: (1) the change in derivatives in certain common and stable glass systems can be small and difficult to observe and (2) the magnitudes of the changes are dependent to a certain extent on the nature and the duration of measurements, this effect is related to relaxation processes. Thus once again a certain amount of uncertainty is folded into T_g. Because of this it has evolved through usages that T_g is now commonly associated with the liquid viscosity, η, of the system. Specifically, T_g is the point at which η attains a value of 10^{13} poise using relatively slow (~min) measurements. Though attempts have been made to correlate T_g to various fundamental thermodynamic parameters of the glassy state, the models which have developed have fallen short of describing the transition in its totality and it is clear that additional developments are required here. Basically owing to the uncertainties connected with T_g it would appear that this quantity is simply a convenient parameter used to describe a complex physical process. Figure 1 illustrates the temperature dependence of the specific heat (C_v) of a sampling of glasses in the vicinity of T_g, a range of behaviors is to be noted in the figure.

The conditions for formation of a glassy state, viz. those which produce nucleation and crystallization, have also been considered by a number of authors beginning with Zachariasen [3, 4]. Which of these two states ensues in cooling a liquid melt depends on an interplay of the nature of the bonding, the kinetics of crystallization and the rate at which thermal energy can be extracted from the liquid. This area has been the subject of various reviews also and readers are referred to these for further details [10–12].

2.2. Terminology and Structure

From the basic definition of an amorphous material it then may be assumed that no local or lattice symmetry nor any periodicity may be expected in a glass. In fact the glassy state has often been described as a liquid which has been frozen in place. The equivalent of a lattice in this state is known as a network or matrix. The network can be built up with single atoms, but the more common glasses generally contain molecular units such as $(SiO_4)^{-2}$ tetrahedra which arrange

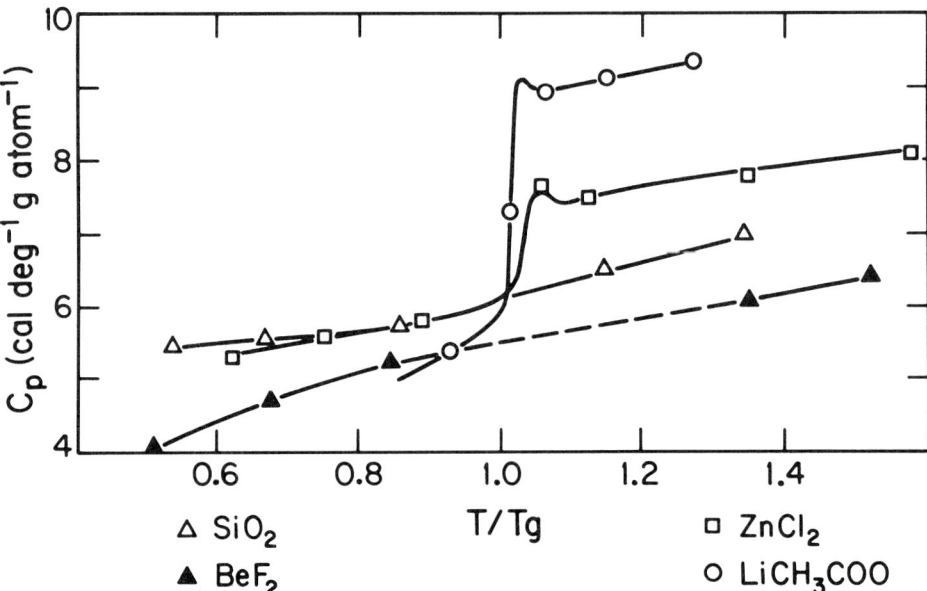

Fig. 1. Changes observed in the vicinity of the glass transition temperature for a number of network glass formers of the MX_2 type. A ionic lithium acetate, $LiCH_3COO$ is also shown for comparison. The glass transition temperatures defined by the temperature at which the viscocity attains a value $\eta = 10^{13}P$ are for SiO_2, $T_g = 1176$ °C, for BeF_2, $T_g = 319$ °C, for $ZnCl_2$, $T_g = 975$ °C and for the acetate, $T_g = 125$ °C. Note the small functional changes in the variation of C_v for the silicate and the fluoride glasses. Data redrafted from Wong and Angell [6].

themselves in ways which parallel their crystalline counterpart and which conform to some simple phenomenological rules. The difference between the glass and the crystalline arrays is that both bond lengths and angles can vary randomly in the former. The cation in these simple glass networks is known as the network former (NWF) [8].

To maintain neutrality the anions in the network will generally connect sufficient number of NWF cations to saturate their valence, if this is the case the anions are said to form bridging bonds. These bonds may be altered to non-bridging type by introducing network modifying cations (NWM) such as alkali or alkaline-earth ions. The NWM bonds to an anion thus modifying the structure by altering the interconnections between the network formers. The type and amount of modifier cations which can be accommodated readily by a simple network depend on the molecular building blocks of the original network. For example, SiO_2 is tetrahedrally coordinated; the addition of a modifier, M_2O, in a ratio of $1:2$ to SiO_2 reduce the average number of bridging bonds to three. Under these conditions, a three-dimensional network may still be formed and many glasses take on this so-called disilicate structure. Addition of the NWM to reach a ratio of $1:1$ with SiO_2 reduces the average number of bonds to two, this metasilicate composition can only produce SiO_2 chains which do not support a three-dimensional structure and no glasses of this type exist in a stable form. Obviously, also, a simple network containing $(SiO_4)^{2-}$ cannot accommodate modifier cations with charge higher than divalent unless additional compensating modifier are added. This is the reason why ternary and more complex glasses exist; typical of such ternary combinations are glasses with Li_2O, SiO_2 and Al_2O_3. Considerations such as those related to bonding plus those concerned with the kinetics of crystallization, delimit the concentrations which ultimately can be used to form a glass; Figure 2 shows the region over which conditions are appropriate for the formation of lithium aluminosilicate glass.

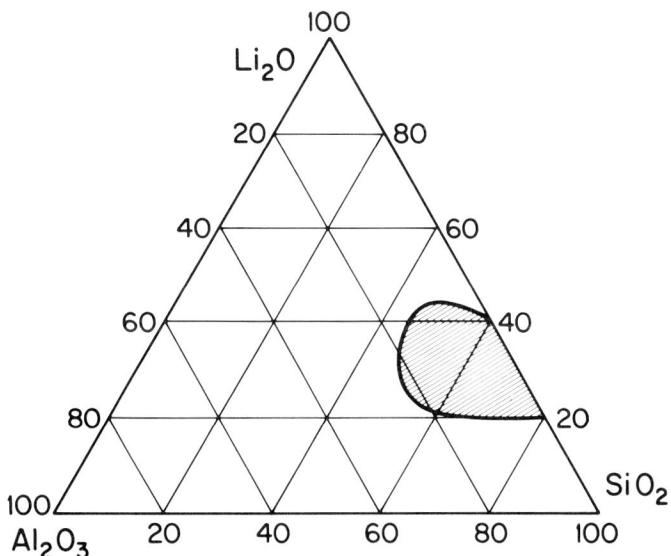

Fig. 2. Glass-forming region for lithium aluminosilicate glass (hatched area). Other compositions are unstable against crystallization/phase separation. Al^{3+} and Li^+ concentrations are approximately the same in these glass compositions [8].

In more complex glass compositions, the distinction between NWF and NWM becomes somewhat fussy. Cations are, as a rule of thumb, classified into one or the other by two parameters, the ionic radius, a, and its effective field strength, eZ/a^2, where eZ is the ion change. If a is small, ~ 0.50 Å, and the effective field is similarly large, then the cation is classified as a NWF; B^{3+}, P^{5+} for example have size and electrostatic properties similar to those of Si^{4+} and are found as formers in many common glasses. Alkali ions, rare earths, etc. on the other hand possess relatively large ionic radii and commensurate lower effective fields thus they generally fall in the modifier category. A number of cations such as Al^{3+}, Zr^{4+}, Mg^{2+} fall in an intermediate region and, depending upon circumstances, can behave either as formers or as modifiers.

Size considerations also yield indications as to the local coordination number of the cations. Smaller ions generally find themselves in four-fold coordinate sites whereas large ions may be surrounded by 6, 8 or more anions. Recent simulations indicate that 5-; 7- and 9-fold coordinations are also possible [13, 14].

Since the glassy state is by definition completely devoid of long-range order, diffraction studies which probe for structural order will not yield sharp interference peaks as would be expected from a crystalline material. However the preceding discussion on average coordination numbers for the cations implies that some degree of short-range order remains in the glassy systems. Indeed, when various types of diffraction experiments are conducted in glass, diffuse structure is observed at low scattering angles [15, 16]. This structure is conjugated to the radial distribution function. Results from such studies, Figure 3, indicate that the distribution of near neighbors in glasses replicates the structure of their crystalline counterparts up to the second near neighbor shell beyond which the distributions become diffuse and approach a structureless spherical distribution. The diffraction patterns are quite similar to those obtained in liquids. Though these studies cannot yield absolute or unique results on the detailed structure of disordered systems, valuable information may be obtained on the average properties of the structure on an atomic scale. The results of Warren, for example, serve to confirm Zachariasen's models of glass structure. These distribution functions have been extensively used for comparison with predictions of simulation and models of disordered systems [17, 18].

As a conclusion for this section, Figure 4 shows a two-dimensional schematic projection of an ordered, a disordered and a modified structure of SiO_2 [19]. The

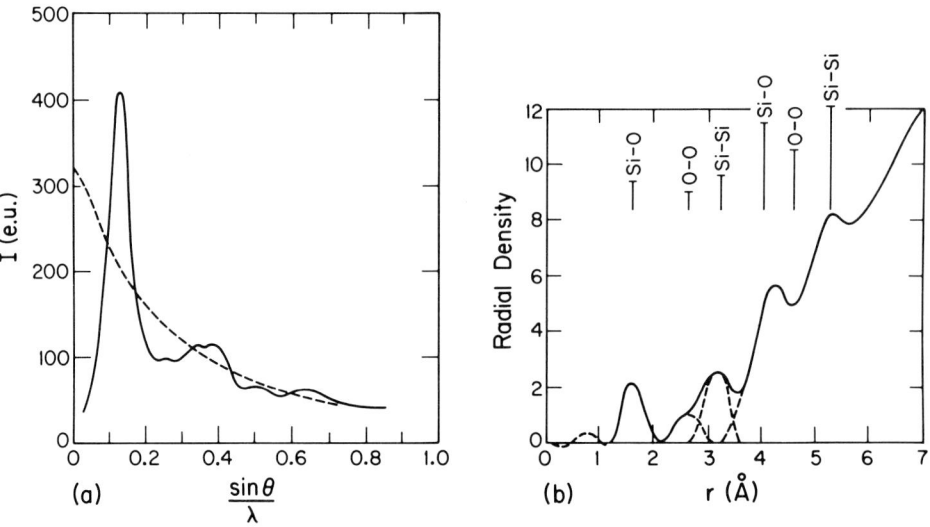

Fig. 3. Radial distribution function for silicate glass as deduced from X-ray scattering data. Different neighbor types are identified in the distribution. Data is from Warren [15] and Warren *et al.* [16] and was viewed as confirmation of the theories of Zachariasen [3]. After Stevels [10] (1962).

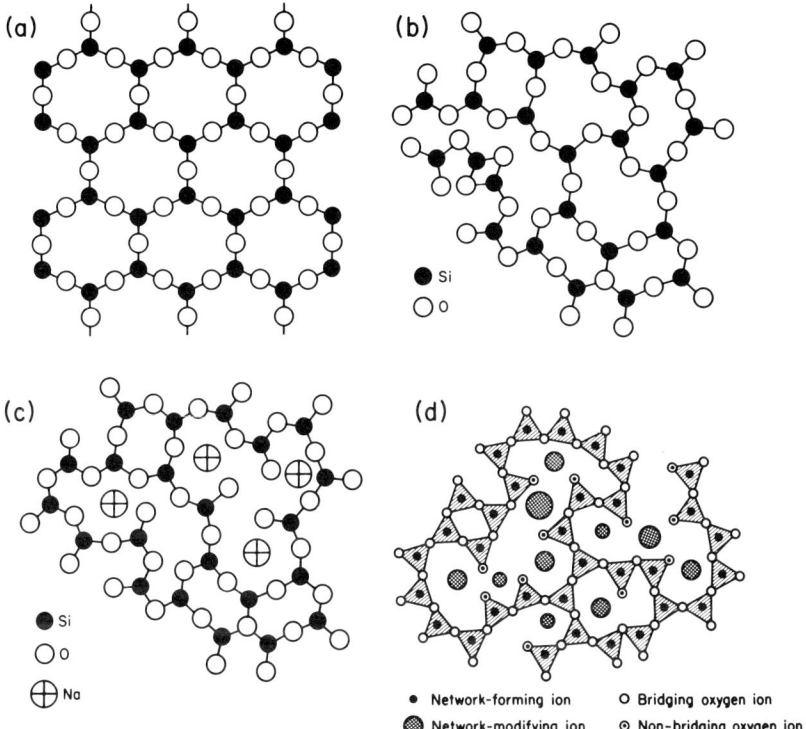

Fig. 4. Two-dimensional representation of (a) α-cristobalite crystal containing SiO$_2$ tetrahedra. In these diagrams the fourth oxygen of the tetrahedra is found outside the plane of the diagram. (b) SiO$_2$ glass which the Si^{4+} coordination remains the same but bond lengths and angles vary to form a disordered network and (c) a sodium oxide modified SiO$_2$ structure in which the O^{2-} coordination varies. From Stookey and Mauer [11]. (d) Zachariasen concept of an oxide glass showing network modifiers (NMW) and bridging and non-bridging O bonds. From Stevels [10].

crystalline phase is known as α-cristobalite. Various definitions introduced above are illustrated in the figure.

2.3. Coloring and activation of glasses

Most glasses of common usage have constituents which show no optical activity in the visible region of the electromagnetic spectrum. This has, of course, been dictated by demand for clear, transparent glasses for various purposes. Much effort in fact was spent, especially in the last century, to develop clear glasses either through the purification of batch raw materials or through the introduction of reagents into the melts to deactivate the offending coloring centers.

As noted earlier, the coloring of glasses developed through unintentional contamination of glass melts. In time, metallic ions of various types began to be correlated with specific glass coloration both when introduced singly or in combinations. Metal ions of the transition metal series, $(3d)^n$, have been the most widely utilized for these purposes. Certain heavier ions have also been used. These include Pb, Tl, Mo and U. The use of rare earth ions, $(4f)^n$, though now widespread, is a relatively recent development. Other elements such as S, Se are also known to color glasses [7, 8].

A number of interesting colored glasses are produced by the incorporation of small metallic particle suspensions into the melts. The coloration and dichroism of the Lycurgus Cup mentioned in the introduction is caused by noble metal particles introduced into the melt. The optical properties of these glasses results from scattering of light from particles rather than from intrinsic electronic transitions of an atomic entity. These classes of ruby glasses, though fascinating, will not be of interest in this review.

The transition metal ions are intermediate in size and following the empirical rules discussed in an earlier section they may enter a glass network either as formers or as modifiers depending upon the circumstances. $(3d)^n$ ions may readily be incorporated in most glasses to ~0.5 wt% without affecting the physical properties of the base glass. This type of solubility is related to the incorporation of the impurity ions into spaces available in the loosely knit glass network. If the $3d$ concentration is increased to a few percent, then these ions begin replacing the basic glass formers and some glasses can manifest changes sufficient at times to lead to crystallization [8].

Heavier and larger ions such as the $4f$ and $5f$ series constituents generally cannot enter as formers in the simple glasses and require the presence of network modifiers. It follows that these ions in most instances do not enter substitutionally but fit in the spaces provided by a modified or compensated network. Certain more complex and heavier glasses have evolved primarily to fulfill optical instrumentation requirements; the lanthanum crown and flint glasses are an example. In these special glasses, rare earth ions can and do serve as network formers.

In this article, an activated glass is defined as one which contains ions, either as dopants (NWM) or as constituents, which are capable of emitting radiation (luminescence) upon excitation. Thus, an activator ion not only will have the general capability of coloring glass through its absorptive transitions but also will contain a state or a set of states capable of emitting light. It is this class of ions which is of greatest interest to us here. The radiative properties of these activators in glasses (as in crystals) are influenced deeply by excitations which are intrinsic to the glass network or to the lattice. These interactions lead to non-radiative relaxation processes and other assisted transitions. Because of these effects, roughly speaking, radiative transitions seldom occur with energies less than, say, 5000 cm^{-1} ($\lambda \approx 2$ μm); this limitation should be kept in mind throughout this discussion.

2.4. COMPOSITION AND PHASE SEPARATION

In view of the discussion above, there is an extremely large set of possibilities for glass compositions which can accommodate a large variety of technical demands. Table I gives a list of nomenclature and composition of sampling of commercially available glasses.

TABLE I
Some typical glass compositions and melt temperatures

Type	Composition (mol%)	Melt T (°C)
Soda lime silicate	SiO_2 — 7.30, Na_2CO_3 — 14.0 $CaCO_3$ — 13.0	1550
Lithium lime silicate	SiO_2 — 60.0, Li_2CO_3 — 27.5 $CaCO_3$ — 10.0, Al_2O_3 — 2.5	1550
Borosilicate	SiO_2 — 65.0, Na_2CO_3 — 20.0 B_2O_3 — 15.0	1440
Lithium borate	B_2O_3 — 66.7, Li_2CO_3 — 32.3, AS_2O_3 — 1.0	1100
Lithium aluminum borate	B_2O_3 — 65.0, Li_2CO_3 — 20.0 Al_2O_3 — 13.9, AS_2O_3 — 1.0	1100
Calcium phosphate	$CaCO_3$ — 33.3, P_2O_5 — 67.7	1100
Fluorophosphate	$Al(PO_3)_3$ — 5.0, AlF_3 — 25.0 NaF — 5.0, LiF — 5.0, MgF_2 — 10.0 CaF_2 — 30.0, SrF_2 — 10.0, BaF_2 — 10.0	850
Tellurite	T_2O_2 — 78.5, Na_2CO_3 — 15.0 K_2O — 5.0, AS_2O_3 — 1.0	800

Recalling that glasses are intrinsically unstable, phase separation of the constitutents in binary and more complex compositions is observed to occur. In appropriate circumstances crystallization or divitrification may occur in one or the other phase. These processes are not fully understood but are known to depend on nucleation and annealing. Exploitation of this phenomena has lead to the development of a variety of ceramics which have had many technical applications. Ceramics constitute a class of materials intermediate between glasses and crystals. The recent development of ceramics which are transparent and which can be activated in either of the ceramic constituent phases presents us with systems which possess potentially very interesting optical properties [20].

3. Optical Properties of Impurity Centers in Inorganic Glass

To fully appreciate the chronology in the evolution of our understanding of the optical properties of colored or activated glasses, it is instructive to peruse a classic book by Weyl originally published in 1951 and reissued in 1959 [7]. The author comprehensively summarizes the status of the field as understood at that time. It is clear from this work that a great deal had already been established by then especially in the chemistry of the composition and formation of colored and fluorescent glasses. However, the optical properties of colored glasses are only loosely related to the actual electronic level structure of the individual coloring ions. One needs to be reminded that the period 1950—60 marks the beginning of increasing activity and interest in understanding the spectroscopy of solids, though the foundations of this understanding had been established some years earlier [21—23]. It is likely that customary delays occurred in the translation of these findings in solid state physics to the glass sciences.

Indeed, the departure point of this section will be the spectroscopic properties of ions in crystals. There is a one to one correlation between crystalline and glassy spectroscopic properties. The principal difference lies in the fact that the spectra of activated glasses are fundamentally affected by the disorder and random variations which define the glass state and hence which produced similar unpredictable changes in the environs of the active center.

In the analysis of the spectra of ions in crystals, the host is initially assumed to simply provide the charges which alter the symmetry or environment of the center and is otherwise inert. This is simply an approximation in which all centers are initially isolated. These assumptions are maintained for glasses in this section, effects arising from the dynamics of the lattice and from intercenter interactions are the subject of later sections. As Figure 5 shows, the transmission properties of some typical glasses are used as a host for coloring and activator dopants. As can be seen from the figure, these glasses are largely inactive in the visible to near infared portions of the spectrum. The near u.v. absorption edge in these glasses of interest can be correlated with charge transfer processes between the constituents, inclusive of conduction. The IR region is dominated by fundamental molecular vibrations of the components as well as contaminants such as water.

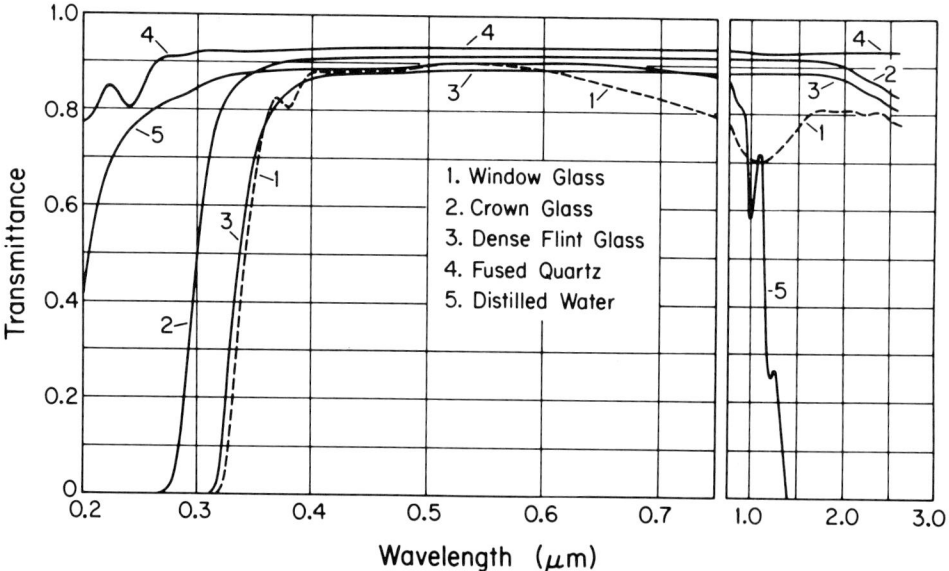

Fig. 5. Transmission characteristics of common glasses prior to activation with coloring ions. From Thomas [133].

3.1. SPECTRA OF IONS IN SOLIDS

Many outstanding reviews have appeared which survey the optical properties of the most important sequences of active ions, i.e. transition, lanthanides and actinides, as they are introduced into ordered crystalline structures [24–29]. Once again space limitations here preclude any lengthy description of the considerable volume of work which has been carried out in this area. Suffice it here to basically outline several approaches commonly used in the interpretation of the spectra of these materials.

Basically, crystal field theory and its extension, i.e. ligand field theory [30] seek to predict the effect of a configuration of charges (the crystal field) on the electronic properties of an optically active ion introduced into this environment. One presupposes that the nature of the wave function of the free ion, i.e. prior to its introduction into the host, is known to some extent though this is often only true in an approximate sense. The result of the perturbation produced on the optically active electrons by the static electric fields of the neighbors depend on the nature of the electrons themselves. The consequence of the perturbations is invariably the shifting of energy levels, the lifting of sundry degeneracies and the alteration of radiative transition probabilities [24].

These theoretical considerations which may be traced to Bethe [21] and Van Vleck [22, 23] have been expanded and refined to include the effects of bonding and the presence of intrinsic excitations of the surrounding complex. These developments and refinements have been applied extensively to principally two series of impurity centers, i.e. the transition metals $(3d)^n$, whence the electrons of interest partially occupy the $3d$ atomic shell and the rare earth series, $(4f)^n$, where partial occupancy of the 14 allowed f states occurs. More recent analyses have dealt with the $(5f)^n$ or actinide series [29, 31] and work exists dealing with the heavier atoms [24, 32].

Though historically efforts to understand the spectra of transition metal complexes were the first to meet success, the spectra of $4f$ ions reveal more readily the influences exerted by the various perturbations on the free ion levels [27, 28]. In these cases, the electrons of the free ion occupy orbitals of the $(4f)^n$ configuration. The inter-electronic Coulomb interaction leads to an initial splitting of states and to the formation of LS terms. The electronic terms are next perturbed by the spin-orbit coupling and under the appropriate conditions the LS terms split further into J-multiplets familiar from the Russell-Saunders approximation. Because of the complexities of the many $4f$ electronic combinations, the Russell-Saunders approximation is often not adequate. The spin-orbit interaction can mix J states from different LS terms; this has required the introduction of intermediate coupling states denoted by

$$|f^n\{\gamma'S'L'\}J\rangle = \sum_{\gamma SL} c(\gamma SL)|f^n \gamma SLJ\rangle. \tag{1}$$

In the above, $|f^n \gamma SLJ\rangle$ are the Russell-Saunders states with γ differentiating states with similar S and L values. The label $\{\gamma'S'L'\}$ usually denotes the dominating Russell-Saunders state in the manifold. The appropriate coefficients c may be obtained from elaborate diagonalizing procedures and accurate representations of the free ion states of trivalent rare earths have been obtained in this way. Typically in $4f$ ions then the states are labelled by $^{2S}L'_J$, e.g. $^3H_4$, $^3P_0$, etc. Deviation from the LS coupling scheme is principally manifested by breakdown of the Lande interval rule.

The free ion states respond to the lowering of symmetry by showing additional splittings. For the lanthanides, because the active electrons are somewhat shielded by higher quantum shells, the effects of external perturbations are relatively weak compared to the Coulomb and $L \cdot S$ interactions. The J-multiplets split into manifolds traceable to their $\{\gamma'S'L'\}$ origins. The nature of the splitting is determined by the symmetry and strength of the crystal field which is parametrized through parameters, B_q^k. The field parameters produce admixing of J values and can alter radiative selection rules.

The majority of the rare earth optical transitions in solids which are observed in the visible occur between states which belong to the same configuration. Because of parity considerations, electric dipole transitions are forbidden in these circumstances unless the crystal field provides an odd component to the perturbation and admixes elements of other configurations into the $4f$ states. Using certain simplifications, Judd [33] and Ofelt [34] have provided a way to conveniently treat the spectral intensities of intraconfigurational transitions of $4f$ ions in solids. In this approximation, the transition strength is written in the form

$$S(aJ:bJ') = e^2 \sum_{t=2,4,6} \Omega_t |\langle f^n \gamma_b S_b L_b J_b || U^t || f^n \gamma_a S_a L_a J_a \rangle| \qquad (2)$$

The Ω_t are known as the Judd–Ofelt parameters, while the reduced matrix elements of the unit tensor operator U^t can be evaluated. Fairly comprehensive tabulations of $|\langle \| U^t \| \rangle|$ are to be found in the literature [35, 36]. Ω_t is a measure of the strength and nature of the odd-parity crystal fields, these parameters are normally derived from fitting experimentally measured absorption and emission strengths.

For transition metal ions the situation differs principally because the active electrons in the $3d$ shells are less shielded and thus are much more sensitive to external perturbations. It is customary to treat these cases in the strong field approximation in which the external crystal field serves as the principal perturbation [25]. There are, of course, five initially degenerate d orbitals in the free transition metal ion. When these free ions are placed in a crystalline field, the orbitals split into two-fold degenerate e_g and three-fold degenerate t_{2g} orbitals to first order. The splitting of the orbitals is assigned a value of $10Dq$, thus Dq is a measure of the field strength; the ordering in energy of the e and t orbitals depends on whether the ion coordination is principally tetrahedral or octahedral.

With the introduction of additional d electrons, the single electron e and t orbitals couple through the Coulomb repulsion and form terms denoted, by $S\Gamma$, similar to the SL terms in the $4f$ case. Γ is now a label used to denote the appropriate group theoretical representation of the symmetry group to which the eigenstate belongs. The $S\Gamma$ term separation is characterized by the Racah parameters A, B and C. A does not affect the term separation whereas the ratio C/B is found to be almost constant for the whole $3d$ series. It follows that two parameters, i.e. B and Dq, then determine the behaviour of the terms of $3d$ levels. This behavior is now commonly represented in Sugano—Tanabe diagrams which plot the energy of the term in units of B, i.e. E/B, versus the ratio Dq/B which is a measure of the field strength [37, 38]. Terms are labelled as $^{2S+1}\Gamma$, Γ comprising the representations A, E and T. A subscript g denoting even parity states is often deleted.

Both the spin-orbit interaction and deviations from high symmetry also affect the energy levels. In the case of $3d$ ions, the $S\Gamma$ terms are split into multiplets analogous to the J multiplets discussed previously. The magnitude of the J-splittings is also of the same order of magnitude as that of $4f$ ions. It is necessary to remember that the source of this splitting differs as it results from a different sequence of perturbations.

Electric dipole transitions are once again forbidden unless some element of odd parity is introduced through the sundry perturbations, the most likely candidate being the crystal field deviations. Lattice coupling can play a role in determining transition strengths, however, spin selection rules are only weakly violated in this series.

By and large, the majority of transitions of interest here may be treated in the spirit of the two approaches sketched above. As they concern transitions between states in the same configuration, in terms of atomic spectroscopy standards the radiative transitions are relatively weak. In the appropriate circumstances this implies long radiative lifetimes and conversely intrinsically very sharp transitions. Owing to this high Q, i.e. $\omega_0/\Delta\omega$, with ω_0 the transition frequency and $\Delta\omega$ the linewidth, these optical transitions may be used to probe for interactions occurring in the solid state. Many examples of this application are to be found in the literature [39].

Actinide ions, $5f$, have been treated in the weak field approximation though certain modifications are required owing to the somewhat less protected nature of the active shells. In addition, $6d$ allowed bands sometimes are lowered sufficiently to interfere with the $5f$ configuration. Though some work has been done to analyze and d bands of $4f$ and $5f$ ions [40], these investigations have not been extensive and likely a thorough analysis will require a molecular orbital approach. Considerations analogous to those applicable to $3d$ ions have also been extended to include $4d$ states [41]; once again these treatments will require additional sophistication.

3.2. INHOMOGENEOUS CONTRIBUTIONS IN THE SPECTRA OF SOLIDS

Even at the lowest temperatures and in the best of crystals, transitions of impurity ions have linewidths which reflect extrinsic contributions [42]. These extrinsic contributions are generically known as inhomogeneous broadening and are distinct from those broadening mechanisms which are fundamental to the radiation process. In the simplest of cases, the latter arises from the widths governed by the Heisenberg principle.

In solids, the source of inhomogeneous broadening can have a complex origin as might be the case when clustering occurs in a doping process. However, for the purposes of this article, it is assumed that broadening processes have a truly random, statistical origin which can be traced to imperfections or random strains in the case of crystalline solids. These strains and imperfections produce random variations of the B_q^k or Dq parameters and thus affect the energy levels of the ions in question. The net result is illustrated in Figure 6, where various types of disorder broadening are also illustrated.

In addition to the above minor perturbation, both compositional disorder and lattice disorder produce inhomogeneously broadened spectra [43]. In the first case, the lattice remains ordered but the constituents and hence the near neighbors to an active center are made up of a randomly constituted admixture. The case of glasses belongs to the latter group in which the lattice is totally disordered and the active ion is exposed to a distribution of environments each with a characteristic set of field parameters. The distribution of fields at individual sites may be large and the energy levels of the ion may show considerable changes from their crystalline counterpart. However, apart from stoichiometric changes which may occur in incorporating active ions into glasses, the principal difference between crystalline and glassine spectra is the magnitude and the nature of the inhomogeneities. Thus it is that the crystal field discussion presented in the previous section is the logical starting point.

3.3. CONVENTIONAL SPECTROSCOPY OF IONS IN GLASSES

3.3.1. Static Properties

Experimental studies of glasses initially were concentrated on transition metal doped systems since these ions played such an important role in the development of colored glasses for various scientific and technical purposes. Invariably these earlier studies simply sought to determine the absorptive optical properties through conventional spectroscopic means. Emissive properties were generally only treated in a cursory way as interest in these properties did not arise until the advent of laser devices [7, 44].

As it has already been noted, Weyl presents a comprehensive summary of optical properties of glasses as they stood in the decade of 1950. A sequence of workers sought to interpret the origin of various absorption bands using models which have now been discredited and much controversy surrounded these interpretations [45–47]. It was not until 1951, that Hartmann and Ilse [48] first interpreted the $(3d)^1$ Ti^{3+} crystalline spectrum successfully in terms of the earlier theories of Bethe and Van Vleck. The considerable activity which followed, of course, culminated in the seminal contributions of Tanabe and Sugano [37, 38] and the subsequent refinements to the interpretation and calculation of $3d$ spectral features [25].

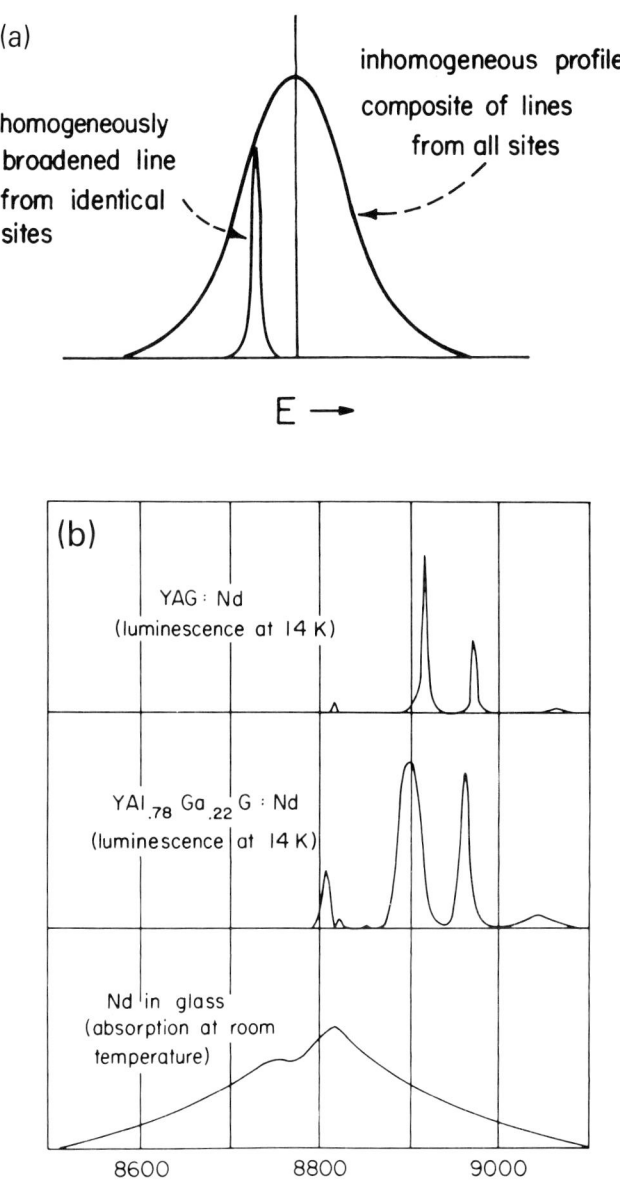

Fig. 6. (a) Schematic representation of an inhomogeneously broadened transition. The constituents of the transition consists of homogeneously broadened levels from similar sites and are some times called monochromats. In glasses the distribution of sites varies greatly and need not be symmetric. (b) Order of magnitude broadening is illustrated for intrinsic, compositional disorder and lattice disorder in the $^4F_{3/2}$–$^4I_{9/2}$ transition in $Y_3Al_5O_{12}$, in a $Y_3Al_5O_{12}$–$Y_3Ga_5O_{12}$ mixed crystal and in glass respectively. After Imbusch and Kopelman [43].

Bates and Douglas [49] first applied crystalline or ligand field theory to the absorption spectra of ions with partially filled $3d$ shells embedded in glasses. This proved to be a logical extension of the developments occurring in other phases of solid state spectroscopy as the absorption spectra of transition metal ions in glasses, aqueous solutions and in crystals were already known to be closely correlated. As it is to be expected, there is a one to one correspondence between the spectra of transition metal ions in crystalline and in amorphous materials. This is most clearly seen whenever the energy levels are well separated or isolated. One such case is illustrated in Figure 7 for $Mn^{3+}(3d^4)$ in a silicate glass and in a crystal [50]. It is apparent from the figure that though the spectrum of the glass is broader, possesses a larger extinction coefficient and is shifted to lower energies, the nature of this state cannot be questioned.

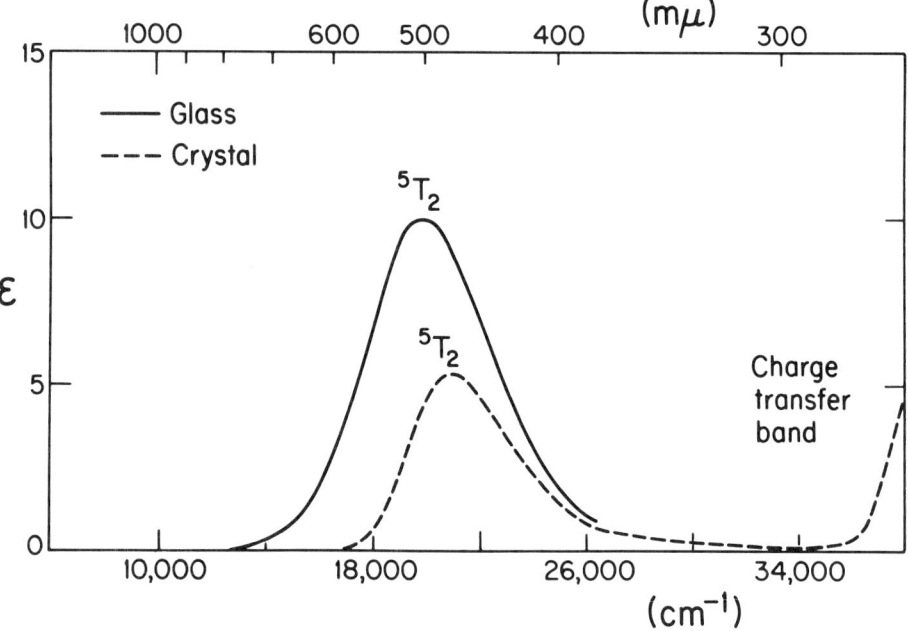

Fig. 7. Comparison of the spectra of $Mn^{3+}(3d^4)$ in a soda—lime—silicate glass and in a similarly coordinated hydrated crystal. A reduction in the crystal fied values for the glass is evinced by the shift of the $^5T_2$ peak towards lower energies. This bond is responsible for the violet-pink coloration in these glasses [50].

Bates [50] was able to deduce factors such as the average coordination numbers of the active ion by carrying out comparisons between crystalline and glass spectra. A measure of the crystalline field strength, i.e. E/B, can also be extracted by fitting peak values of absorption bands. In common glasses, it is found that the E/B values are generally 5—10% lower than those found in similarly coordinated ions in crystals. It has been pointed out that given the limited number of bands observed in $3d$ glass systems and the loss of fine structure because of disorder broadening, the interpretation of spectral features in the above terms is at best a crude approximation to average bulk properties of the glass system.

There has been a hiatus in the experimental activities concerned with $3d$ glasses which encompasses the late sixties and the decade of the seventies. This dearth of interest can be correlated to the fact that only a few of the $3d$ ions in glass show useful emission properties. It is only very recently that a comprehensive survey of the optical properties of $3d$ glasses has been initiated [51]. Representative results along with the Sugano—Tanabe diagrams indicating the nature of the bands are shown in the sequence of Figures 8(a)—(c).

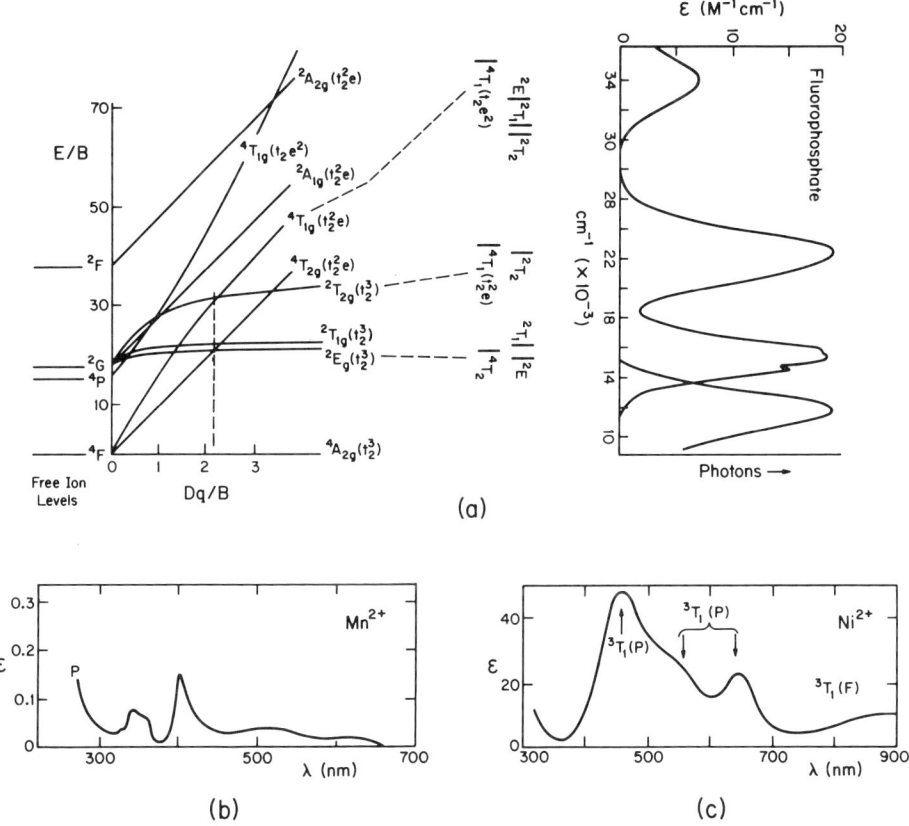

Fig. 8. (a) Absorption and emission spectra of Cr^{3+} doped fluorophosphate glass, adjacent Sugano—Tanabe diagram and bars in figure identify the appropriate octahedral states. Note that the $^4T_2$ and $^2T_1$, ^2E overelap in this case; structure appearing in the absorption band have been identified as a Fano antiresonance [57]. Figure redrafted from Andrews et al. [54] (b) and (c). Absorption spectra of Mn^{2+} in a phosphate glass and Ni^{2+} in a silicate glass respectively. The Mn^{2+} is weak as it originates from spin forbidden sextet to quartet transition. It is thought that the Ni^{2+} is the composite of 4- and 6-fold coordinated Ni, tentative assignments are shown. Redrafted from Patek [8].

Regardless, it would appear that among the $3d$ series the properties of Cr^{3+} doped glasses have been studied in the greatest detail. In most instances, [50, 52—54], the crystalline field strengths are such that the familiar ^2E and $^4T_2$ states are not distinctly separated from each other as is the case in ruby, for example. Analysis of the resulting spectra requires the incorporation of the disorder inheritant in glass structures, i.e. the random and large variation of fields. The luminescence spectrum of Cr^{3+} glasses may be deconvoluted into a 'narrow' band and a broad band component which result from ^2E and Stokes shifted $^4T_2$ emission from individual sites of higher and lower fields strengths, respectively. Similarly the distribution in field strengths results in a site to site variation in transition or oscillator strengths which in turn are reflected in the non-exponential decay of the total luminescence of Cr^{3+} glasses when excited. The ^2E, $^4T_2$ crossover also affects the non-radiative relaxation processes in these glasses in a fundamental way by providing additional channels for the dissipation of optical energy [55]. The net result is that for systems reported on to date, the quantum efficiency of Cr^{3+} glasses has generally been poor. Some of these aspects will be discussed in greater detail in a later section. It is noteworthy that certain structure appearing in the $^4T_2$ band has been identified as resulting from a Fano antiresonance [56—58].

There has been in the past decade intense activity in the study of the optical properties of rare earth glasses. This activity is related to efforts to develop high power laser devices and has resulted in an extremely comprehensive data base on these properties especially in those ions which have been found to be most useful for stimulated devices such as Nd^{3+}, Er^{3+} [59, 61]. The properties of some of these ions in a wide ranging set of glass compositions have been surveyed using conventional spectroscopic methods and results can be found in tabulations [59, 61].

Historically, the spectra of rare earth ions in solids was not fully unraveled until the 1950s, consequently no realistic effort was made to understand the properties of rare earth glasses until that time [27]. Representative traces of rare earth spectra in some common glasses are shown in the sequence Figure 9(a)—(c). Apart from the effects of inhomogeneous or disorder broadening which washes out details of the J-manifold structure, it can be seen from the figures that transitions in the glass can clearly be traced to their $L \cdot S$ origins. The correspondence between spectra in crystalline and amorphous materials is immediately obvious. This is, of course, simply a consequence of the better shielding the $4f$ electrons experience.

As noted previously, rare earth ions enter into glass as modifiers (NWM) and are thus not generally simply coordinated. Levin and Block [62] using immiscibility arguments deduced 6, 7 or 9 as the coordination numbers, though evidence also exists that the smaller trivalent rare earths can be octahedrally coordinated [63, 64]. These possibilities, coupled with variations in bond angles and near neighbor distances between cations and the center, once again result in a random spread of crystalline field values, B_q^k in this case. The vanishing J-manifold structure noted above may thus be attributed to the site to site difference in the

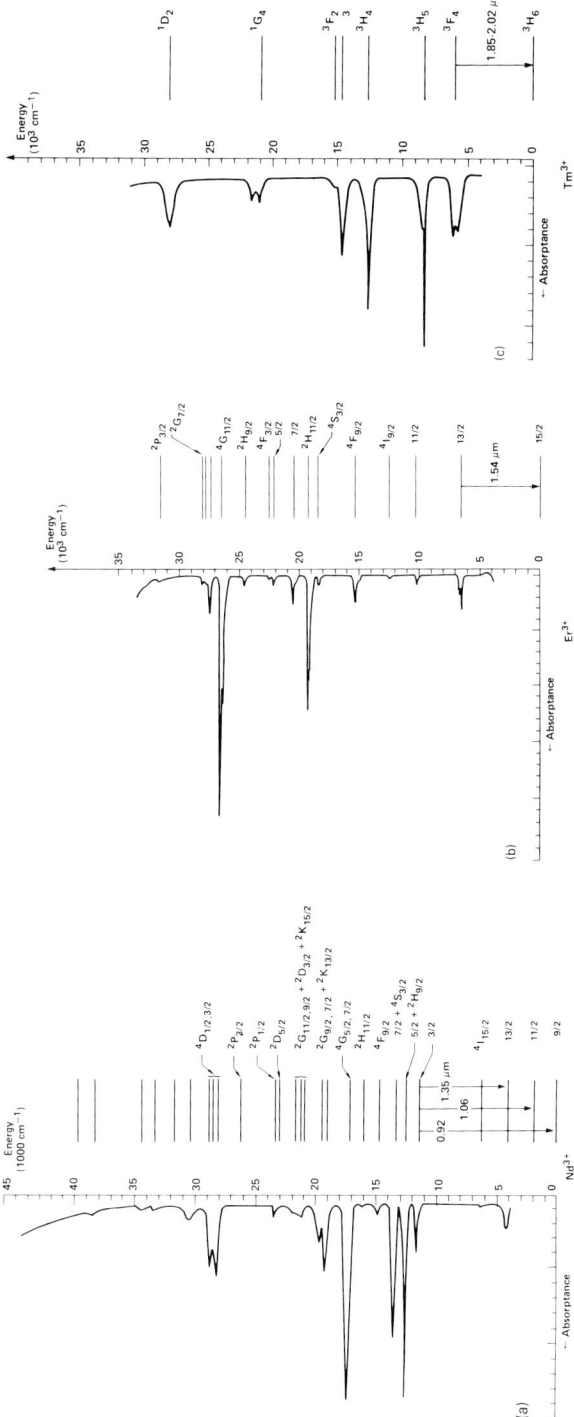

Fig. 9. Absorption spectra and energy levels for some of the more common 4f ions in glass. (a) Nd^{3+} has been the best studied 4f dopant in glass. Diagram shows the principal laser transitions (b) Er^{3+} and (c) Tm^{3+}. The Stark level structure is not shown in the level diagram but can be discerned in the inhomogeneously broadened transitions of the majority of 4f doped glasses [59].

intramanifold splittings. Fluorescence lifetimes of emitting states in rare earth doped glasses are then also highly non-exponential as field parameters determine oscillator strengths.

Noticeable but not spectacular changes of the spectrum of $4f$ glasses are observed as the composition of the glass is changed. Typical changes in the Nd^{3+} fluorescence and absorption are shown in Figure 10 for a set of glasses with different compositions [59]. These spectroscopic changes are totally consistent with the weak field approximation used in the case of rare earth ions.

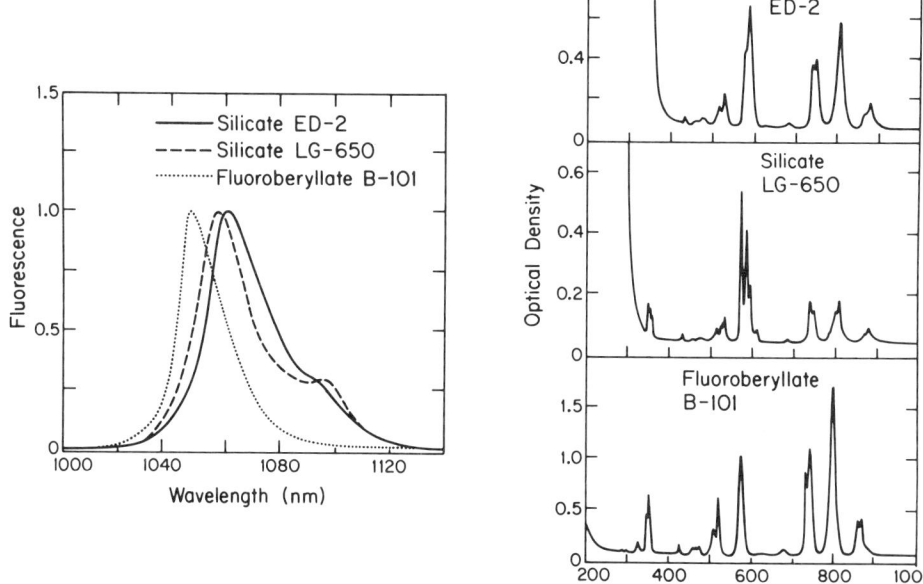

Fig. 10. Compositional dependent changes in the fluorescence ($\sim 1.06\,\mu$) Nd^{3+} transition and in the absorption spectrum of Nd^{3+} in a standard laser glass, ED-2, a low cross section silicate glass. LG-650, and a fluoroberyllate glass. Redrafted from Stokowski [59].

Analysis of the spectra of rare earth ions in glasses once again seek some descriptive average of the field parameters. Attempts have been made to fit the spectrum by assuming a dominant coordination around the ion and accounting for deviations from this coordination by adding an inhomogeneous breadth to each state [65, 66]. This approach has been proven to be fallacious by later work as the random field variations actually produce variations in the J-splittings which can be very large.

A common and useful approach for the purpose of describing cross sections for emission in laser glasses has been the Judd—Ofelt treatment. Making reasonable assumptions for the tensor operator elements, $|\langle\|U\|\rangle|$, appearing in Equation (2) from values calculated for crystals, the Judd—Ofelt parameters, Ω_j, may be derived from the total absorption/emission intensity per ion of the glass. Though once again by integrating transition strengths, one ends up with some average of the $4f$ ion properties, nevertheless the cross sections calculated through the expression

$$\sigma(\lambda_p) = \frac{8\pi^3 e^2}{27he(2J+1)} \cdot \frac{\lambda_p}{\Delta\lambda_{\text{eff}}} \cdot \frac{(n^2+2)^2}{n} S(aJ:bJ'), \tag{3}$$

where λ_p is the peak fluorescence wavelength, $\Delta\lambda_{\text{eff}}$ the effective linewidth and n the glass refractive, index, have proven to be surprisingly accurate. These cross sections can in turn be used as a convenient index or summary of the gross optical behavior of a given glass [61].

The preponderant majority of work has concentrated on the trivalent rare earths, specifically involving the forbidden f—f transitions. Spectra are available on certain divalent rare earths, the most common of which is Eu^{2+}. Dy^{2+} and Sm^{2+} are found only in a limited number of oxygen free glasses [8]. The first set of electric dipole allowed transitions in the trivalent series comprise states of the $5d$ configuration and are normally beyond the u.v. absorption edge of the glass host. The single exception is the case of Ce^{3+} ($4f^1$) which can show a broad absorption and Stokes shifted emission in the near ultraviolet corresponding to a $4f \rightleftarrows 5d$ transition. The presence of an observable transition depends on the nature of the host to a very large extent for it has been shown that the $5d$ states which are no longer shielded are extremely sensitive to the field strengths. Ce^{3+} finds use as a general u.v. sensitizer for other $4f$ optical transitions [67, 68] and as a inhibiting agent for solarization in glasses [69].

Some experimental work exists on actinides, principally uranium complexes [31] and other heavy ions such as Bi^{3+} and Pb^{2+} in glass. The majority of these spectra consist of relatively featureless broad absorption and emission bands; summaries of some of these spectral properties have appeared in the literature [70]. Analysis of the spectra has not been extensive and the heavier ions are of interest predominantly in the context of sensitization for $4f$ ions [32].

3.3.2. Dynamical Properties

Conventional spectroscopic studies of glasses have also helped to establish that processes extraneous to the electronic structure of the active ion greatly affect the optical properties of these systems. Once again there is a one to one correspondence between glassy and crystalline systems.

The Raman spectrum of a fluorophosphate glass is shown in Figure 11(a) [71]. It shows that insulating glasses have excitations which are comparable to those of crystalline phonons up to a frequency of ~ 3–500 cm^{-1}. In addition, relatively sharp vibrational modes are observed at higher frequencies. The latter correspond to the normal modes of the molecular building blocks of the glass, SiO_2 in this case. The presence of both bulk and localized vibrational excitations in the glass allow interactions similar to the phonon—ion (lattice—orbit) processes in crystals

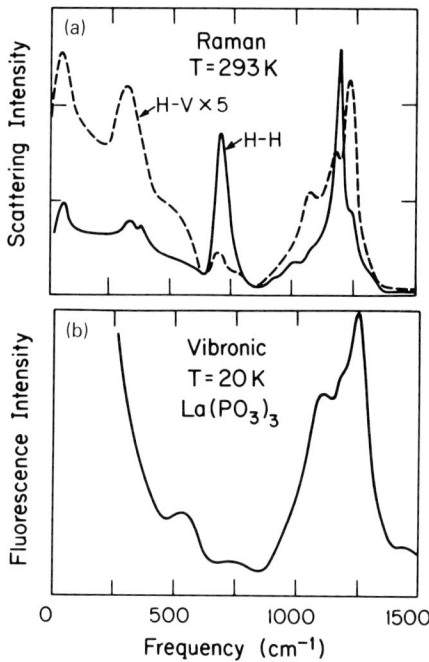

Fig. 11. (a) Polarized Raman spectra of a metaphosphate glass showing broad phonon scattering bands and molecular vibrations from the phosphate groups. (b) Vibronic sideband accompanying a FLN fluorescence of the $^6P_{7/2} \rightarrow {}^8S_{7/2}$ transition of Gd^{3+} when the glass is doped with this ion. From Hall et al. [71].

to occur. Thermalization within level distributions, various forms of non-radiative relaxation and quenching and energy transferral and diffusion are all observed in glasses and may be attributed to some form of phonon/vibrational mediation. These processes have been reviewed in a number of places for ions in crystals and in glasses [36, 72, 73].

The number of states of a given ion in a solid which are metastable or which fluoresce are limited because of non-radiative relaxation which can efficiently dissipate the energy between adjacent levels. It has been long established that this relaxation rate follows a gap law [74, 75]

$$W_{nr} = W_0(T)\exp(-\alpha\,\Delta E) \tag{4}$$

in the case of 4f ions in crystals. In Equation (4), $W_0(T)$ depends on the host crystal and on temperature, α is a parameter related to the maximum phonon energy available in the lattice and ΔE is the gap between two levels [76, 77]. An analogous expression has also been derived for the case of stronger lattice coupling. Figure 12 shows that the non-radiative relaxation rates in glasses also obey this relation, except that in the case of glasses the high-frequency molecular vibrations appear to play the principal role in producing relaxation [78]. As the non-radiative rates are inversely dependent on the number of vibrations which

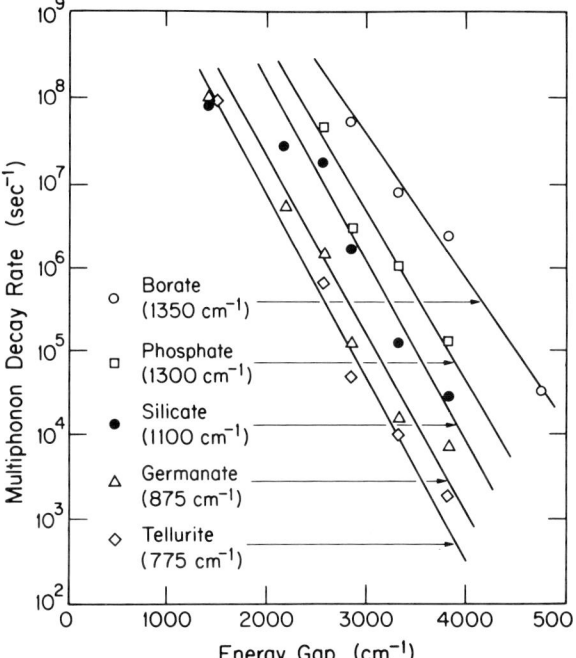

Fig. 12. Multiphonon or non-radiative relaxation rates as a function of intermanifold energy gap for a series of 4f states in a sequence of glass compositions. The decay rates are seen to obey an exponential gap law as discussed in the text. Highest optical phonon/vibrational energy are shown in the figure; generally because of the existance of these high energy excitations, these decay rates of ions in glasses are faster. From Layne et al. [78].

bridge the gap and as the molecular vibrations are of higher energy than ordinary crystalline phonons, non-radiative rates are faster in glasses than in crystals. It follows also that the number of emitting states are fewer in a glass than they are in a crystal, given the same ion of course.

In the case of transition metal ions additional factors must be taken into consideration. In these cases the coupling strength of the phonons to the various electronic states may vary and phonon overlap factors resulting from the configurational coordinate models used are affected strongly by anharmonicities in the lattice [54, 76]. Similar considerations have not been extended to the relaxation of transition metal ions in glasses. It is possible for example that such factors may be responsible for the notoriously low quantum efficiency of Cr^{3+} glasses [55, 79, 80].

Since molecular vibrations of constituents have been shown to be effective in producing relaxation, it should then come as no surprise that other molecular contaminants in a glass matrix also help to decrease the overall quantum efficiency of emitting ions in these systems. The OH^- molecular vibrations have been found to be the most common source of this additional de-excitation channel.

As the concentration of optically active ions is increased or as other ion species are introduced into the glass, then the possibility of two like or unlike ions find themselves near enough to each other to interact increases. In certain cases the interaction may be strong enough that distinct pairs or higher complexes may be formed. Though the appearance of dimer and more complex spectra is known in crystals as well as in organic solids, no such spectra has been reported in inorganic glasses. This is most likely because interionic coupling is generally weak compared to the inhomogeneous broadening. The presence of interactions leads to the non-radiative transfer of excitation from an excited (donor) to an unexcited center. Transfer of this type is responsible for a host of phenomena observed in crystals as well as in disordered materials.

Figure 13 for example illustrates the common phenomena of concentration quenching in Nd^{3+} in various glasses. In such cases, the excitation residing in one ion ends up being shared at some intermediate states with an originally unexcited neighbor. The intermediate states then dissipate their energy non-radiatively reducing the luminescence efficiency of the material [81].

A similar process is involved in sensitization of a secondary radiation center. In this case, optical energy is introduced into an active ion. This energy is transferred to the radiation center and the latter emits its own characteristic luminescence

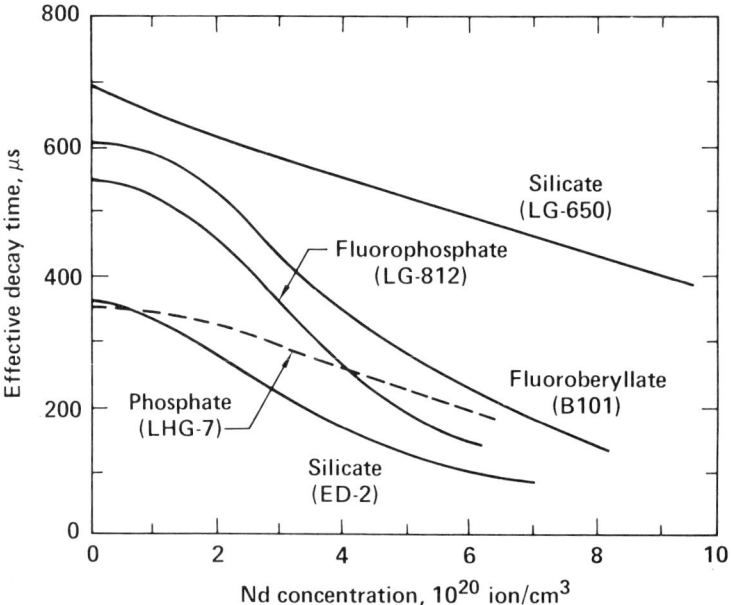

Fig. 13. First e-folding time behavior of the $^4F_{3/2}$ lasering state of Nd^{3+} in glasses of different compositions. The observed decay contains contributions from radiative decay and from a non-radiative transfer component arising from cross-relaxation to other Nd^{3+}. Approximately $1/\tau_{obs} = 1/\tau_{rad} + 1/\tau_{nr}$. Rapid decay here is known as concentration quenching and scales in most instances as the square of the Nd concentration. From Stokowski [59].

[82, 83]. This scheme is generally used to increase the pumping range of solid state lasers in the ultraviolet, for example. An example of sensitized luminescence of Eu^{3+} is shown in Figure 14, where both non-radiative and radiative transfers have been observed to occur [84].

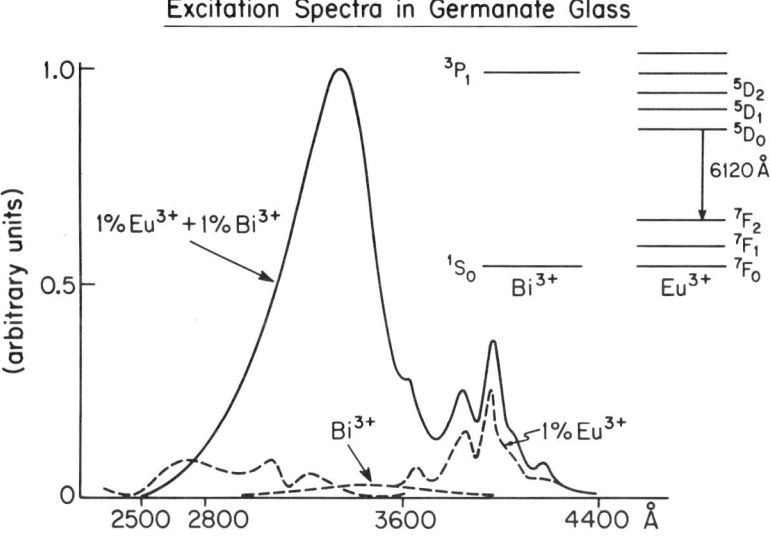

Fig. 14. An example of sensitized luminescence in a Eu^{3+} doped Germanate glass. In these experiments the 612.0 nm $^5D_0 \to {}^7F_2$ fluorescence of Eu^{3+} is monitored as a function of excitation energy in a sample doped with Eu and with Bi^{3+}. The absorption bands of the singly doped glasses are given by the dashed lines; transfer from Bi^{3+} to Eu^{3+} is evident as pumping into the Bi^{3+} absorption produces $^5D_0$ fluorescence. This system also shows radiative transfer. From Boulon et al. [84].

In insulating solids the interaction leading to the diffusion of energy is understood to arise from the fields produced by the excited ion or by an exchange interaction [85, 86]. In most cases these interactions have to be mediated by lattice excitations in order to preserve energy conservation [87, 88]. Various procedures and models exist in order to translate the parametric dependence of the interionic interaction to the observables in the system. The latter entail measurements such as the manner and the rate at which the excited state populations decay. Invariably the translation from microscopic to macroscopic entails averaging procedures often over random variables. In addition the macroscopic behavior depends on the relative rates between donor—donor and donor—acceptor transfer. In general this is a complex problem which has only begun to be resolved recently [89].

The occurrence of transfer of various types is relatively easy to observe in activated glasses. There is a large volume of literature evincing transfer in single and codoped systems with a myriad of compositions [8, 70, 72]. The bulk of these studies has been conducted using conventional spectroscopic techniques which in most instances probe the whole distribution of sites in the glass. As in the case of the interpretation of the static spectra of glasses, such inhomogeneous measurements yield some bulk averaged result which at times carries questionable weight in detailing the actual process taking place.

4. Laser Spectroscopy of Ions in Glasses

The advent of tunable lasers has had an immediate and significant impact of various aspects of optical spectroscopy [39]. These lasers have provided us with a versatile and convenient, and in certain cases a unique source of radiation with which to conduct experiments. The properties of high-frequency and high-temporal resolution as well as the properties of high power and coherence of laser radiation can all be utilized advantageously for spectroscopic purposes. Of special interest to us here are the laser spectroscopic techniques which have evolved and which allow us to suppress purely inhomogeneous contributions from the spectrum [90]. Fluorescence line narrowing (FLN) and its time-resolved pulsed version have proven to be the most common techniques in the study of ions in glasses. More detailed reviews of the techniques as well as their application to glass studies are also to be found in the literature [9].

4.1. STATIC SPECTROSCOPIC STUDIES AND STRUCTURE

The suppression of the large inhomogeneous contributions arising from the disordered glass structure implies that one has the means to obtain site specific information and thus the ability to probe into the microscopic structure of the glass in the vicinity of the active center. To demonstrate that suppression of inhomogeneously broadened spectra or spectral narrowing is achievable simply requires an excitation source with a frequency width narrower than the inhomogeneous width produced by the disordered glass distributions. This is not a very stringent requirement and indeed the first demonstration of what has become known as FLN was achieved using incoherent sources [91]. Tunable lasers, however, because of the higher resolution and power have made laser spectroscopic techniques applicable to a wide range of glass systems [92, 93].

In FLN, a narrowband source is simply used to excite a selected portion of an inhomogeneous distribution. In the limit of no interactions other than with the radiation field, the fluorescence in this case will arise only from that subset of ions which is in resonance with the source, the emission spectrum is said to be narrowed when the fluorescence and excitation are in resonance, the FLN signal will have a spectral width which is a convolution of the intrinsic homogeneous width of an average ion in the excited subset and the instrumental width, i.e. source, detection system etc. [94]. For FLN in glasses some additional factors must be taken into account as the disorder broadening is such that different subsets may be in resonance at once [9]. Further, extreme narrowing is observed only in the resonant fluorescence, emission involving intermediate state show varying degrees of narrowing which depend on the 'accidental coincidence' effect [95].

FLN using lasers was first demonstrated by Riseberg [92] and by Motegi and Shionoya [93]. Since then these techniques have been applied extensively to the study of principally $4f$ activated glasses. The most comprehensive studies have occurred in glasses containing Eu^{3+} and Yb^{3+} [96, 97] because of their simple energy level structure and Nd^{3+} doped materials because of their importance as laser sources [98].

In experiments on Eu^{3+} containing glasses, for example, the $^7J_0 \rightarrow {}^5D_0$ inhomogeneously broadened transition is selective excited and the resulting narrowed fluorescence to the ground state ($^7F_0$, weak) and to the $^7F_1$ and $^7F_2$ excited manifolds are observed [93, 99]. In the low glass symmetries, the $^7F_1$ and $^7F_2$ split into the three and five components respectively. As Figure 15(a) shows for a Eu^{3+} doped silicate glass, dramatic changes occur in the FLN spectra as the laser is tuned across the inhomogeneous $^5D_0$ absorption. These spectral changes are obviously connected with large site to site variations in the local electrostatic fields at the Eu^{3+} sites. Since the manifold splittings are functions of the crystalline parameters, the energy level diagrams obtained through FLN may be used to derive insights on the microscopic structure and bonding of the glass.

Brecher and Riseberg [99] have derived level assignments and the appropriate crystalline field parameters for the glass shown in Figure 15. These are shown in Figure 15(b). The variations in the even B_q^k required to describe the sites in these glasses encompass the whole range of values which are encountered in crystals. Using these parameters and placing reasonable physical constraints such as packing densities, they have derived geometrical models of the nearest neighbor environment of the Eu^{3+} ion in oxide and fluoride glasses. The coordination derived from above is shown in Figure 16. Calculated values of the parameters B_q^k for this structure are shown as dashes in Figure 15. Though only the first shell of neighbors is considered and the cell cannot be shown to be unique, these studies indicate that it is possible to get at these microscopic details of disordered structures.

Parallel efforts in computer simulation of the atomic structure of glasses have been made by Brawer [100–103] and others [104, 105] using Monte Carlo and molecular dynamics techniques. Molecular dynamics simulations are carried out by assuming a finite number of glass constituent molecules having a characteristic temperature dependent velocity and which interact with each with an assumed set of potentials. The motion and trajectories of this collection of particles are followed first by increasing the temperature to randomize the system and then by reducing it to quench the particles into a disordered structure. Each gas or fluid starting 'melt' results in a different glass configuration. Many such simulations are carried out, the collection of which is then assumed to represent the range of local environments present in an actual glass. Three examples of simulated structures for a system of a single Eu^{3+} ion in a BeF_2 glass are shown in Figure 17. These structures can then be tested against measurements such as X-ray or neutron diffraction to determine, for example, average anion—cation separations and

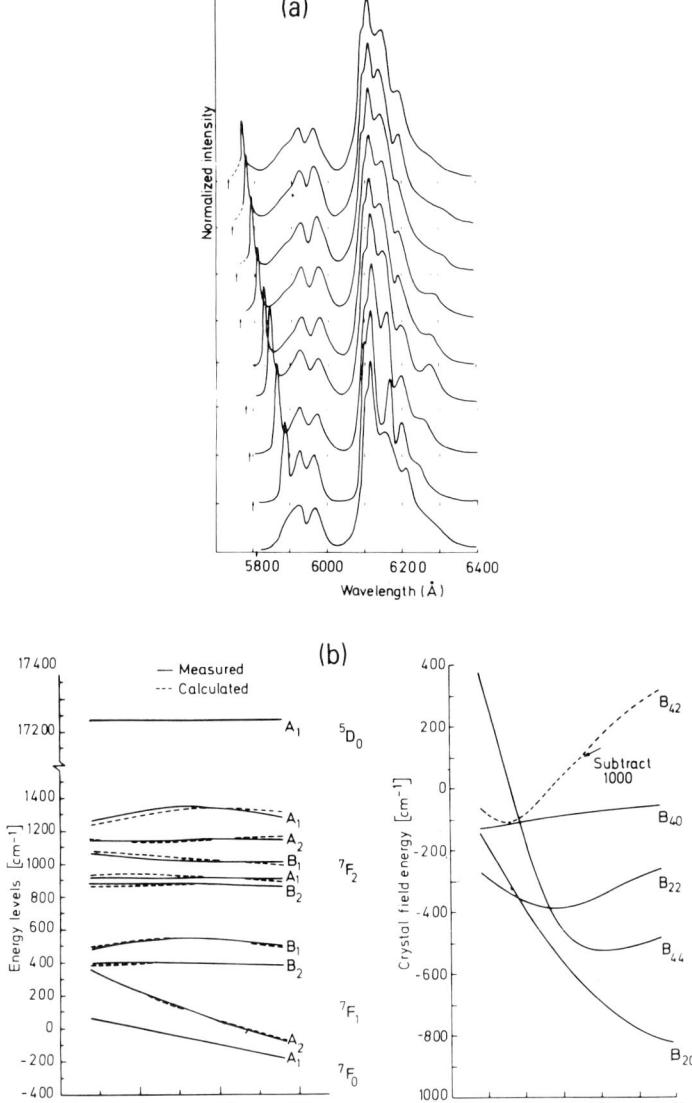

Fig. 15. (a) FLN spectra of Eu^{3+} in a silicate glass showing the large distributions of fields affecting the $^5D_0 \to {}^7F_1$ transitions. The arrow points at the laster excitation wavelengths within the inhomogeneously broadened absorption $^7F_0 \to {}^5D_0$ in this glass. (b) Variations of the $^7F_0$, $^7F_1$ and $^7F_2$ observed in the silicate glass with $^5D_0$ energy held as a constant. B_q^k calculated to yield the ^7F variations are shown to the right. The variation of the A_2 level of $^7F_1$ (C_{2v} symmetry) is due to the large B_{44} variation as a function of $^5D_0$ pumping energy. From Brecher and Riseberg [13].

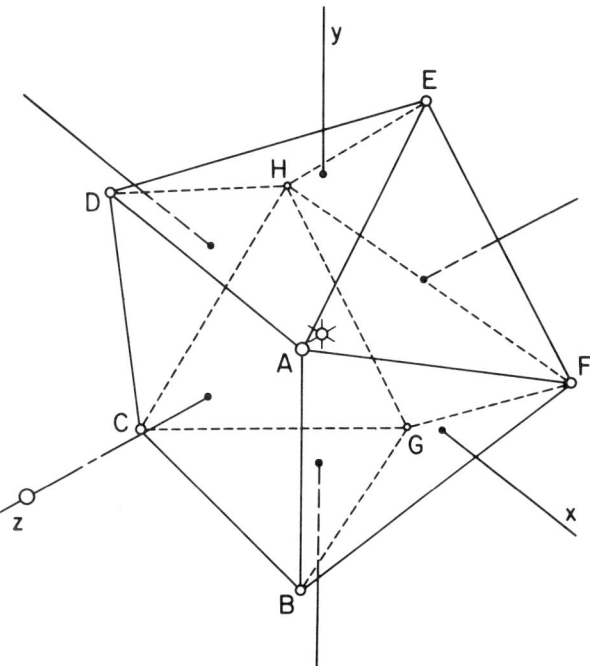

Fig. 16. Geometric model derived from field diagrams as those shown in Figure 15(b) and from some reasonable glass packing and density requirements. Result shown above is for an oxide such as the silicate glass. The Eu^{3+} sits at the center of this structure with a principal coordination of eight equidistant oxygens. A ninth oxygen (I) introduced along the z-axis distorts this structure by enlarging the *ABVD* area and by streching the *EFGH* plane towards negative z-values. From Brecher and Riseberg [99] and Weber [9].

coordination numbers. Though some discrepancies exist in details, the overall agreement between simulation and measurements is quite impressive specially in view of the limitations inherent in the finite size of the set and the two body potentials used. The splitting of the Eu^{3+} ion in BeF_2 glass have been calculated for a set of 159 different configurations, the results are shown in Figure 18. Comparing these with the FLN shown in Figure 15 indicates that the $^7F_1$ is in this case behaving in roughly the correct manner [14].

Once again the structural determinations which derive from simulations have well-understood limitations and once again the distributions are not unique. Nevertheless the information which the simulations can generate potentially can impact various aspects of disordered structures including the dynamics or excitations of these structures [102, 103].

In this area the rare earths have served us well principally because of the less pronounced effects due to the crystalline field. Likely because the opposite is true in $3d$ and heavier ion spectra, the FLN studies carried out in the latter systems

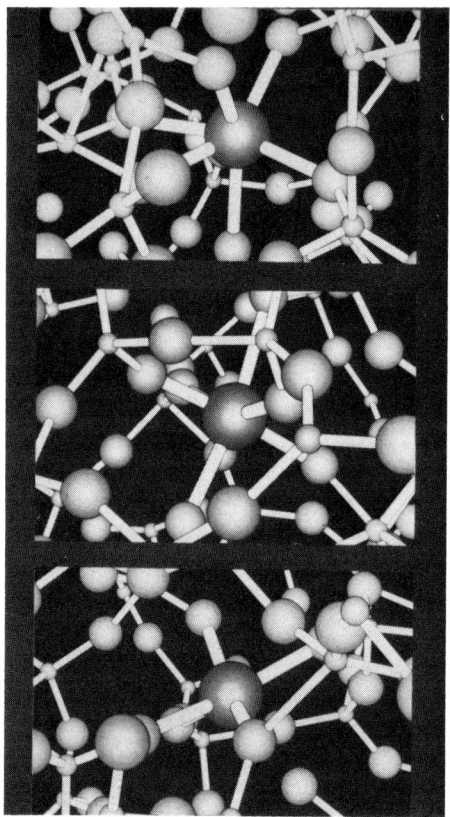

Fig. 17. Computer graphic display of three $4f$ ion sites in a BeF_2 glass obtained through a molecular dynamic simulation. The large spheres in the center represents a rare earth ion, the smallest spheres are beryllium ions coordinated with fluorine ions, 5, 6 and 7 coordinated $4f$ ions are shown in the sequence. [18].

have not been as comprehensive nor have they yielded the wealth of detail achieved in the former or $4f$ systems. Nevertheless, a recent study of Cr^{3+} doped glasses has allowed the deconvolution of sites with fields above and below the crossover between the ^2E and $^4T_2$ states [55]. As can be seen from Figure 19, states above the crossover emit a sharp line appropriate for a $^2E \rightarrow {}^4A_1$ transition. Results of this type will be important in resolving the behavior of these systems.

Complementary methods of probing into inhomogeneously broadened distributions are those techniques which are generically known as 'hole-burning' [106–108]. These methods have not been used extensively in glass studies except in the context of depopulating by stimulation a portion of an inverted line and in the context of persistent spectral hole-burning. Both have been restricted to rare earth glasses. In some ways, these absorptive methods are simpler to interpret as generally only two levels are involved.

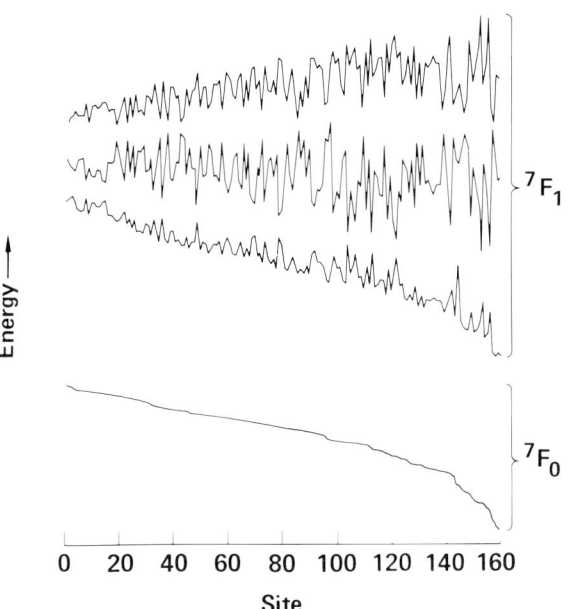

Fig. 18. $^7F_0$ and $^7F_1$ states of Eu^{3+} obtained through 159 different simulated BeF_2 glass configurations using a point charge model. The general behavior calculated here is to be compared to those shown in Figure 15(b) [18].

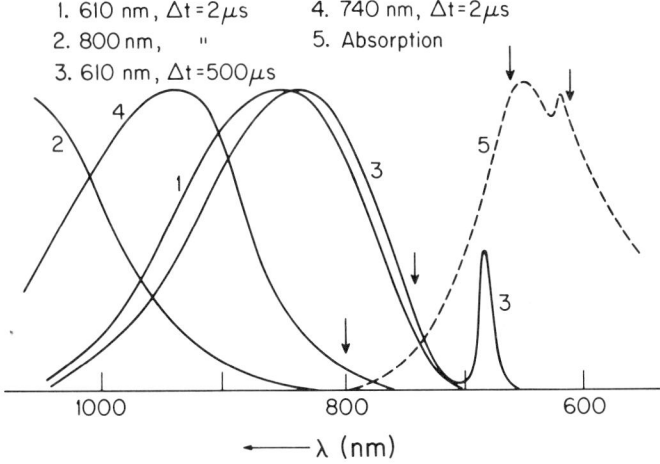

Fig. 19. FLN of Cr^{3+} in a phosphate glass, absorption band is shown as a dashed line with higher energy pumping (610 nm) sites with ^2E states lowest, i.e. higher field configurations, are selectively excited and a resonably sharp ^2E emission is observed, in this case resulting from pumping of a vibronic sideband. As lower field configurations are excited only broad band Stokes shifted $^4T_2$ emission is observed. From Glynn et al. [55].

4.2. RADIATIVE AND NON-RADIATIVE TRANSITIONS

As noted in the previous section, the radiative decay of metastable states of ions in glasses is invariably highly non-exponential. This results not only from the intrinsic changes which are caused by changes of the crystalline field parameters within the distribution of sites but also because the coupling with the lattice may vary, and in the case of higher concentrations, transfer may also be site dependent. In the weak concentration limit, FLN experiments may be used to determine the radiative decay rates of selectively chosen subsets of ions within the inhomogeneous envelopes. When this is done, the lifetimes become pure exponentials or nearly so and additional parametric information may be derived from the transition strengths. Variations of lifetimes of factors of 2 to 5 across the broadened profile are encountered in $4f$ glass systems.

Results of a study on Nd^{3+} glass by Brecher et al. [98] are shown in Figure 20. These workers used FLN and quantum efficiency measurements to deconvolute the effects due to nonradiative relaxation. As can be seen from the figure, the latter is also site dependent as one would have expected. The site os site variation of the phonon—ion coupling is much more pronounced in the d-series of ions consistent with our earlier discussions. In the case of Mo^{3+}, a $4d$ ion equivalent to Cr_{3+}, the FLN decays remain non-exponential and vary in their first e-folding by almost one order of magnitude across the ^2E inhomogeneous width [41]. Both the non-exponential decays and the observed lifetime variations may be accounted for by variations in the ion to lattice coupling. Similar variations are thought to affect the ^2E, $^4T_2$ state of Cr^{3+} glasses [54].

It has been generally accepted that stronger ion to lattice coupling leads to stronger vibronic spectra. This is certainly the case in many features of transition metal spectra in crystal and in glasses. The coupling a more pronounced in the heavier elements and the Mo^{3+} transitions mentioned above are thought to be mostly vibronic. A consequence of phonon—ion coupling is the formation of vibronic sidebands. Some of the structure appearing in the FLN Cr^{3+} spectra of Figure 19 may be attributed to sidebands of the relatively sharp ^2E transitions when they are selectively excited. The $4f$ ions are then, by the same reasoning, expected to produce much weaker sidebands, such sidebands have been observed in Gd^{3+} in a fluorophosphate glass and are shown in Figure 11(b), these are to be compared with the Raman spectra for the same glass shown in Figure 11(a) [71].

4.3. THERMALIZATION AND HOMOGENEOUS LINEWIDTHS

Given the appropriate experimental circumstances, intrinsic properties of representative ions within a preselected monochromat of an inhomogeneously broadened line may be extracted through FLN and other laser spectroscopic techniques. It has been long established that transition widths of ions in crystals are determined by ion—lattice interactions which in turn lead to thermalization [109]. This is, of course, in the absence of interionic interactions.

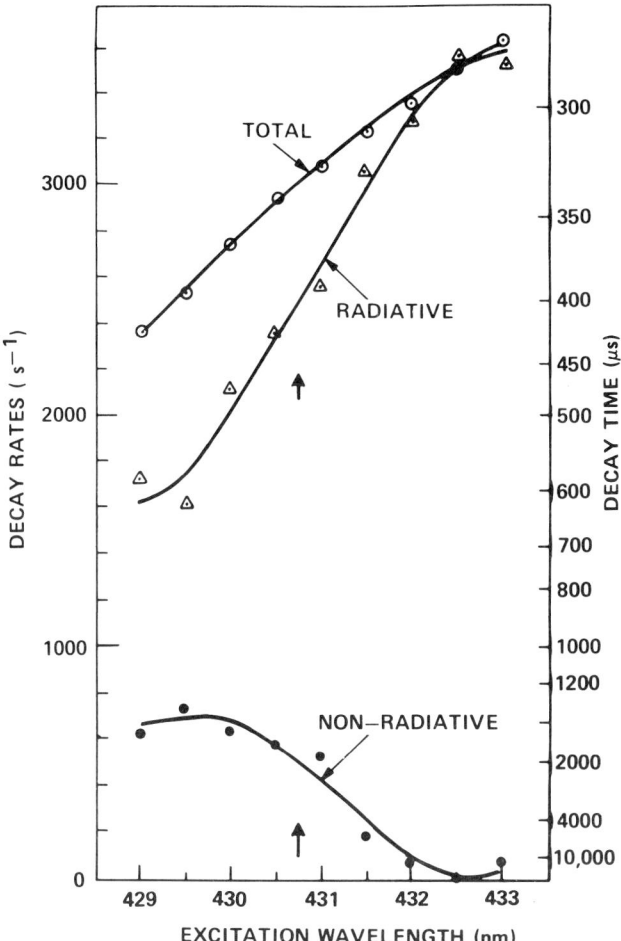

Fig. 20. Observed variation of the lifetimes of the $^4F_{3/2}$ state in a Nd^{3+} doped silicate glass upper trace when pumping across the inhomogeneously broadened absorption. Radiative and non-radiative contributions were deconvoluted using quantum efficiency measurements and indicate the effects of the field distributions and the changing phonon—ion coupling strengths respectively. From Brecher et al. [98].

The investigation of the linewidth of ions in glasses through FLN principally in rare earth systems produced some surprising results which have led to considerable theoretical activity and which have been responsible for the interest in amorphous organic systems [110, 111]. Most of this work is being reviewed in this volume and hence is not repeated here. Suffice it once again to delve briefly into the salient features of the doped glasses.

In FLN, studies in a variety of systems containing rare earths yielded linewidth results all of which evinced a T^n dependence, where T is the glass temperature and n is a fractional power ~ 2. In addition, at moderately low temperatures

(~5 K) there appears to be an anomalously large dephasing rate as expressed in the linewidth of the ions in glass and compared to ions in similarly coordinated crystals. In correlated work, Pellegrino et al. [112] also found that the homogeneous linewidths scaled as $\bar{v}_s^{-2.6}$, where $\bar{v}_s$ is the average velocity of sound in the glass, this is similar to the behavior of ions in crystals [113]. Though the disordered excitations or two levels systems (TLS) were suspected to play a part in the relaxation process [110, 114], it became difficult to differentiate the observed temperature dependence from normal Raman scattering of phonons provided a low effective Debye temperature is assumed [115]. The task was then reduced to seeking a crossover temperature dependence, from $T^7 \to T^2$ if the normal Raman process is effective or from $T^{n'} \to T^2$, where n' approximates 1 if TLS played a role in the relaxation.

Recently two types of studies have been conducted which shed additional light on the subject and which also indicate that additional work is required to resolve this problem in disordered systems. Macfarlane and Shelby [116] and Shelby [117] have measured the $Eu^{3+}[^7F_0 \leftrightarrow {}^5D_0]$ and the $Pr^{3+}[^3H_6 \leftrightarrow {}^1D_2]$ linewidths in glass compositions used in the FLN measurements. These workers used the techniques of absorptive hole burning and of cumulative photon echoes [118]. In the first instance, they obtained a measure of the linewidth of Eu^{3+} at 1.6 K which is consistent with the value extrapolated from the higher temperature values of Selzer et al. [110]. This result in conjunction with others implies that the $^5D_0$ width scales as $T^{1.8}$ over almost three decades in the temperature. On the other hand, the $^1D_2$ state of Pr^{3+} showed a T^1 dependence in contrast to the T^2 dependence reported by Hegarty and Yen [111] in the $^3P_0$ state. The results of Macfarlane would seem to imply that a cross over behavior to T^2 might be expected in this state of Pr^{3+}, unfortunately, it appears to be difficult to burn a hole at higher temperatures than those reported. Additionally in this case it is not possible to conduct analogous FLN in the $^1D_2$ state because this state shows a very rapid temperature dependence in fluorescence quenching. It is of some physical as well as technical interest that the holes burned in these measurements were persistent and were accompanied by 'antiholes', implying some configuration rearrangement of the active centers. This phenomena in inorganic systems is not thoroughly understood.

The linewidths in Nd^{3+} silicate glass fibers similar in composition to one studied by Pellegrino et al. [112] were studied by Hegarty et al. [119] at very low temperatures using a very unique geometry for photon echoes. Their results are summarized in Figure 21 along with studies of the same matrix using phonon echoes and phonon saturation [120]. the dephasing time is seen to scale as $T^{1.3}$, this implied that a crossover behavior of the linewidths between 1 and ~20 K, the lowest temperatures used by Pellegrino. The dephasing dependence has been analyzed by Huber et al. [121] and it is tentatively concluded that electric dipole–dipole interactions between rare earth ions and TLS can satisfactorily account for the observed T dependence in this region. A model invoking fractons has also been proposed by Lyo (1982), and though fractals seem to be appropriate for certain etched glasses [122] it is presently not clear whether they impact common glasses as were used in these studies.

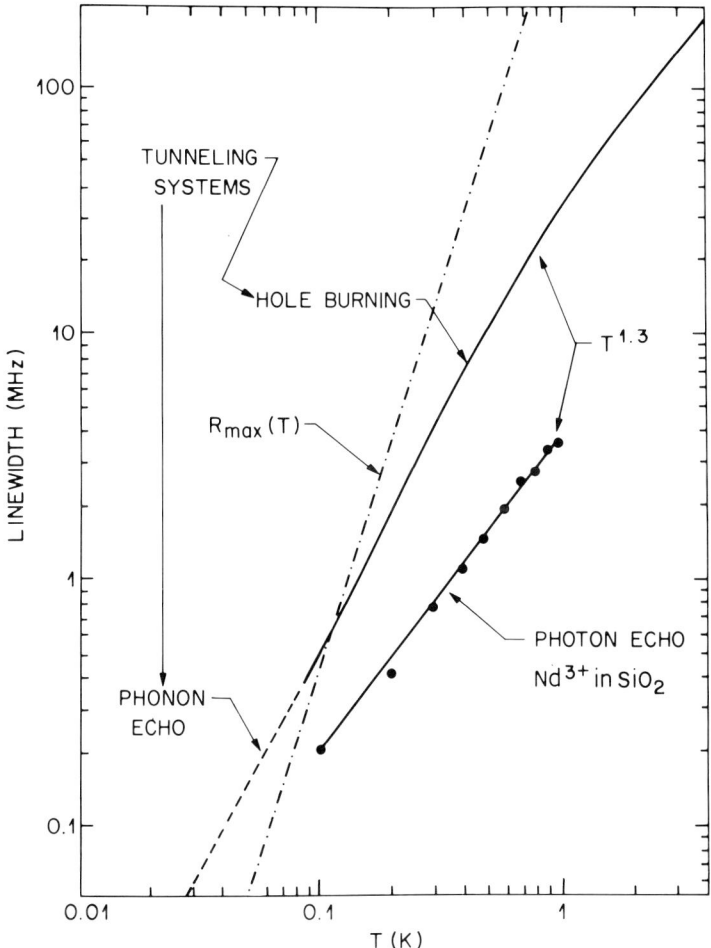

Fig. 21. Temperature dependence of the $^4F_{3/2}$ state linewidths of Nd^{3+} in a silicate glass at ultralow temperatures as measured by photon echoes. Composition of glass used here was analogous to those used by Pellegrino et al. [112] (1980) and the $T^{1.3}$ dependence shown implies there is a turn over to the T^2 linewidth dependence observed at higher temperatures. Optical linewidth results are shown alongside those obtained through ultrasonic methods. The line R_{max} defines the maximum T_1's attainable through TLS and is discussed by Huber et al. [121]. See also Hegarty et al. [119].

Though clearly steady progress has been made in this area, a number of puzzles still remain. It is apparent that a comprehensive study of the temperature dependence in the complete temperature range using a single sample and complementary measuring techniques is required at this time. Among others such a study would confirm whether coherent and FLN methods yield equivalent linewidths [133].

We note in passing that the understanding of this specific linewidth problem is of considerable interest to laser technology for this quantity ultimately plays a role in the energy extraction of inverted glass laser/amplifier systems [59, 123].

4.4. ENERGY TRANSFER OF OPTICAL EXCITATION IN GLASSES

As already noted, if the concentration of ions is increased additional dynamical effects occur which originate in the interionic interactions. Invariably these interactions result in the transferral of optical energy as reflected in the richness of phenomena observed in optically active solids. All these effects have of course been well established in crystals and their experimental observation in glasses is well documented [70, 83].

The preponderant majority of transfer studies in glasses including some comprehensive studies have been done using conventional as distinct from laser techniques. It has been pointed out that such measurements suffer from certain shortcomings connected with the inability of conventional techniques in providing a measure of processes occurring in like ion systems easily [124]. The time-resolved version of FLN, TRFLN, done within an inhomogeneously broadened line, has remedied this situation [125]. Hegarty et al. [126, 127] have shown that a combination of TRFLN within the donor state coupled with conventional measurements of the $D-A$ dynamics is required to resolve the problem completely.

Though beginning with Motegi and Shionoya [93], several studies of TRFLN in the inhomogeneously broadened transitions in rare earth systems have been conducted, the analysis of the dynamics occurring in the spectral evolution have not been done in the same detail as those carried out in similarly doped crystals [97–128]. This is because in order to obtain an appropriate theoretical model for transfer within a line, a myriad of factors such as number densities within the distribution, variations in transition strength, etc., connected with the random glass distributions must be folded into the averages which determine the macroscopic behavior of the donor–acceptor system. Such a complete treatment remains to be done [9].

Figure 22 shows the TRFLN dynamics observed in Yb^{3+} in silicate glass at low temperatures by Brundage and Yen [97]. The meaningful quantity to be analyzed in such traces is the ratio

$$R(t) = \frac{\text{FLN intensity at } t}{\text{Total intensity at } t};$$

perhaps reasonably this quantity has been analyzed in terms of the Inokuti–Hirayama [129] model and tentative conclusions have been drawn from the behavior regarding $D-D$ dynamics. As the majority of the conclusions remain unconfirmed, we will only summarize the experimental observations. It is apparent that the multipolar interactions as well as exchange effective in energy transfer in crystals are also responsible for the transfer of optical excitation in glasses; in addition, phonon mediation of these interactions is invariably present. From the asymmetry of the transfer rates, i.e. downward and upwards transfer in energy, it may be concluded that a one-phonon direct process is effective in the case of Yb^{3+} intraline transfer. As Holstein et al. [88] have pointed out, all of these phonon-mediated processes have characteristic temperature dependences, unfor-

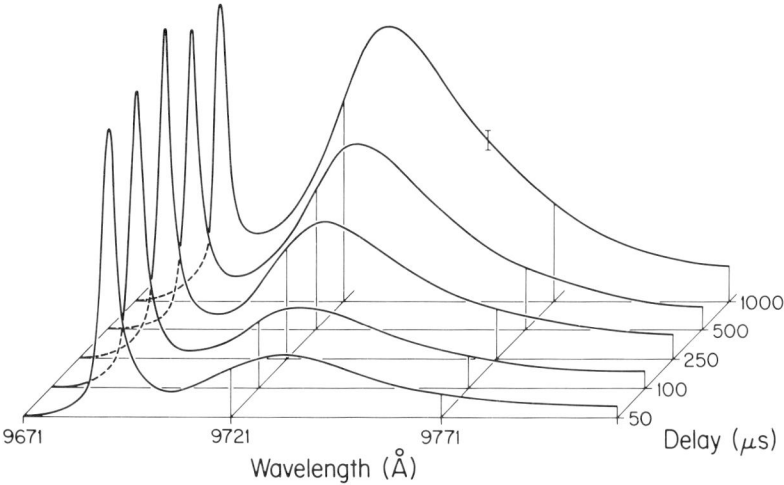

Fig. 22. TRFLN in the $^2F_{7/2} \to {}^2F_{5/2}$ transition of Yb^{3+} in a silicate glass. The laser excitation is a 969.4 nm, the temperature is 15 K. Fluorescence intensity is in number of photons and plots are given as a function of delay; results have been compensated for the radiative decay of the $^2F_{7/2}$ state. From Brundage and Yen [97].

tunately, it is only recently that a comprehensive study of this dependence has been attempted and the study is still in progress though preliminary conclusions have appeared [130].

Once again, sensitization, quenching and optical trapping have all been shown to occur in glasses. It also may be safely concluded that various phonon assisted multipolar and exchange interactions are responsible for the interionic transfer. Again here the analysis have been oversimplified as generally they have not taken into account the details of the variations inherent in the random distributions in glasses. As in the case of crystals, the $D-D$ process as reflected in TRFLN of an inhomogeneously broadened line must be unravelled first before the difficult yet interesting problem of unlike ion transfer in disordered systems can hope to be totally resolved.

5. Concluding Remarks

The diversity of glass systems is such that clearly no review other than an encylopedic one would do justice to this subject. Consequently the intention here has simply been to touch upon various subjects concerning the optical properties of inorganic glasses to impart a feeling of this area of studies to the reader. Basically, this review sought to illustrate that spectroscopic studies of glasses can be carried out conveniently and that the results obtained in these studies may be generalized and applied to other systems. In this, rare earth ions in particular have played an important role; as their interactions with external factors are weak but controllable, they have been used as probes for effects connected with disorder such as structure and dynamics. A comprehensive understanding of the latter, for example, in terms of energy transfer and relaxation would certainly have an impact on the general problem of transport in all amorphous systems.

There has been, in addition, recent interest in certain transparent ceramics which may be activated by ions [131]. In these materials, the separation of phases is such that crystalline domains are surrounded by glassy material. However the activators may be dissolved into the crystalline or into the glassy phases. Comprehensive studies of the optical properties of these ceramics are likely to be very interesting as they will elucidate the behavior of optical energy as it crosses an ordered to a disordered phase, or vice versa.

Understanding of the general properties of optically active glasses, finally, have immediate technical consequences. Proof of this lies in the fact that the activity in this whole area has been fueled in the past decade principally by the requirements of the technical community concerned with lasers. Though this motivation is expected to wane, the properties of glasses from a scientific point of view will remain an interesting and fundamental area to pursue as a host of problems still require full resolution.

Acknowledgments

The author's interest in this area may be traced to a period of interaction with the laser effort at Lawrence Livermore National Laboratories. As a result, a number of workers there nurtured my interest and influenced my outlook in this subject — these include Dr M. J. Weber, Drs. Stokowski and Dr S. Brawer. An expression of gratitude is due to them. Preparation of this review was performed under the auspices of the Materials Science Program of the U.S. Department of Energy. The author has also benefited from a A. v. Humboldt Foundation Senior U.S. Scientist Award.

References

1. R. W. Douglas and S. Frank, *A History of Glassmaking*, Fowles, Oxforshire (1972).
2. Anita Engle, *Readings in Glass History*, Vols 6/7, Phoenix Publications, Jerusalem (1967).
3. W. H. Zachariasen, *J. Amer. Chem. Soc.* **54**, 3844 (1932).
4. W. H. Zachariasen, *J. Chem. Phys.* **3**, 162 (1935).
5. L. M. Mollenauer, Ph.D. Thesis, M. L. Report, No. 1325, Stanford Univ. (1965).
6. J. Wong and C. A. Angell, *Glass: Structure by Spectroscopy*, Dekker, New York (1976).
7. W. A. Weyl, *Colored Glasses*, Dawson's of Pall Mall, London (1959).
8. K. Patek, *Glass Lasers*, Iliffe (Butterworth), London (1970).
9. M. J. Weber, 'Laser Excited Fluorescence Spectroscopy in Glass' in *Laser Spectroscopy of Solids, Topics in Applied Physics* (eds W. M. Yen and P. M. Selzer), Chap. 6, Springer-Verlag, Berlin (1981).
10. J. M. Stevels, 'The Structure and Physical Properties of Glass' in *Handbuch der Physik* (ed. S. Flügge), Vol. XIII, Springer-Verlag, Berlin (1962).

11. S. D. Stookey and R. D. Mauser, 'Glass' in *Handbook of Physics* (eds. E. U. Condon and H. Odishaw), McGraw-Hill, New York, pp. 8—85 *et seq.* (1961).
12. D. R. Uhlmann and H. Yinnon, 'The Formation of Glasses' in *Glass-Science and Technology* (eds D. R. Uhlmann and N. J. Kreidl), Vol. 1, Chap. 1, Academic Press, New York, (1983).
13. C. Brecher and L. A. Riseberg, *Phys. Rev.* **B21**, 2607 (1980).
14. M. J. Weber, *J. Non-Cryst. Solids* **47**, 117 (1982).
15. B. E. Warren, *Z. Kristallog.* **86**, 349 (1933).
16. B. E. Warren, H. Krutter, and O. Morningstar, *J. Amer. Ceram. Soc.* **19**, 202 (1936).
17. A. Bienestock, 'Radial Distribution Functions and EXAFS Studies of Amorphous Materials' in *The Structure of Non-Crystalline Materials* (ed. P. H. Gaskell), p. 5, Taylor and Francis, London (1977).
18. M. J. Weber, 'Computer Simulation of Rare Earth Sites in Glass' in *Proc. Int. Conf. on Defects in Insulating Crystals* (ed. F. Lüthi) (to be published).
19. F. Oberlies and A. Dietzel, *Glastechn. Ber.* **30**, 37 (1957).
20. G. H. Beall and D. A. Duke, 'Glass Ceramics Technology' in *Glass-Science and Technology*, V-1, (eds D. R. Uhlmann and N. J. Kreidl), Academic Press, New York (1983).
21. H. Bethe, *Ann. Physik.* **3**, 133 (1929).
22. J. H. Van Vleck, *Phys. Rev.* **41**, 208 (1932).
23. J. H. Van Vleck, *J. Chem. Phys.* **41**, 67 (1937).
24. D. S. McClure, 'Electronic Spectra of Molecules and Ions in Crystals', *Solid State Reprints*, Academic Press, New York (1959).
25. S. Sugano, Y. Tanabe, and H. Kamimura, *Multiplets of Transition Metal Ions in Crystals*, Academic Press, New York (1970).
26. B. R. Judd, *Operator Techniques in Atomic Spectroscopy*, McGraw-Hill, New York (1963).
27. G. H. Dieke, *Spectra and Energy Levels of Rare Earth Ions in Crystals*, Wiley-Interscience, New York (1968).
28. S. Hüfner, *Optical Spectra of Transparent Rare Earth Compounds*, Plenum Press, New York (1978).
29. J. Hessler and W. T. Carnall, 'Energy Levels of Trivalent Actinides in Crystals' in *Lanthanide and Actinide Chemistry and Spectroscopy* (ed. N. M. Edelstein). American Chemical Society, Washington, D.C. (1980).
30. C. J. Ballhausen, *Introduction to Liquid Field Theory*, McGraw-Hill, New York (1962).
31. A. R. C. Rodriguez, C. W. Parmelee, and A. E. Badger, *J. Am. Ceram. Soc.* **26**, 136 (1943).
32. G. Boulon, *Proc. Colloque Inter. CNRS* **255**, 295 (1977).
33. B. R. Judd, *Phys. Rev.* **127**, 750 (1962).
34. G. S. Ofelt, *J. Chem. Phys.* **37**, 511 (1962).
35. L. A. Riseberg and M. J. Weber, *Prog. Opt.* **14**, 91 (1976).
36. W. T. Carnall, H. Crosswhite, and H. M. Crosswhite, 'Energy Level Structure and Transition Probabilities of Trivalent Lanthanides in LaF_3', Argonne National Laboratory Report No. 60439, Argonne (unpublished) (1979).
37. Y. Tanabe and S. Sugano, *J. Phys. Soc. Japan* **9**, 766 (1954).
38. S. Sugano and Y. Tanabe, *J. Phys. Soc. Japan* **13**, 880 (1958).
39. W. M. Yen and P. M. Selzer, 'High Resolution Laser Spectroscopy of Ions in Crystals' in *Laser Spectroscopy of Solids, Topics in Applied Physics* (eds W. M. Yen and P. M. Selzer), Chap. 5, Springer-Verlag, Berlin (1981).
40. W. S. Heaps, L. R. Elias, and W. M. Yen, *Phys. Rev.* **B13**, 94 (1976) and references therein.
41. M. J. Weber, S. A. Brawer, and A. J. DeGrott, *Phys. Rev.* **B23**, 11(1981).
42. A. L. Schawlow, *Advances in Quantum Electronics* (ed. J. R. Singer), p. 50, Columbia Univ. Press, New York (1961).
43. G. F. Imbusch and R. Kopelman, 'Optical Spectroscopy of Electronic Centers in Solids' in *Laser Spectroscopy of Solids, Topics in Applied Physics* (eds W. M. Yen and P. M. Selzer), Chapter 1, Springer-Verlag, Berlin, (1981).
44. G. E. Rindone, 'Luminescence in the Glassy State' *Luminescence of Inorganic Solids* (ed. P. Goldberg), Academic Press, New York (1962).

45. K. Fajans, *Naturwiss.* **11**, 165 (1923).
46. S. Kato, *Sci. Pap. Inst. Phys. Chem. Res., Tokyo* **13**, 49 (1930).
47. A. Mead, *Trans. Faraday Soc.* **30**, 1052 (1934).
48. H. Hartmann and F. E. Ilse, *Z. Phys. Chem.* **197**, 239 (1951).
49. T. Bates and R. W. Douglas, *Trans. Soc. Glass Tech.* **43**, 289 (1959).
50. T. Bates, 'Ligan Field Theory and Absorption Spectra of Transition Metal Ions in Glasses' in *Modern Aspects of the Vitreous State* (ed. J. D. Macken), Vol. 2, Chap. 5, Butterworth, London (1962).
51. S. E. Stokowski, Private communication (1985).
52. C. R. Bamford, *Phys. Chem. Glasses* **3**, 189 (1962).
53. S. A. Brawer and W. B. White, *J. Chem. Phys.* **67**, 2043 (1977).
54. L. J. Andrews, A. Lempicki, and B. C. McCollum, *J. Chem. Phys.* **74** 5526 (1981).
55. T. J. Glynn, V. P. Gapontsev, and W. M. Yen, *J. Opt. Soc. Am.* **73**, 1391 (1983).
56. U. Fano, *Phys. Rev.* **124**, 1866 (1961).
57. M. D. Sturge, H. J. Guggenheim, and M. H. L. Pryce, *Phys. Rev.* **B2**, 2459 (1970).
58. A. L. Lempicki, S. J. Nettel, and B. C. McCollum, *Phys. Rev. Lett.* **44**, 1234 (1980).
59. S. E. Stokowski, 'Glass Lasers' in *Handbook of Laser Science and Technology* (ed. M. J. Weber), Vol. 1, p. 215, CRC Press, Boca Raton (1982).
60. V. P. Gapontsev, S. M. Matitsiu, A. A. Isineev, and V. B. Kravchenko, *Optics and Laser Technology*, p. 189, Butterworth, London (1982).
61. S. E. Stokowski, R. A. Saroyan, and M. J. Weber, 'Neodynium Doped Glass Spectroscopic and Physical Properties', Lawrence Livermore National Laboratory Report No. M-095 (2nd revision), Livermore (unpublished) (1981).
62. E. M. Levin and S. Block, *J. Am. Ceram. Soc.* **40**, 95; **40**, 113 (1957).
63. C. C. Robinson and J. T. Fournier, *Chem. Phys. Lett.* **3**, 517 (1969).
64. C. C. Robinson and J. T. Fournier, *J. Phys. Chem. Solids* **31**, 895 (1970).
65. D. K. Rice and L. G. DeShazer, *Phys. Rev.* **186**, 387 (1969).
66. M. M. Mann and L. G. DeShazer, *J. Appl. Phys.* **41**, 2951 (1970).
67. R. Reisfeld and U. Eckstein, *Appl. Phys. Lett.* **26**, 253 (1975).
68. R. R. Jacobs, C. B. Layne, M. J. Weber, and C. Rapp, *J. Appl. Phys.* **47**, 2020 (1976).
69. I. M. Buzhinskii, E. I. Koryagina, and S. K. Mamonov, *J. Appl. Spectrosc.* **22**, 250 (1975).
70. R. Reisfeld, 'Spectra and Energy Transfer of Rare Earths in Inorganic Glass' in *Structure and Bonding*, Vol. 13, p. 53, Springer-Verlag, Berlin (1973).
71. D. W. Hall, S. A. Brawer, and M. J. Weber, *Phys. Rev.* **B25**, 2828 (1981).
72. R. Reisfeld, 'Radiative and Non-radiative Transitions of Rare Earth Ions in Glasses' in *Structure and Bonding*, Vol. 22, p. 123, Springer, Verlag, Berlin (1975).
73. R. Reisfeld, 'Excited States and Energy Transfer from Donor Cations to Rare Earths in Condensed Phases' in *Structure and Bonding*, Vol. 30, p. 65, Springer-Verlag, Berlin (1976).
74. H. W. Moos, *J. Lumin.* **1/2**, 106 (1970).
75. M. J. Weber, *Phys. Rev.* **B8**, 54 (1973).
76. M. D. Sturge, *Phys. Rev.* **B8**, 6 (1973).
77. W. H. Fonger and C. W. Struck, *Phys. Rev.* **B11**, 3251 (1975).
78. C. B. Layne, W. H. Lowdermilk, and M. J. Weber, *Phys. Rev.* **B16**, 10 (1977).
79. E. Strauss and W. Seelert, *J. Lumin.* **31/32**, 191 (1985).
80. G. F. Imbusch, Private communication (1984).
81. B. I. Denker, V. V. Osiko, A. M. Prokhorov, and I. A. Sheherbakov, *Sov. J. Quantum. Electron.*, 485 (1978).
82. J. C. Wright, 'Radiationless Processes in Molecules and Condensed Phases' in *Topics in Applied Physics* (ed. F. K. Fonj), Chap. 4, Springer-Verlag, Berlin (1976).
83. W. M. Yen, 'Experimental Studies of Energy Transfer Glasses' in *Coherence and Energy Transfer in Glasses* (eds P. A. Fleury and B. Golding), Plenum Press, New York (1984).
84. G. Boulon, B. Moline, and Y. Kalisky, *Proc. Int. Conf. on Lasers '80* (ed. C. B. Collins), p. 365, STS Press, McClean (1981).
85. T. Föster, *Ann. Physik* **2**, 55 (1948).

86. D. L. Dexter, *J. Chem. Phys.* **21**, 836 (1953).
87. T. Miyakawa and D. L. Dexter, *Phys. Rev.* **B1**, 2961 (1970).
88. T. Holstein, S. K. Lyo, and R. Orbach, 'Excitation Transfer in Disordered Systems' in *Laser Spectroscopy of Solids, Topics in Applied Physics* (eds W. M. Yen and P. M. Selzer), Chap. 2, Springer-Verlag, Berlin (1981).
89. D. L. Huber, 'Dynamics of Incoherent Transfer' in *Laser Spectroscopy of Solids, Topics in Applied Physics* (eds W. M. Yen and P. M. Selzer), Chap. 3, Springer-Verlag, Berlin (1981).
90. P. M. Seltzer, 'General Techniques and Experimental Methods in Laser Spectroscopy of Solids' in *Laser Spectroscopy of Solids, Topics in Applied Physics* (eds W. M. Yen and P. M. Selzer), Chap. 4, Springer-Verlag, Berlin (1981).
91. Y. V. Denisov and V. A. Kizel, *Opt. Spectrosc.* **23**, 251 (1967).
92. L. A. Riseberg, *Phys. Rev. Lett.* **28**, 789 (1972).
93. N. Motegi and S. Shionoya, *J. Lumin.* **8**, 1 (1973).
94. J. Hegarty, R. T. Brundage, and W. M. Yen, *App. Opt.* **19**, 1889 (1980).
95. R. Flach, D. S. Hamilton, P. M. Selzer, and W. M. Yen: *Phys. Rev.* **B15**, 1248 (1977).
96. J. A. Paisner, S. S. Sussman, W. M. Yen, and M. J. Weber, *Bull. Am. Phys. Soc.* **20**, 447 (1975).
97. R. T. Brundage and W. M. Yen, *J. Lumin.* **31/32**, 827 (1985).
98. C. Brecher, L. A. Risberg, and M. J. Weber, *Phys. Rev.* **B18**, 5799 (1978).
99. C. Brecher and L. A. Riseberg, *Phys. Rev.* **B13**, 81 (1976).
100. S. A. Brawer, *J. Chem. Phys.* **72**, 4264 (1980).
101. S. A. Brawer, *J. Chem. Phys.* **75**, 3516 (1981).
102. S. A. Brawer, *Phys. Rev. Lett.* **46**, 778 (1981).
103. S. A. Brawer, *J. Chem. Phys.* **79**, 4539 (1983).
104. S. A. Brawer and M. J. Weber, *Phys. Rev. Lett.* **45**, 460 (1980).
105. S. A. Brawer and M. J. Weber, *J. Chem. Phys.* **75**, 3572 (1981).
106. S. A. Brawer and M. J. Weber, *App. Phys. Lett.* **35**, 31 (1979).
107. V. I. Nikitin, M. S. Soskin, and A. I. Khizhnyak, *Sov. Tech. Phys. Lett.* **2**, 64 (1976).
108. V. I. Nikitin, M. S. Soskin, and A. I. Khizhnyak, *Sov. Tech. Phys. Lett.* **3**, 5 (1977).
109. W. M. Yen, W. C. Scott, and A. L. Schawlow: *Phys. Rev.* **136**, A271 (1964).
110. P. M. Selzer, D. L. Huber, D. S. Hamilton, W. M. Yen, and M. J. Weber, *Phys. Rev. Lett.* **36**, 813 (1976).
111. J. Hegarty and W. M. Yen, *Phys. Rev. Lett.* **43**, 1126 (1979).
112. J. M. Pellegrino, W. M. Yen, and M. J. Weber, *J. Appl. Phys.* **51**, 6332 (1980).
113. J. M. Pellegrino and W. M. Yen; *Phys. Rev.* **B24**, 6719 (1981).
114. S. K. Lyo, *Phys. Rev. Lett.* **48**, 688 (1982).
115. D. L. Huber, *J. Non-Cryst. Solids* **51**, 241 (1982).
116. R. M. Macfarlane and R. M. Shelby, *Opt. Comm.* **45**, 46 (1983).
117. R. M. Shelby, *Opt. Letters* **8**, 88 (1983).
118. W. H. Hesseling and D. A. Wiersma, *Phys. Rev. Lett.* **43**, 991 (1979).
119. J. Hegarty, M. M. Broer, B. Golding, J. R. Simpson, and J. B. MacChesney, *Phys. Rev. Lett.* **51**, 2033 (1983).
120. B. Golding and J. E. Graebner, *Amorphous Solids* (ed. W. A. Phillips), Chap. 7, Springer-Verlag, Berlin (1981).
121. D. L. Huber, M. M. Broer, and B. Golding, *Phys. Rev. Lett.* **52**, 2281 (1984).
122. U. Even, K. Radamann, J. Jortner, M. Manor, and R. Reisfeld, *Phys. Rev. Lett.* **52**, 2164 (1984).
123. D. W. Hall, M. J. Weber, and R. T. Brundage, *J. Appl. Phys.* **55**, 2642 (1984).
124. W. M. Yen, 'Experimental Studies of Energy Transfer in 4f Ions in Crystals' in *Spectroscopy of Rare Earth Ions in Crystals* (eds R. M. Macfarlane and A. A. Kaplyanskii), North Holland, Amsterdam (1985).
125. W. M. Yen, *J. Lumin.* **18/19**, 639 (1979).
126. J. Hegarty, D. L. Huber, and W. M. Yen, *Phys. Rev.* **B23**, 6271 (1981).
127. J. Hegarty, D. L. Huber, and W. M. Yen, *Phys. Rev.* **B25**, 5638 (1982).

128. M. J. Weber, J. A. Paisner, S. S. Sussman, W. M. Yen, L. A. Risberg, and C. Brecher, *J. Lumin.* **12/13**, 737 (1976).
129. M. Inokuti and H. Hirayama, *J. Chem. Phys.* **43**, 1978 (1965).
130. R. T. Brundage, Ph.D. Thesis, Univ. of Wisconsin, unpublished (1985).
131. L. J. Andrews, B. C. McCollum, S. Stone, D. E. Guenther, G. J. Murphy, and A. Lempicki, 'Development of Materials for a Luminescent Solar Report', G.T.E. Laboratories Inc. Final Report, DOE/ER/04996—4, Waltham (unpublished) 1983.
132. Woodlief Thomas, Jr, *SPSE Handbook of Photographic Science and Engineering*, Wiley-Interscience, New York (1973).
133. R. T. Brundage and W. M. Yen, *Phys. Rev.* **B33**, 4436 (1986).

MODEL CALCULATION OF OPTICAL DEPHASING IN GLASSES*

P. REINEKER and K. KASSNER

Abteilung für Theoretische Physik, Universität Ulm, 7900 Ulm, Germany

1. Introduction

1.1. ANOMALOUS LOW-TEMPERATURE PROPERTIES OF AMORPHOUS SOLIDS — EXPERIMENTS AND INTERPRETATIONS

In 1971, Zeller and Pohl [2] presented experimental evidence that thermal conductivity and heat capacity of several insulating glasses show a behavior differing strikingly from that of their crystalline counterparts. Earlier measurements [3—5] which had already pointed at these facts had not found much notice. Before 1970, it was believed by most scientists that the low-temperature properties of glasses were similar to those of crystals, because the microscopic structure of a material should become unimportant with increasing phonon wavelength. The results of Zeller and Pohl — a linear temperature dependence of the (phonon part of the) heat capacity and a T^2 dependence of the thermal conductivity with an order of magnitude almost independent of materials — were therefore extremely interesting and triggered off a large number of experimental studies and theoretical explanations.

Within few years about ten different models [6—17] trying to explain the anomalies discovered were proposed. The experimental demonstration that ultrasonic absorption can be saturated in glasses [18, 19] basically decided the matter in favour of the tunneling model put forward by Anderson, Halperin, and Varma [9] and, independently, by Phillips [10]. This saturation could be explained through scattering of acoustic phonons from additional degrees of freedom in the glass, provided they were described in terms of two-level systems (TLSs). The tunneling systems postulated in [9, 10] just were such two-level systems. Since then most experimentalists have tried to interpret their data in the framework of this model. More details on the AHV—P model will be given in Section 2. In this place, we merely mention a few experiments, which either directly supported the model or else could be most successfully explained by it, and some theoretical developments.

Besides the linear temperature dependence [20, 21], a time dependence of the heat capacity, postulated by the tunneling model, was finally discovered after [23, 24] several fruitless attempts. Deviations from the linear temperature dependence demonstrated by Lasjaunias *et al.* [22] could be attributed to this time dependence. A logarithmic temperature dependence of the velocity of sound

---

* Based on a Ph.D. Thesis by K. Kassner [1].

(below 1.5 K) was interpreted in terms of resonant one-phonon processes whereas a maximum in the temperature dependence was explained by relaxation absorption resulting from the modulation of the TLS energy splitting by the phonon field. The same effects showed up in polymer glasses [27].

Experimental evidence for an interaction between the TLSs was found [28]. This interaction is a consequence of the elastic dipole nature of the TLSs; the influence of the dipole deformation field on the density of states was investigated theoretically [29].

Phonon echo measurements [30] yielded direct information on the strength of the TLS coupling to acoustic phonons. Because heat capacities and thermal conductivities calculated from acoustic measurements did not agree with direct experimental results, Black and Halperin [31] postulated additional anomalous two-level systems coupling weakly to the phonons but contributing to the specific heat. Dielectric measurements [26, 32] and thermal expansion data [33] were consistent with the model as well.

Whether the TLSs are intrinsic to glasses in the sense that their existence follows from the lack of long-range order, is still an open question. Early acoustic measurements on amorphous silicon [34] seemed to defeat this idea — no TLSs were found. In the meantime, however, even in this material — despite its four-fold covalent bond structure — two-level systems could be detected, although in minute concentration [35]. Thus it seems that the influence of chemical composition on the TLS density of states is larger than that of structure. The microscopic nature of the two state tunneling systems is not yet clear.

1.2. Optical low-temperature properties of glasses

The main problem with optical spectra of amorphous materials is their large inhomogeneous broadening. Several novel laser techniques allowed this problem to be overcome.

Two Russian groups [36, 37] were the first to gain detailed information from the inhomogeneous spectrum by **optical hole-burning**. In Figure 1, a schematic representation of the process is given. The upper part of the figure shows the inhomogeneous line which is a superposition of many homogeneous lines with different central frequencies. That these frequencies are shifted with respect to each other results from structural differences in the environment of the optically absorbing guest molecules. Irradiation of the material with a narrow-band laser leads to a selective excitation of guest molecules the central frequencies of which are (almost) at resonance. If the optical properties of the guest molecules change after the excitation, their contribution is removed from the optical spectrum and a 'hole' remains which may be scanned with a low-power laser.

There are three mechanisms of hole-burning [38]. Photochemical hole-burning [37, 39] means that the spectrum of excited guest molecules is modified in consequence of a chemical reaction triggered by the excitation. Non-photochemical hole-burning [40] which is characteristic for amorphous systems is based on

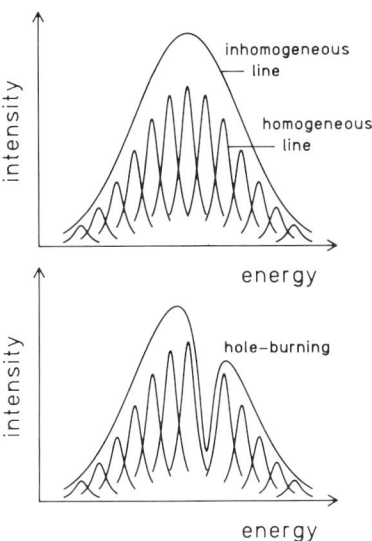

Fig. 1. Inhomogeneous line originating from homogeneous lines with different central frequencies. Schematic representation of a hole burnt into the inhomogeneous line.

reorientation processes (producing frequency shifts) in the environment of a molecule. In the case of transient hole-burning [41, 42] the excited molecules go into a metastable (e.g. triplet) state; the lifetime of holes is much smaller (microseconds to seconds) than in the other two cases (hours to weeks). Details of the mechanisms becoming effective during hole-burning have been investigated [43—48].

In addition to hole-burning experiments [49, 50] the technique of fluorescence line narrowing (FLN) is used to study the temperature dependence of homogeneous linewidths. FLN was first applied to glasses by Selzer et al. [51]. Experiments of this kind on rare earth ions doped into inorganic glasses yielded a quadratic temperature dependence between 8 and 300 K [52] in contrast to a T^7 power law (up to 50 K) in a crystal. Homogeneous linewidths in glasses [53—55] were one to two orders of magnitude larger than in crystals at low temperatures.

Shelby [56] was the first to measure homogeneous linewidths in glasses with a photon echo technique. The temperature dependences obtained in these experiments varied between $T^{1.0}$ and $T^{2.2}$ [49, 50, 55, 56] and usually were valid in the whole range of temperature considered.

Since 1982 a series of hole-burning experiments have been performed by Völker et al. [42, 57—61] in which a power broadening of the holes could be most certainly avoided by use of sufficiently low laser powers. These measurements on organic glasses resulted in extremely small linewidths (the Heisenberg-limit was reached in several cases) and a rather universal temperature exponent of 1.3 describing the homogeneous linewidth down to below 1 K. On the other hand, for the first time [61] a crossover between a $T^{1.0}$ dependence below 1 K and a $T^{1.3}$

dependence above was directly observed. With a photon echo technique, the $T^{1.3}$ behavior was also found in an inorganic glass [62].

Whereas these experiments are interpreted as describing the homogeneous linewidth of guest molecules in glasses, another series of hole-burning experiments, resulting in holewidths which depend logarithmically on time, are interpreted in terms of spectral diffusion [63—65]. Furthermore, it was shown that the holewidth varies appreciably with the applied pressure [66]. Until now, there is no completely satisfactory theory of homogeneous linewidths of impurities in glasses. A first calculation by Lyo and Orbach [67], in which the guest molecule was coupled electrostatically to a single TLS, could not explain the T^2 dependence solely known at that time, as was shown in [68]. Reinecke [69] was the first to apply the concept of spectral diffusion [31, 70] to the optical problem. He assumed a coupling of the guest molecule to the deformation fields of the TLSs via the orbit-lattice interaction and found a linear temperature dependence.

More recent models in part are extensions of the first [71—73], in part of the second approach [74, 75], in part they introduce additional dephasing mechanisms [76, 77]. A discussion of the successes and problems of some of these models can be found in [38].

1.3. Organization of the Paper

The aim of this contribution is to give a detailed study of a model for the description of homogeneous optical linewidths in glasses the basic features of which go back to [67]. As we shall see, however, the model is not restricted to electrostatic but can be specified to any coupling.

The paper is organized as follows. In Section 2, the ideas underlying a theoretical description of glasses are given and used to establish the model Hamiltonian containing the energies and interactions of the guest molecule, the TLS ensemble, and the phonon bath. The system Hamiltonian describing the system without phonons is diagonalized exactly. Section 3 contains a derivation of general equations of motion for the correlation functions needed in the line shape calculation. This calculation is performed within Mori's formalism. In Section 4 we examine a system consisting of one guest molecule and a single TLS. Applying the formalism of Section 3 we obtain a 4×4 eigenvalue problem which is solved exactly. The functional dependence of the eigenvalues on several system parameters and some analytical approximations are discussed. In Section 5 we calculate the line shape of the single-TLS system. The physical meaning of two different averaging procedures is discussed in Section 6. An average of the linewidth over asymmetry and overlap parameters is performed numerically exactly as well as analytically approximately. An additional average over ΔV is carried through. Furthermore, averaged line shapes of the single-TLS system are given.

From Section 6 we learn that it is essential to treat the system containing many TLSs. This is done in Section 7. A line shape formula is derived justifying one of the averaging procedures of Section 6. Linewidths and line shapes are calculated numerically. The former are compared to experimental results. Section 8 contains a discussion of the possible mechanisms by which observed experimental linewidths can be explained within the model.

2. The Model and its Hamiltonian

In this paper we give a contribution to the theory of the homogeneous optical linewidth of guest molecules in glasses. Therefore, our model must include the essential features of the glass, of the guest molecules and of the interaction between both. The aim of the following subsections is to give a physical justification of the Hamiltonians which are used to model these components and their interactions.

2.1. THEORETICAL DESCRIPTION OF THE DYNAMICS OF GLASSES

2.1.1. *Models for the Structure of Glasses and its Vibrations*

Glass has been used for 6000 years for various purposes. Despite this very long history it is probably one of those materials which are least understood. First considerations about the microscopic structure of glasses have been published by Zachariasen [78] in 1932. Broader activities in this field have arisen, however, only since about 1960 [79]. Most of the developments since then have been based on two rather different models of amorphous materials. One of these models, the **random close-packed structure**, originally devised by Bernal [80] for mono-atomic liquids, proved to be an excellent description of amorphous metals [81]. Glasses with covalent and ionic bonds were described in the **continuous random network** model, going back to the ideas developed in [78]. Organic glasses, in which also van der Waals forces play an important role, are more closely related to the second model than to the first. The following considerations refer exclusively to non-metallic glasses. Our picture of these glasses therefore is that of a random network of chemical bonds connecting the atoms or molecules of the host material. Starting from this structural picture the dynamics of the glass may, in lowest-order approximation, be described by small vibrations about the (mechanical) equilibrium state of the random network. Expanding, as usual, the potential of the atoms or molecules up to second order in the displacements, the Hamiltonian H_{ph} of the vibrations is given by

$$H_{\text{ph}} = \frac{1}{2} \sum_{\mathfrak{n}} M_{\mathfrak{n}} \dot{\mathfrak{u}}_{\mathfrak{n}}^2 + \frac{1}{2} \sum_{\mathfrak{n}} \sum_{\mathfrak{m}} \mathfrak{u}_{\mathfrak{m}} \Phi_{\mathfrak{m}\mathfrak{n}} \mathfrak{u}_{\mathfrak{n}}. \qquad (2.1)$$

$M_{\mathfrak{n}}$ is the mass of the atom or molecule at site $\mathfrak{n}$, $\Phi_{\mathfrak{m}\mathfrak{n}}$ the second derivative of the potential with respect to the site vectors $\mathfrak{m}$ and $\mathfrak{n}$. Equation (2.1) has the same form as the Hamiltonian of a crystal; $\mathfrak{m}$ and $\mathfrak{n}$, however, do not describe a translationally invariant lattice. It is well known from theoretical mechanics that a system described by the Hamiltonian (2.1) can be diagonalized by transformation to normal coordinates Q_α and their conjugate momenta $P_\alpha = \dot{Q}_\alpha$. (For a macroscopic system of N_a atoms the practical accomplishment of this diagonalization is impossible in general.) In terms of the normal coordinates the Hamiltonian reads

$$H_{\text{ph}} = \frac{1}{2} \sum_{a=1}^{3N_a} (P_a^2 + \omega_a^2 Q_a^2). \tag{2.2}$$

The quantization of this system of $3N_a$ harmonic oscillators is carried through by requiring the usual commutation relations for the coordinates Q_a and their conjugate momenta P_a. Introducing creation and annihilation operators b_a^+ and b_a, respectively, we obtain from (2.2)

$$H_{\text{ph}} = \sum_{a=1}^{3N_a} \hbar \omega_a b_a^+ b_a. \tag{2.3}$$

Equation (2.3) shows that in an amorphous network the vibrations can be described by phonons as well as in a crystal. However, the index a characterizing the various vibrational modes cannot be interpreted in the same simple manner as in a crystal, where it can be chosen as wavevector and polarization. At very low temperatures only the normal virbrations with the lowest energies are excited. In the case of a crystal we know that these low-energy excitations are acoustic phonons with wavelengths of several lattice constants. On account of the large wavelengths these phonons do not see details of the lattice structure, and the crystal can be described as an elastic continuum. Because the structure of the network — whether crystalline or amorphous — does not enter this description, it should also be appropriate for glasses. In the elastic continuum approximation of the glass the index a characterizing eigenfrequencies and phonon operators can be replaced by a double index $(\mathbf{q}, \sigma)$, denoting wave vector and polarization. In this approximation, the vibrations in the glass are described by the Hamiltonian

$$H_{\text{ph}} = \sum_{\mathbf{q}, \sigma} \hbar \omega_{\mathbf{q}\sigma} b_{\mathbf{q}\sigma}^+ b_{\mathbf{q}\sigma} \tag{2.4}$$

with a linear dispersion relation

$$\omega_{\mathbf{q}\sigma} = c_\sigma |\mathbf{q}|, \tag{2.5}$$

and c_σ being the velocity of sound of branch σ.

These considerations suggest that at low temperatures crystals and glasses will behave in the same manner as regards their thermal properties. This, however, contradicts experimental results that we know from the introductory section.

Rivier [82] showed that this simple continuum picture of the glass has to be modified for topological reasons. In the structural network of a glass there exist rings with an odd number of bonds. These odd rings are connected by disclination lines. On the basis of group theoretical arguments Rivier concluded [83] that each disclination line corresponds to two configurations with different energies. Furthermore, the macroscopic nature of these disclinations in the continuum limit suggests that they can couple to long wavelength phonons.

Somewhat less abstract is the picture that Anderson et al. [9] and Phillips [10] have given of these additional degrees of freedom, which are necessary to describe the thermal and acoustic properties of glasses. In amorphous materials there exist atoms or groups of atoms which can take on different configurations which are energetically almost equivalent. Using probability theory one can conclude that in most cases the number of such configurations is two. Figure 2 shows some possibilities for SiO_2.

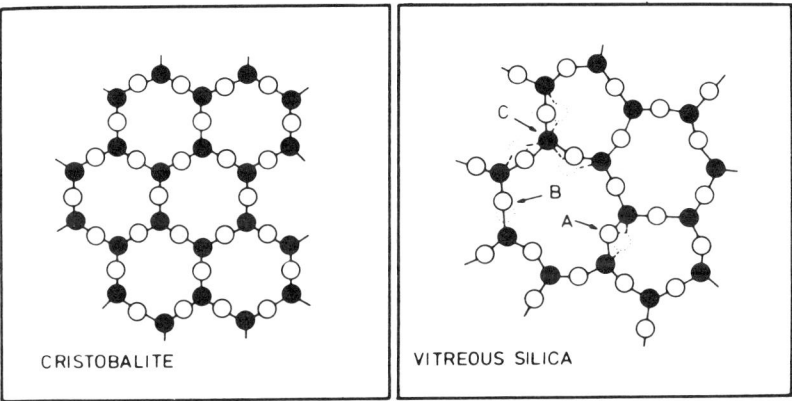

Fig. 2. Structure of cristobalite, a crystalline modification of SiO_2, and of vitreous silica (schematic picture), taken from [84]. Full circles represent silicon atoms, open ones oxygen atoms. Three possible types of defects are indicated.

From both pictures two-state tunneling systems emerge. However, up to now it has not been shown how (or whether at all) the two pictures are connected which demonstrates once more that the physical nature of the tunneling two-level systems (TLS) is not clear. However, the results of thermal and acoustic experiments necessarily require additional degrees of freedom with the properties of TLSs. For this reason in our theory of the optical linewidth of guest molecules in glasses we shall take into account these TLSs, considering them as a phenomenological — however microscopically treated — part of the glass.

2.1.2. Quantum Mechanical Description of the TLSs

The TLSs have been introduced for a microscopic description of the additional degrees of freedom of the glassy state and have been visualized since then [9, 10] by a quantum particle moving in a double-well potential (see Figure 3). Such a system may be described approximately by two quantities, the asymmetry parameter Δ and the tunneling matrix element $W/2$. Δ is the energy difference of the two localized states, i.e. the states with $W = 0$, and W describes the overlap of the wave functions in the two states and may be written in the following way

$$W = 2\hbar\omega_0 e^{-\lambda}. \tag{2.6}$$

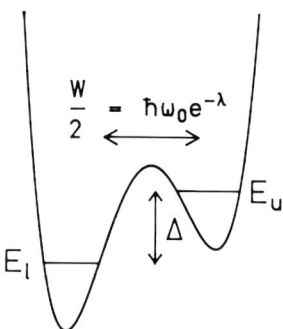

Fig. 3. Double well potential in configuration space and energy levels of a TLS. Several relevant parameters are pictured.

$\hbar\omega_0$ is on the order of magnitude of the zero point vibrational energy in the minima of the potential curve [9] and the overlap parameter λ is connected to the height V_0 of the potential wall between the minima and to their distance Δx via [9, 85]

$$\lambda = \frac{1}{2}\sqrt{\frac{2mV_0}{\hbar^2}}\,\Delta x. \tag{2.7}$$

In the (orthogonalized) localized basis $\{|l\rangle, |u\rangle\}$, denoting the lower and upper state in the potential the Hamiltonian of a single TLS (index s) is

$$H_{\text{TLS},s} = E_l|l\rangle\langle l| + E_u|u\rangle\langle u| + W/2(|l\rangle\langle u| + |u\rangle\langle l|). \tag{2.8}$$

The diagonalization of this Hamiltonian is carried out easily with the following result

$$H_{\text{TLS},s} = \sum_{n=1}^{2} E_n|\psi_n\rangle\langle\psi_n|, \tag{2.9}$$

where

$$E_n = \tfrac{1}{2}(E_u + E_l) + \tfrac{1}{2}(-1)^n E \tag{2.10}$$

with

$$E = \sqrt{\Delta^2 + W^2}. \tag{2.11}$$

The eigenvectors are given by

$$\begin{pmatrix}|\psi_1\rangle \\ |\psi_2\rangle\end{pmatrix} = \begin{pmatrix}\cos\phi & -\sin\phi \\ \sin\phi & \cos\phi\end{pmatrix}\begin{pmatrix}|l\rangle \\ |u\rangle\end{pmatrix}, \tag{2.12}$$

where

$$\cos\phi = \frac{1}{\sqrt{2}}\sqrt{1 + \frac{\Delta}{E}}, \quad \sin\phi = \frac{1}{\sqrt{2}}\sqrt{1 - \frac{\Delta}{E}}. \tag{2.13}$$

The notation used in (2.8)–(2.13) is the same as in [73] except for the argument in the trigonometric functions. The connection between the angle ϕ of this paper and Φ of [73] is

$$\phi = \frac{\pi}{2} - \Phi. \quad (2.14)$$

In a large part of the literature (see e.g. [84]) the analogy of the TLSs with a spin 1/2 system is exploited. The connection between our Hamiltonian (2.9) and the spin Hamiltonian is simply

$$H_{TLS,s} = \tfrac{1}{2}(E_1 + E_2)(|\psi_1\rangle\langle\psi_1| + |\psi_2\rangle\langle\psi_2|) - \tfrac{1}{2}E(|\psi_1\rangle\langle\psi_1| - |\psi_2\rangle\langle\psi_2|)$$
$$= \tfrac{1}{2}(E_1 + E_2) - \tfrac{1}{2}E\sigma_z, \quad (2.15)$$

where the Pauli matrix σ_z is defined with respect to the diagonal basis $\{|\psi_1\rangle, |\psi_2\rangle\}$. The term proportional to the unit operator does not contribute to the dynamics of the system and may therefore be dropped.

In a glass, there are many TLSs which in direct generalization of the results given above are described by

$$H_{TLS} = \sum_{k=1}^{N} \sum_{n=1}^{2} E_n^k |\psi_n^k\rangle\langle\psi_n^k| = \sum_{k=1}^{N} (\tfrac{1}{2}(E_1^k + E_2^k) - \tfrac{1}{2}E^k\sigma_z^k). \quad (2.16)$$

k is a TLS index, N the total number of TLSs. An important feature of the TLSs in a glass is the broad distribution of asymmetry parameters Δ_k and tunneling parameters λ_k for the various TLSs. Transitions between the two states of a TLS, either localized or energy states, are mediated by the phonons of the glass matrix. Therefore, the following subsection discusses the interaction between TLSs and vibrations.

2.1.3. Interaction between TLSs and Phonons

To derive an expression for the interaction energy of a TLS coupled to the phonons we start from the Hamiltonian of a particle in a double well potential in its localized form (2.8). We assume that the elastic wave gives rise to modulation of the energies E_l and E_u and of the tunneling matrix element W of the TLS. In linear approximation we write for the interaction Hamiltonian

$$H_{T-ph,s} = \sum_{jk} \varepsilon_{jk}(\mathfrak{x}_l) \frac{\partial E_l}{\partial \varepsilon_{jk}} |l\rangle\langle l| + \varepsilon_{jk}(\mathfrak{x}_u) \frac{\partial E_u}{\partial \varepsilon_{jk}} |u\rangle\langle u| +$$
$$+ \frac{1}{2} \varepsilon_{jk}(\mathfrak{x}_{lu}) \frac{\partial W}{\partial \varepsilon_{jk}} (|l\rangle\langle u| + |u\rangle\langle l|). \quad (2.17)$$

In this expression ε_{jk} is the (classical) deformation tensor, $\mathfrak{x}_l$ and $\mathfrak{x}_u$ are the sites of the minima of the double well potential, and $\mathfrak{x}_{lu}$ is an average of both. The partial derivatives have to be evaluated for $\varepsilon_{jk} = 0$. The deformation tensor is calculated from the displacement field, the quantized form of which is given by [86, 87]

$$\mathfrak{R}(\mathfrak{x}) = \sum_{q\sigma} \sqrt{\frac{\hbar}{2\omega_{q\sigma}\rho V}} \, \mathfrak{e}_{q\sigma}(b_{q\sigma} + b^+_{-q\sigma}) \, e^{iq\mathfrak{x}}. \tag{2.18}$$

In this expression $\mathfrak{e}_{q\sigma}$ is the polarization vector of phonons of wavevector $\mathfrak{q}$ and branch σ. ρ and V are density and volume, respectively.

According to the definition of the deformation tensor we obtain from (2.18)

$$\varepsilon_{jk} = \frac{1}{2} \left(\frac{\partial R_j}{\partial x_k} + \frac{\partial R_k}{\partial x_j} \right)$$

$$= \frac{1}{2} \sum_{q\sigma} i \sqrt{\frac{\hbar\omega_{q\sigma}}{2\rho V c_\sigma^2}} \, (e^j_{q\sigma} e^k_q + e^k_{q\sigma} e^j_q)(b_{q\sigma} + b^+_{-q\sigma}) \, e^{iq\mathfrak{x}}. \tag{2.19}$$

e^k_q is the kth component of the vector $\mathfrak{e}_q$. Furthermore, we have used dispersion relation (2.5). Inserting (2.19) in (2.17) we obtain

$$H_{T-\text{ph},s} = \sum_{q\sigma} \sum_{r,s=l}^{u} h^{rs}_{q\sigma} |r\rangle\langle s| (b_{q\sigma} + b^+_{-q\sigma}), \tag{2.20}$$

where

$$h^{rs}_{q\sigma} = i \sqrt{\frac{\hbar\omega_{q\sigma}}{2\rho V c_\sigma^2}} \, g^{rs}_\sigma \exp(iq\mathfrak{r}_{rs}) \tag{2.21}$$

and

$$g^{ll}_\sigma = \sum_{jk} \frac{\partial E_l}{\partial \varepsilon_{jk}} \, e^j_{q\sigma} e^k_q, \tag{2.22a}$$

$$g^{uu}_\sigma = \sum_{jk} \frac{\partial E_u}{\partial \varepsilon_{jk}} \, e^j_{q\sigma} e^k_q, \tag{2.22b}$$

$$g^{lu}_\sigma = g^{ul}_\sigma = \frac{1}{2} \sum_{jk} \frac{\partial W}{\partial \varepsilon_{jk}} \, e^j_{q\sigma} e^k_q. \tag{2.22c}$$

In deriving the expressions (2.22) for the deformation potentials g^{ll}_σ, g^{uu}_σ the symmetry of the deformation tensor was used. If the quantities E_l, E_u and W are functions of the deformation tensor only, the g^{rs}_σ are frequency independent. In this case $h^{rs}_{q\sigma}$ is proportional to $\sqrt{\omega_{q\sigma}}$ for small values of q.

If the parameters of the TLS, however, depend not only on the displacements, but also on their velocities, the g^{rs}_σ will become frequency dependent. For an explicit calculation of such a dependence, a microscopic model for the TLS should be used which, as discussed above, is not yet known. To allow for a frequency dependence of the g^{rs}_σ, we shall write (which is a phenomenological approach) $h_{q\sigma}^{rs} \propto \sqrt{\omega_{q\sigma}^{\kappa}}$. In the long-wavelength approximation ($\exp(i\mathbf{q}\mathbf{r}_{rs}) = 1$) we obtain

$$h_{q\sigma}^{rs} = i\sqrt{\frac{\hbar\omega_{q\sigma}}{2\rho V c_{\sigma}^2}} \sqrt{\left(\frac{\omega_{q\sigma}}{\omega_c}\right)^{\kappa-1}} f_{\sigma}^{rs}. \qquad (2.23)$$

The parameters f_{σ}^{rs} are defined by (2.23) and ω_c is a characteristic frequency of the system which is introduced for dimensional reasons. The value $\kappa = 1$ of the phenomenological exponent leads back to the original case. The main advantage of the introduction of the parameter κ — as we shall see later on — is that it allows to discuss the influence of several physical quantities, such as density of TLSs or density of states of the phonons, on the optical linewidth in a convenient way.

Because the Hamiltonian is hermitean the coupling parameters have to fulfill the following relation:

$$h_{q\sigma}^{rs*} = h_{-q\sigma}^{rs}, \qquad f_{\sigma}^{rs*} = -f_{\sigma}^{rs}. \qquad (2.24)$$

An estimate of the deformation potentials shows [88, 89] that the non-diagonal coupling parameters f_{σ}^{lu} and f_{σ}^{ul} can be neglected as compared to the diagonal ones $f_{\sigma}^{ll}, f_{\sigma}^{uu}$ whose order of magnitude is about 1 eV. The non-diagonal elements are estimated by

$$\frac{\partial W}{\partial \varepsilon_{jk}} \approx \frac{\partial W}{\partial \lambda} \frac{\partial \lambda}{\partial \varepsilon_{jk}} = -W \frac{\partial \lambda}{\partial \varepsilon_{jk}}. \qquad (2.25)$$

Here $\partial \lambda/\partial \varepsilon_{jk} \approx 1$ [88] and W is of the order of magnitude of phonon energies and thus smaller by a factor of 100 than the diagonal elements. Therefore, the interaction Hamiltonian becomes

$$H_{T-ph,s} = \sum_{q\sigma} \sum_{r} h_{q\sigma}^{r} |r\rangle\langle r| (b_{q\sigma} + b_{-q\sigma}^{+})$$

$$= \sum_{q\sigma} \tfrac{1}{2} \{h_{q\sigma}^{l} + h_{q\sigma}^{u} + (h_{q\sigma}^{l} - h_{q\sigma}^{u})(|l\rangle\langle l| - |u\rangle\langle u|)\} (b_{q\sigma} + b_{-q\sigma}^{+}), \qquad (2.26)$$

where for simplicity we have replaced $h_{q\sigma}^{rr}$ by $h_{q\sigma}^{r}$.

To transform (2.26) into the diagonal representation of the TLS, we use (2.12)

$$|l\rangle\langle l| - |u\rangle\langle u| = \cos 2\phi (|\psi_1\rangle\langle\psi_1| - |\psi_2\rangle\langle\psi_2|) +$$
$$+ \sin 2\phi (|\psi_1\rangle\langle\psi_2| + |\psi_2\rangle\langle\psi_1|) \qquad (2.27a)$$

with

$$\cos 2\phi = \cos^2 \phi - \sin^2 \phi = \frac{\Delta}{E}, \qquad (2.27b)$$

$$\sin 2\phi = 2\sin\phi\cos\phi = \sqrt{1 - \left(\frac{\Delta}{E}\right)^2} = \frac{W}{E}. \qquad (2.27c)$$

Using the abbreviations

$$S_{q\sigma} = \tfrac{1}{2}(h^l_{q\sigma} + h^u_{q\sigma}), \quad D_{q\sigma} = \tfrac{1}{2}(h^l_{q\sigma} - h^u_{q\sigma}) \tag{2.28}$$

we obtain from (2.26)

$$H_{T-\text{ph},s} = \sum_{q\sigma} \sum_{n=1}^{2} \left\{ \left(S_{q\sigma} - (-1)^n D_{q\sigma} \frac{\Delta}{E} \right) |\psi_n\rangle\langle\psi_n| + \right.$$

$$\left. + D_{q\sigma} \frac{W}{E} |\psi_n\rangle\langle\psi_{\tilde{n}}| \right\} (b_{q\sigma} + b^+_{-q\sigma}), \tag{2.29a}$$

where

$$\tilde{n} \equiv \begin{cases} 1, & \text{if } n = 2 \\ 2, & \text{if } n = 1 \end{cases}. \tag{2.29b}$$

Taking into account all TLSs in the glass the interaction Hamiltonian is generalized to

$$H_{T-\text{ph}} = \sum_{k=1}^{N} \sum_{n=1}^{2} \sum_{q\sigma} \left\{ \left(S^k_{q\sigma} - (-1)^n D^k_{q\sigma} \frac{\Delta^k}{E^k} \right) |\psi^k_n\rangle\langle\psi^k_n| + \right.$$

$$\left. + D^k_{q\sigma} \frac{W^k}{E^k} |\psi^k_n\rangle\langle\psi^k_{\tilde{n}}| \right\} (b_{q\sigma} + b^+_{-q\sigma})$$

$$= \sum_{k=1}^{N} \sum_{q\sigma} \left\{ S^k_{q\sigma} + D^k_{q\sigma} \left(\frac{\Delta^k}{E^k} \sigma^k_z + \frac{W^k}{E^k} \sigma^k_x \right) \right\} (b_{q\sigma} + b^+_{-q\sigma}), \tag{2.30}$$

where the k sum runs over all TLSs.

This form of the interaction Hamiltonian has been used in [73, 90–92]. The versions used in [82, 89, 93] are given in a semiclassical form. A comparison of matrix elements, however, shows that these versions are essentially equivalent to our Hamiltonian (for $\kappa = 1$).

2.1.4. Interaction between TLSs

The TLSs interact with each other on account of the distortion fields generated by each of them. Quantum mechanically this interaction is described by the exchange of virtual phonons [89, 94]. Using the spin picture for the TLSs it is represented by the following Hamiltonian

$$H_{TT} = H_{zz} + H_{xx} = \sum_{j<k} J^{jk}_{zz} \sigma^j_z \sigma^k_z + \sum_{j<k} J^{jk}_{xx} \sigma^j_x \sigma^k_x. \qquad (2.31)$$

It can be shown [31, 89] that for phonons of large wavelength $J^{jk} \propto r^{-3}$, which is characteristic for dipolar interaction. Furthermore, J^{jk}_{xx} is different from zero only for pairs of TLSs which are at resonance [89]. Because of the broad distribution of TLS parameters there are only few TLSs which are at resonance and therefore the influence of J^{jk}_{xx} can be neglected.

The neglect of the terms containing J^{jk}_{zz} is harder to justify. We are mainly interested in the influence of the TLSs on the temperature dependence of the optical linewidth of the guest molecule. This influence is taken into account by a direct interaction between a guest molecule and the TLSs. Via this interaction changes in the TLSs are seen by the guest molecule. In the following we shall consider explicitly the influence of the phonons on the TLS ensemble. The Hamiltonian describing this interaction has qualitatively the same structure as the TLS—TLS interaction, at least if one of the spins in (2.31) is replaced by a mean field. Therefore, the TLS—phonon interaction and the TLS—TLS interaction will create similar dephasing processes in the guest molecule or ion. The essential difference of the two interactions is that one of them is temperature dependent whereas the other is not. Because we are mainly interested in temperature dependent processes, the influence of the interaction between TLSs is no longer considered in the following.

2.2. INTERACTION BETWEEN A GUEST MOLECULE AND THE TLSs

In actual experiments [49, 62, 95] the concentration of guest ions or molecules usually is rather low. Therefore, we may neglect the interaction between the guests and in the following we shall consider a single guest molecule only. This guest is described in the simplest possible way by only two states, the ground state $|\alpha\rangle$ and relevant excited state $|\beta\rangle$ with energies E_α and E_β, respectively. The Hamiltonian of the guest is given by

$$H_g = \sum_{\rho=\alpha}^{\beta} E_\rho |\rho\rangle\langle\rho|. \qquad (2.32)$$

The distance between the electronic states is of the order of a hundred Debye energies. Therefore, higher excited electronic states could be neglected in (2.32) and, furthermore, for temperatures of about 1 K, at which many experiments are carried out, only the lower of the two electronic states is occupied.

The interaction between the TLSs and the guest molecule considered is written in the following way:

$$H_{gT} = \sum_{\rho=\alpha}^{\beta} \sum_{k=1}^{N} (V_{l\rho}^{k}|l^{k}\rangle\langle l^{k}| + V_{u\rho}^{k}|u^{k}\rangle\langle u^{k}|)|\rho\rangle\langle\rho|. \quad (2.34)$$

This Hamiltonian is diagonal in the states of the guest, i.e. it cannot cause transitions between the electronic states. The physical reason for this ansatz is that for such a transition energy could not be conserved because of the large difference of excitation energies in the guest and in the TLSs. The $V_{l\rho}^{k}$ and $V_{u\rho}^{k}$ are the coupling matrix elements of the guest in the state ρ with the kth TLS in the states $|l\rangle$ and $|u\rangle$, respectively. These two matrix elements have to be different from each other to create dephasing in the guest molecule. They are also different from each other for $\rho = \alpha$ and $\rho = \beta$. We transform H_{gT} from (2.33) to the diagonal bases of the TLSs (2.12) and use

$$S_{\rho}^{k} = \tfrac{1}{2}(V_{l\rho}^{k} + V_{u\rho}^{k}), \quad (2.34a)$$

$$V_{\rho}^{k} = \tfrac{1}{2}(V_{l\rho}^{k} - V_{u\rho}^{k}). \quad (2.34b)$$

Now the interaction Hamiltonian between the guest considered and the TLSs of the glass reads

$$H_{gT} = \sum_{\rho, k, n} |\rho\rangle\langle\rho| \left\{ \left(S_{\rho}^{k} - (-1)^{n} V_{\rho}^{k} \frac{\Delta^{k}}{E^{k}} \right) |\psi_{n}^{k}\rangle\langle\psi_{n}^{k}| + V_{\rho}^{k} \frac{W^{k}}{E^{k}} |\psi_{n}^{k}\rangle\langle\psi_{\bar{n}}^{k}| \right\}$$

$$= \sum_{\rho, k} |\rho\rangle\langle\rho| \left\{ S_{\rho}^{k} + V_{\rho}^{k} \frac{\Delta^{k}}{E^{k}} \sigma_{z}^{k} + V_{\rho}^{k} \frac{W^{k}}{E^{k}} \sigma_{x}^{k} \right\}. \quad (2.35)$$

2.3. THE TOTAL MODEL

2.3.1. *The Hamiltonian*

We wish to give a microscopic description of the temperature dependence of the line shape and linewidth of guest molecules in a glassy matrix. In the preceding subsections we have discussed the various components, i.e. the glass with its vibrational degrees of freedom and the TLSs, the guest molecule and the interactions between TLSs and vibrations and between guest molecule and TLSs. We have not considered the direct interaction between the guest and the phonons of the glass for the following reason. This interaction should give similar contributions to the linewidth in crystals and in glasses. The optical linewidth in crystals, however, is much smaller than in glasses. Therefore, we conclude that this direct interaction is of minor importance only for the linewidth in glasses, and it is neglected further on. Collecting now the various contributions we obtain the total Hamiltonian:

$$H = H_{1} + H_{12} + H_{2} + H_{23} + H_{3}, \quad (2.36)$$

where

$$H_1 \equiv H_g = \sum_\rho E_\rho |\rho\rangle\langle\rho|, \tag{2.37}$$

$$H_2 \equiv H_{TLS} = \sum_{k,n} E_n^k |\psi_n^k\rangle\langle\psi_n^k|, \tag{2.38}$$

$$H_{12} \equiv H_{gT} = \sum_{\rho,k,n} |\rho\rangle\langle\rho| \left\{ \left(S_\rho^k - (-1)^n V_\rho^k \frac{\Delta^k}{E^k}\right) |\psi_n^k\rangle\langle\psi_n^k| \right.$$
$$\left. + V_\rho^k \frac{W^k}{E^k} |\psi_n^k\rangle\langle\psi_{\bar{n}}^k| \right\}, \tag{2.39}$$

$$H_3 \equiv H_{ph} = \sum_q \omega_q b_q^+ b_q, \tag{2.40}$$

$$H_{23} \equiv H_{T-ph} = \sum_q \sum_{k,n} \left\{ \left(S_q^k - (-1)^n D_q^k \frac{\Delta^k}{E^k}\right) |\psi_n^k\rangle\langle\psi_n^k| \right.$$
$$\left. + D_q^k \frac{W^k}{E^k} |\psi_n^k\rangle\langle\psi_{\bar{n}}^k| \right\} (b_q + b_{-q}^+). \tag{2.41}$$

In the above expressions we have used $\hbar = 1$ and abbreviated (q, σ) by q and $(-q, \sigma)$ by $-q$.

Figure 4 shows the different components of the total system together with their interactions. As indicated in the figure we shall decompose the total system in two parts. The part we are mainly interested in contains the guest and the TLSs and will be denoted as **system** in the following. The second part contains the phonons which are considered a heat **bath** and will be eliminated from our equations by use of a projection technique. The influence of the phonons finally shows up in temperature dependent coefficients.

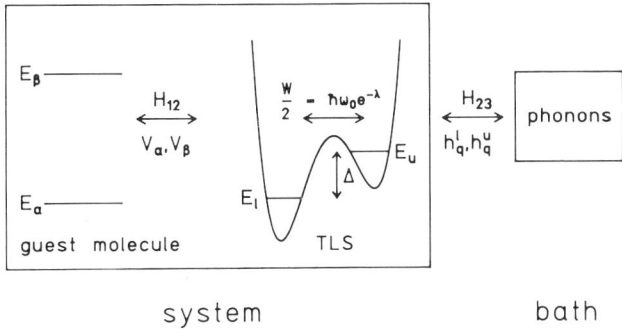

Fig. 4. The components of the model and their interactions.

2.3.2. Investigation of the System Part

In a first step of our investigation we consider the system part only and neglect the influence of the bath. We are interested in the energy level scheme and in transition energies. These quantities are obtained by diagonalizing the system Hamiltonian

$$H_S \equiv H_1 + H_{12} + H_2. \tag{2.42}$$

To make the discussion clearer we shall first consider the interaction of the guest molecule with a single TLS. The influence of several TLSs is discussed subsequently.

2.3.2.1. Interaction of the guest molecule with a single TLS.
The state vectors $\{|\alpha\rangle, |\beta\rangle\}$ of the guest and $\{|\psi_1\rangle, |\psi_2\rangle\}$ of a single TLS form a complete set for the guest and the TLS, respectively, with completeness relations

$$\sum_{\rho=\alpha}^{\beta} |\rho\rangle\langle\rho| = 1, \tag{2.43}$$

$$\sum_{n=1}^{2} |\psi_n\rangle\langle\psi_n| = 1. \tag{2.44}$$

The system consisting of a guest atom or molecule interacting with a single TLS is described by $H_{S,s} = H_1 + H_2 + H_{12}$ from (2.37)–(2.39) with $k = 1$ in (2.38, 2.39). A basis for this system is given by the four product states $|\rho\rangle|\psi_j\rangle$, $\rho = \alpha, \beta$, $j = 1, 2$. With these product states, $H_{S,s}$ reads

$$H_{S,s} = \sum_{\rho, n} \left\{ \left(E_\rho + E_n + S_\rho - (-1)^n V_\rho \frac{\Delta}{E} \right) |\psi_n\rangle\langle\psi_n| + \right.$$

$$\left. + V_\rho \frac{W}{E} |\psi_n\rangle\langle\psi_{\tilde{n}}| \right\} |\rho\rangle\langle\rho|. \tag{2.45}$$

This Hamiltonian is diagonal in the guest states. Diagonalization with respect to the TLS states gives the eigenvalues and eigenvectors of $H_{S,s}$. The eigenvalues read

$$\varepsilon_{\rho n} = E_\rho + S_\rho + \tfrac{1}{2}(E_1 + E_2) + (-1)^n \tfrac{1}{2}\varepsilon_\rho \tag{2.46}$$

with

$$\varepsilon_\rho = \sqrt{W^2 + (\Delta - 2V_\rho)^2}. \tag{2.47}$$

The eigenvectors are $|\psi_n^\rho\rangle|\rho\rangle$, where

$$|\psi_n^\rho\rangle = \cos(\rho_s)|\psi_n\rangle + (-1)^n \sin(\rho_s)|\psi_{\tilde{n}}\rangle \tag{2.48}$$

with

$$\cos(\rho_s) \equiv \frac{1}{\sqrt{2}} \sqrt{1 + \frac{E - 2V_p \Delta/E}{\varepsilon_p}}, \qquad (2.49a)$$

$$\sin(\rho_s) \equiv \frac{1}{\sqrt{2}} \operatorname{sign}(V_p) \sqrt{1 + \frac{E - 2V_p \Delta/E}{\varepsilon_p}}. \qquad (2.49b)$$

We arrange the eigenvalues according to their magnitude, i.e. $\varepsilon_1 = \varepsilon_{\alpha 1}$, $\varepsilon_2 = \varepsilon_{\alpha 2}$, $\varepsilon_3 = \varepsilon_{\beta 1}$, and $\varepsilon_4 = \varepsilon_{\beta 2}$.

The four eigenvectors are then given by

$$|1\rangle = |\alpha 1\rangle \equiv |\alpha\rangle |\psi_1^\alpha\rangle, \qquad |2\rangle = |\alpha 2\rangle \equiv |\alpha\rangle |\psi_2^\alpha\rangle,$$
$$|3\rangle = |\beta 1\rangle \equiv |\beta\rangle |\psi_1^\beta\rangle, \qquad |4\rangle = |\beta 2\rangle \equiv |\beta\rangle |\psi_2^\beta\rangle, \qquad (2.50)$$

and $H_{S,s}$ from (2.45) becomes

$$H_{S,s} = \sum_{j=1}^{4} \varepsilon_j |j\rangle\langle j|. \qquad (2.51)$$

The energy level scheme of (2.51) is represented in Figure 5. It consists of two pairs of closely spaced energy levels. The energy splitting within a pair is on the order of magnitude of the energy separation in the TLS, whereas the energy difference between the upper and lower pair corresponds to the energy splitting of the guest. The full arrows R_1 to R_4 correspond to optical transitions. In the following we shall see that the dashed arrows describe relaxational transitions induced by the coupling of the system to phonons. These transitions reduce the lifetime of the eigenstates of the system, resulting in a broadening of the levels and hence of the optical lines. The spectral position of the four lines is approximately determined by the energy differences

$$\varepsilon_{jk} \equiv \varepsilon_j - \varepsilon_k, \qquad (2.52)$$

where $j = 3, 4$ and $k = 1, 2$. The width of the transition denoted by R_1 goes to

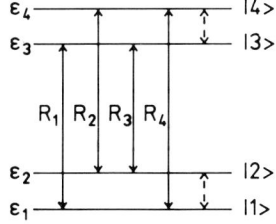

Fig. 5. Energy level scheme of the system.

zero with decreasing temperature, because the transitions $1 \to 2$ and $3 \to 4$ can only occur when a phonon is absorbed and the probability for this process vanishes with disappearing temperature. The width of the transitions R_2 to R_4, however, is influenced by phonon emission. On account of spontaneous emission processes, which are still possible at $T = 0$, the widths of these transitions remain finite. Phonon induced transitions can also lead to a narrowing of lines, if the scattering rate between two optical transitions is larger than their frequency difference. This process is well known from electron paramagnetic resonance [96]. The optical line shape which is actually seen in an experiment is, of course, determined by the relative intensities of the four lines which depend on the occupation probabilities of the four energy levels and on the dipole moments of the transitions.

Figure 5(a) shows a typical behavior of the transition energies ε_{jk} (2.52) as a function of the asymmetry parameter Δ. The parameters in the figure are

$$E_0 \equiv E_\beta - E_\alpha + S_\beta - S_\alpha \tag{2.53}$$

and $\Delta V \equiv V_\beta - V_\alpha.$ (2.54)

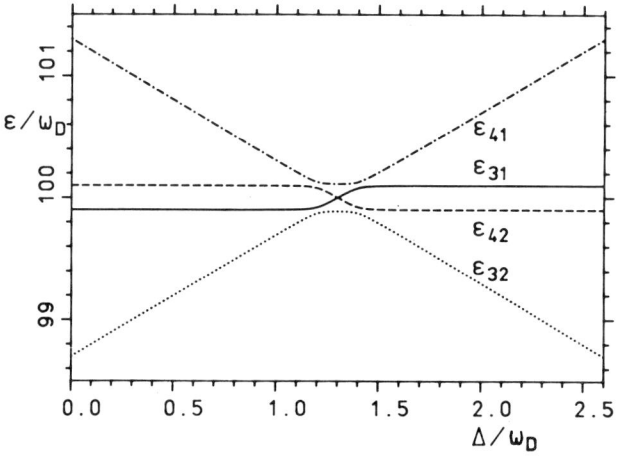

Fig. 5(a). Energies of the optical transitions as a function of the asymmetry parameter ($E_0 = 100\omega_D$, $V_\alpha = 0.6\omega_D$, $\Delta V = 0.1\omega_D$, $\lambda = 3$).

ω_D is the Debye energy and λ the overlap parameter defined in (2.6, 2.7). We have chosen this representation of the transition energies, because for different values of V_α and V_β the graph is only shifted parallel to the Δ axis as long as ΔV remains constant. This is immediately seen by considering (2.46, 2.47, 2.51): ε_α depends on $\Delta - 2V_\alpha$ and ε_β on $\Delta - 2V_\beta = \Delta - 2V_\alpha - 2\Delta V$; hence for given values of ΔV and the other parameters the graph depends only on $\Delta - 2V_\alpha$. At this point it might be interesting to mention that as a result of the interaction between guest and TLS we have arrived at an energy level scheme which can be interpreted as superposition of two TLSs with renormalized asymmetry parameters $\Delta_{\alpha/\beta} = \Delta - 2V_{\alpha/\beta}$. The two TLSs are not independent but connected by the energy difference of the guest.

With regard to the discussion of the linewidth in Section 6 it should be mentioned that ΔV determines the difference of the transition energies ε_{31} and ε_{42} for $|\Delta - 2V_{\alpha/\beta}| \gg W$:

$$\varepsilon_{42} - \varepsilon_{31} = \varepsilon_{43} - \varepsilon_{21} \approx |\Delta - 2V_\beta| - |\Delta - 2V_\alpha|$$

$$= \begin{cases} +2\Delta V \text{ for } \Delta < 2V_\alpha, 2V_\beta \\ -2\Delta V \text{ for } \Delta > 2V_\alpha, 2V_\beta \\ +2(\Delta - (V_\alpha + V_\beta)) \text{ for } 2V_\beta < \Delta < 2V_\alpha \\ -2(\Delta - (V_\alpha + V_\beta)) \text{ for } 2V_\alpha < \Delta < 2V_\beta. \end{cases} \quad (2.55)$$

Furthermore, ΔV seems to be important also from a different point of view. We know that it is characteristic for glasses that the parameters of the TLSs have a broad distribution. Therefore, $\Delta - 2V_\alpha$ and $\Delta - 2V_\beta$ are described by distributions shifted by $2V_\alpha$ and $2V_\beta$, respectively. As long as the distribution for Δ is broad this shift for itself may not be too important. However, the distributions for $\Delta - 2V_\alpha$ and $\Delta - 2V_\beta$ are shifted by different amounts, and the important quantity is their overlap which depends on $\Delta V = V_\beta - V_\alpha$.

Finally, we wish to mention that the energy splitting for $\Delta \approx 2V_\alpha$ of the pairs of transition energies ε_{42} and ε_{41} and of ε_{31} and ε_{32} (repelling each other) is determined by λ. A similar statement holds at $\Delta \approx 2V_\beta$. In the figure we have chosen $\lambda = 3$ in order to render this splitting visible (for $\lambda \to \infty$ it vanishes). In [9] it is assumed that λ is equidistributed between $\lambda = 5$ and $\lambda = 20$. However, even for $\lambda = 3$ $W = \omega_D e^{-\lambda}$ is so small that for most Δ values (2.55) is well satisfied.

2.3.2.2. Interaction of the guest molecule with several TLSs.
In this subsection we diagonalize the system Hamiltonian $H_S = H_1 + H_2 + H_{12}$ from (2.37)–(2.39) where the sum over k now runs over all N TLSs interacting with the guest. The diagonalization is carried through immediately, if we exploit the fact that H_S can be split into a sum of N commuting parts

$$H_S = \sum_{k=1}^{N} H_S^k, \quad (2.56)$$

$$H_S^k = \sum_{\rho,n} \left\{ \left(E_\rho^k + E_n^k + S_\rho^k - (-1)^n V_\rho^k \frac{\Delta^k}{E^k} \right) |\psi_n^k\rangle\langle\psi_n^k| + \right.$$

$$\left. + V_\rho^k \frac{W^k}{E^k} |\psi_n^k\rangle\langle\psi_{\bar{n}}^k| \right\} |\rho\rangle\langle\rho|, \quad (2.57)$$

where

$$\sum_{k=1}^{N} E_\rho^k = E_\rho \tag{2.58}$$

has to be fulfilled. From (2.57) it is immediately seen that

$$[H_S^k, H_S^l] \qquad \text{for } k, l \in \{1, 2, \ldots, N\}. \tag{2.59}$$

Because of the property (2.59) which is essentially caused by H_S being diagonal in $|\alpha\rangle, |\beta\rangle$, all H_S^k can be diagonalized simultaneously. All of these operators have the same algebraic structure and therefore the diagonalization of one member of the sum solves the problem for all terms. Furthermore, the structure of (2.57) is the same as that of $H_{S,s}$ in (2.45). Thus we can immediately take over the results obtained from the diagonalization of (2.45). The energies E_ρ^k are arbitrary except for the condition (2.58) and do not correspond to a quantum mechanical observable. Decomposition (2.56) is just a mathematical trick which reduces the diagonalization of the 'many-particle system' described by H_S to that of a 'single-particle system' described by $H_{S,s}$. (Of course, we could achieve this diagonalization as well using a product ansatz for the many-particle wave function. By this procedure, however, the simple form (2.60) given below for H_S is not obtained as easily as by our method.)

The physical reasons for this simplification of the problem are firstly that the TLSs do not interact with each other but only with the guest atom and secondly that the coupling of the TLSs to the guest in the excited state is not influenced by the coupling to the ground state — mathematically speaking the coupling is diagonal in the states of the guest. The effect of this coupling is that we finally arrive at a system of N dynamically non-interacting four-level systems or pairs of non-interacting two-level systems, each of which is denoted by the quantum numbers α and β of the guest molecule. The many-particle nature of the system shows up in a correlation effect: either only α state TLSs are occupied by the system or only β state ones; system states consisting of mixed α and β TLS states are not allowed.

The diagonalized form of the system Hamiltonian reads

$$H_S = \sum_{\rho, k, n} \varepsilon_{\rho k n} |\rho k n\rangle\langle \rho k n|, \tag{2.60}$$

where

$$\varepsilon_{\rho k n} = E_\rho^k + S_\rho^k + \tfrac{1}{2}(E_1^k + E_2^k) + (-1)^n \tfrac{1}{2}\varepsilon_\rho^k, \tag{2.61}$$

$$\varepsilon_\rho^k = \sqrt{(W^k)^2 + (\Delta^k - 2V_\rho^k)^2}. \tag{2.62}$$

The eigenfunctions are given by

$$|\rho k n\rangle = |\rho\rangle|\psi_n^{\rho k}\rangle = \cos(\rho_k)|\rho\rangle|\psi_n^k\rangle + (-1)^n \sin(\rho_k)|\rho\rangle|\psi_{\bar{n}}^k\rangle \tag{2.63}$$

with

$$\cos(\rho_k) \equiv \frac{1}{\sqrt{2}} \sqrt{1 + \frac{E^k - 2V_\rho^k \Delta^k/E^k}{\varepsilon_\rho^k}}$$

$$= \frac{1}{\sqrt{2}} \sqrt{1 + \frac{(W^k)^2 + \Delta^k(\Delta^k - 2V_\rho^k)}{E^k \varepsilon_\rho^k}}$$

$$= \frac{1}{\sqrt{2}} \sqrt{1 + \frac{\varepsilon_\rho^k + 2V_\rho^k(\Delta^k - 2V_\rho^k)/\varepsilon_\rho^k}{E^k}}, \quad (2.64a)$$

$$\sin(\rho_k) \equiv \frac{1}{\sqrt{2}} \operatorname{sign}(V_\rho^k) \sqrt{1 - \frac{E^k - 2V_\rho^k \Delta^k/E^k}{\varepsilon_\rho^k}}. \quad (2.64b)$$

The E^k are defined in (2.11, 2.16).

2.3.3. Transformation of the Model Hamiltonian H to the Basis System of H_S

In Section 2.3.1 we have introduced the splitting of the Hamiltonian of the total system into a system part, a bath part, and the interaction between both. In the following sections we shall eliminate the bath part from our equations using a projection technique. This is most conveniently carried out, if the total Hamiltonian is represented in terms of eigenstates $|\rho k n\rangle$ (2.63) of H_S. H_S and $H_B = H_3$ are already written in the desired representation (see (2.40, 2.60)). The transformation of the interaction operator $H_{SB} \equiv H_{23}$ is given in the following. To that end we multiply H_{23} from (2.41) by the completeness relation (2.43). In this way we obtain the system operators $|\rho\rangle\langle\rho| |\psi_n^k\rangle\langle\psi_n^k|$ and $|\rho\rangle\langle\rho| |\psi_n^k\rangle\langle\psi_{\tilde{n}}^k|$; $\tilde{n}$ is defined in (2.29b). Using (2.63) we get

$$|\rho\rangle\langle\rho| |\psi_n^k\rangle\langle\psi_m^k|$$
$$= \cos^2(\rho_k) |\rho k n\rangle\langle\rho k m| + (-1)^n(-1)^m \sin^2(\rho_k) |\rho k \tilde{n}\rangle\langle\rho k \tilde{m}| -$$
$$- \sin(\rho_k)\cos(\rho_k)((-1)^m |\rho k n\rangle\langle\rho k \tilde{m}| + (-1)^n |\rho k \tilde{n}\rangle\langle\rho k m|). \quad (2.65)$$

H_{23} then becomes

$$H_{23} = \sum_q \sum_k \left\{ S_q^k + \sum_\rho D_q^k \sum_{n=1}^{2} \left\{ (-1)^n \left[-\frac{\Delta^k}{E^k} \cos(2\rho_k) + \frac{W^k}{E^k} \sin(2\rho_k) \right] \times \right.\right.$$
$$\times |\rho k n\rangle\langle\rho k n| +$$
$$\left.\left. + \left[\frac{\Delta^k}{E^k} \sin(2\rho_k) + \frac{W^k}{E^k} \cos(2\rho_k) \right] |\rho k n\rangle\langle\rho k \tilde{n}| \right\}\right\} (b_q + b_{-q}^+). \quad (2.66)$$

The first bracket in (2.66) is denoted by A_ρ^k, the second by B_ρ^k. From (2.64) we get

$$\cos(2\rho_k) = \frac{E^k}{\varepsilon_\rho^k} - \frac{2V_\rho^k \Delta^k}{E^k \varepsilon_\rho^k}, \tag{2.67}$$

$$\sin(2\rho_k) = \frac{W^k}{E^k} \frac{2V_\rho^k}{\varepsilon_\rho^k}, \tag{2.68}$$

and using these relations we obtain for A_ρ^k and B_ρ^k

$$A_\rho^k = -\frac{\Delta_\rho^k}{\varepsilon_\rho^k}, \tag{2.69}$$

$$B_\rho^k = \frac{W^k}{\varepsilon_\rho^k}. \tag{2.70}$$

The interaction between system and bath is then written in the following way

$$\begin{aligned} H_{SB} &\equiv H_{23} \\ &= \sum_{q,\rho,k,n} \left\{ \left(S_q^k - (-1)^n D_q^k \frac{\Delta_\rho^k}{\varepsilon_\rho^k} \right) |\rho k n\rangle\langle \rho k n| + \right. \\ &\quad \left. + D_q^k \frac{W^k}{\varepsilon_\rho^k} |\rho k n\rangle\langle \rho k \tilde{n}| \right\} (b_q + b_{-q}^+) \\ &\equiv \sum_{q,k} \Lambda_q^k \varepsilon_q. \end{aligned} \tag{2.71}$$

The last equality introduces abbreviations which will prove useful in later calculations. ε_q is the sum of the phonon operators, the definition of Λ_q then becomes obvious.

The total Hamiltonian, describing the interaction between system and bath and being the starting point of the following calculations, then reads

$$\begin{aligned} H &= H_S + H_B + H_{SB} \\ &= \sum_{\rho,k,n} \varepsilon_{\rho k n} |\rho k n\rangle\langle \rho k n| + \\ &\quad + \sum_q \omega_q b_q^+ b_q + \sum_{q,\rho,k,n} \left\{ \left(S_q^k + (-1)^n D_q^k A_\rho^k \right) |\rho k n\rangle\langle \rho k n| + \right. \\ &\quad \left. + D_q^k B_\rho^k |\rho k n\rangle\langle \rho k \tilde{n}| \right\} (b_q + b_{-q}^+). \end{aligned} \tag{2.72}$$

3. Optical Line Shape Calculated with Mori's Formalism

3.1. CORRELATION FUNCTIONS FOR THE OPTICAL LINE SHAPE

It is well known that the optical absorption is determined by the imaginary part $\chi''(\omega)$ of the dielectric susceptibility. Using linear response theory $\chi''(\omega)$ is calculated according to [73, 97, 98]

$$\chi''(\omega) = \frac{1}{2} \int_0^\infty d\tau (e^{i\omega\tau} - e^{-i\omega\tau}) \mathrm{Tr}([p(\tau), p]\rho). \tag{3.1}$$

In (3.1) we have used the long-wavelength approximation for the light field which renders the susceptibility independent of the wavevector of the light field. $p(\tau)$ is the operator of the electric dipole moment in the Heisenberg representation

$$p(\tau) = e^{iH\tau} p\, e^{-iH\tau}, \tag{3.2a}$$

and $p = p(0)$. ρ is the equilibrium density operator of the total system

$$\rho = \frac{e^{-\beta H}}{\mathrm{Tr}(e^{-\beta H})}, \qquad \beta = \frac{1}{kT}, \tag{3.2b}$$

where H is the Hamiltonian (2.72).

The correlation functions in (3.1) cannot be calculated exactly. Assuming that the interaction between system and bath is sufficiently weak, we factorize the density operator (3.2b)

$$\rho \approx \rho_0 = \rho_S \rho_B, \tag{3.3}$$

where

$$\rho_S = \frac{e^{-\beta H_S}}{\mathrm{Tr}_S(e^{-\beta H_S})} \equiv \frac{1}{Z_S} e^{-\beta H_S}, \tag{3.4a}$$

$$\rho_B = \frac{e^{-\beta H_B}}{\mathrm{Tr}_B(e^{-\beta H_B})} \equiv \frac{1}{Z_B} e^{-\beta H_B}, \tag{3.4b}$$

are the equilibrium density operators of system and bath without coupling, and Z_S and Z_B are their partition functions. $\chi''(\omega)$ is then rewritten as

$$\chi''(\omega) = \frac{1}{2} \int_0^\infty d\tau (e^{i\omega\tau} - e^{-i\omega\tau}) \mathrm{Tr}_S(\langle p(\tau)[p, \rho_S]\rangle), \tag{3.5}$$

where the angular brackets denote the thermal average

$$\langle \ldots \rangle \equiv \text{Tr}_B(\ldots \rho_B). \tag{3.6}$$

We assume that the dipole moment operator $p = p(0)$ at the initial time and thus $[p, \rho_S]$ as well are pure system operators which can be expanded into a complete set $\{O_j\}$ of system operators. If $\{|l\rangle\}$ is a basis of the Hilbert space of the system, such a complete set is given, e.g., by

$$\{O_\mathbf{j}\} = \{O_{lm}\} = \{|l\rangle\langle m|\} \tag{3.7}$$

($\mathbf{j}$ is an abbreviation of the double index lm and printed boldface.) Using a complete operator set we may write p and $[p, \rho_S]$ as

$$p = \sum_j v_j O_j = \sum_j v_j^* O_j^+, \tag{3.8a}$$

$$[p, \rho_S] = \sum_j \gamma_j O_j. \tag{3.8b}$$

Explicit expressions for the expansion coefficients v_j and γ_j will be given in later sections. Inserting (3.8) into the correlation function of (3.5) we obtain

$$\text{Tr}_S(\langle p(\tau)[p, \rho_S]\rangle) = \sum_{j,k} v_j^* \gamma_k \text{Tr}_S(\langle O_j^+(\tau) O_k \rangle)$$

$$= \sum_{j,k} v_j^* \gamma_k \text{Tr}(O_j^+(\tau) O_k \rho_B). \tag{3.9}$$

From (3.5, 3.9) it is obvious that the line shape can be calculated if the correlation functions

$$S_{jk}(\tau) \equiv \text{Tr}(O_j^+(\tau) O_k \rho_B) \tag{3.10}$$

are known. There are several ways of determining such correlation functions (see e.g. [1]); the procedure we have used is Mori's formalism which will be summarized in the following subsection.

3.2. CALCULATION OF CORRELATION FUNCTIONS WITH MORI'S FORMALISM

The Mori formalism [99] is a general method to derive equations of motion for the projections of dynamical variables on a subspace of the super Hilbert space spanned by them. These equations have the structure of generalized Langevin equations [99, 100] and the **exact**. In most cases they can be evaluated approximately only and the stochastic nature of the fluctuation term remains a hypothesis [100]. This aspect, however, is unimportant if the formalism is applied to calculate correlation functions.

The choice of the subspace is to some extent arbitrary and guided by physical intuition. The equations of motion can be derived for any subspace. If, however, the subspace is chosen too small, the integral kernel becomes too complex to allow for a good approximation; if it is too large, the number of final equations becomes intractable. In our case the relevant subspace is spanned by all operators of the system, the irrelevant subspace by the bath operators. Later on we shall see that not all possible system operators are relevant to the calculation of the optical line shape.

In Mori's formalism the projection is carried out with the help of a scalar product defined on the space of dynamical variables. Many authors including Mori himself [99, 101–103] prefer a scalar product which goes back to Kubo. For our purpose, the calculation of the correlation functions S_{jk} defined in (3.10), another scalar product is more convenient. We define

$$(A|B) = \mathrm{Tr}(\rho_B A^+ B), \tag{3.11}$$

(which is identical to the so-called Hilbert–Schmidt scalar product [104] if A and B are system operators); the correlation function $S_{jk}(t)$ may then be written

$$S_{jk}(t) = (O_j(t)|O_k). \tag{3.12}$$

Using such a scalar product, whose explicit form does not enter the general derivation of the equations of motion, the projection operator is defined by

$$\bar{\mathbf{P}} = \sum_{j,k} (\underline{\mu}^{-1})_{jk} |O_j)(O_k|. \tag{3.13}$$

Each of the same over $\mathbf{j}$ and $\mathbf{k}$ runs over a complete set of vectors $|O_j)$, $(O_k|$, and the matrix $\underline{\mu}$ is calculated from

$$\mu_{jk} = (O_j|O_k). \tag{3.14}$$

In the special case we have in mind the projector is evaluated from

$$\bar{\mathbf{P}} \odot = \sum_{j,k} (\underline{\mu}^{-1})_{jk} O_j \mathrm{Tr}\, \rho_B O_k^+ \odot, \tag{3.15}$$

where $\odot$ denotes any operator acting on the Hilbert space of the total system. It is easy to show that $\bar{\mathbf{P}}$ satisfies

$$\bar{\mathbf{P}}^2 = \bar{\mathbf{P}} \tag{3.16}$$

and is hermitean with respect to the scalar product:

$$(\bar{\mathbf{P}}A|B) = (A|\bar{\mathbf{P}}B). \tag{3.17}$$

Our aim in this section is to derive equations of motion for the correlation functions $S_{jk}(t)$. To that end we multiply the disentanglement theorem [105], written as

$$e^{i(1-\tilde{P})Lt} = e^{iLt} + \int_0^t d\tau\, e^{iL(t-\tau)}(-i\tilde{P}L)\, e^{i(1-\tilde{P})L\tau}, \qquad (3.18)$$

where $\mathbf{L} = \mathbf{L_S} + \mathbf{L_B} + \mathbf{L_{SB}}$ is the Liouvillian of the total system, by $(1-\tilde{P})\dot{O}_l(0)$ from the right-hand side and obtain

$$e^{i(1-\tilde{P})Lt}(1-\tilde{P})\dot{O}_l(0) = f(t) + \int_0^t d\tau\, g(t,\tau) \qquad (3.19\mathrm{a})$$

with

$$f(t) = e^{iLt}(1-\tilde{P})\dot{O}_l(0), \qquad (3.19\mathrm{b})$$

$$g(t,\tau) = e^{iL(t-\tau)}(-i\tilde{P}L)\, e^{i(1-\tilde{P})L\tau}(1-\tilde{P})\dot{O}_l(0). \qquad (3.19\mathrm{c})$$

By use of the definition (3.13) of $\tilde{P}$, $f(t)$ and $g(t,\tau)$ become

$$f(t) = \dot{O}_l(t) - i \sum_j \Omega_{lj}^* O_j(t), \qquad (3.20)$$

$$g(t,\tau) = \sum_j \Phi_{lj}^*(\tau) O_j(t-\tau). \qquad (3.21)$$

Ω_{lj}^* is the complex conjugate of

$$\Omega_{lj} = \sum_k (LO_l \mid O_k)(\underline{\underline{\mu}}^{-1})_{kj} \qquad (3.22)$$

and $\Phi_{lj}^*(\tau)$ that of

$$\Phi_{lj}(\tau) = \sum_k (L\, e^{i(1-\tilde{P})L\tau}(1-\tilde{P})LO_l \mid O_k)(\underline{\underline{\mu}}^{-1})_{kj}. \qquad (3.23)$$

With the help of (3.20, 3.21) we can cast (3.19a) into the following form

$$\dot{O}_l(t) = i \sum_j \Omega_{lj}^* O_j(t) - \int_0^t d\tau \sum_j \Phi_{lj}^*(\tau) O_j(t-\tau) + e^{i(1-\tilde{P})Lt}(1-\tilde{P})\dot{O}_l(0). \qquad (3.24)$$

This equation describes the time evolution of operator O_l. The frequency matrix Ω_{lj}^* is connected with collective oscillations of the system, the memory functions $\Phi_{lj}^*(\tau)$ are related to its damping. Equation (3.24) can be considered a stochastic model of the open system, if the last term on the right hand side is a **stochastic operator** [100]. Whether or not this hypothesis holds is unimportant if we are only interested in equations for the correlation functions instead of those for the operators themselves. This is immediately seen if we insert (3.24) into the scalar product (describing a correlation function) $(O_m | \dot{O}_l(t))$. The fluctuation term vanishes and we get

$$(O_m | \dot{O}_l(t)) = i \sum_j \Omega_{lj}^*(O_m | O_j(t)) - \int_0^t d\tau \sum_j \Phi_{lj}^*(\tau)(O_m | O_j(t - \tau)). \qquad (3.25)$$

The quantity we want to calculate is $S_{jk}(t)$ which is just the complex conjugate of $(O_k | O_j(t))$. Forming the complex conjugate of (3.25) we end up with the following set of equations for the correlation functions

$$\dot{S}_{jk}(t) = -i \sum_l \Omega_{jl} S_{lk}(t) - \int_0^t d\tau \sum_l \Phi_{jl}(\tau) S_{lk}(t - \tau). \qquad (3.26)$$

In matrix notation this set of equations reads

$$\dot{\underline{\underline{\mathfrak{S}}}}(t) = -i\underline{\underline{\Omega}}\,\underline{\underline{\mathfrak{S}}}(t) - \int_0^t d\tau\, \underline{\underline{\Phi}}(\tau)\underline{\underline{\mathfrak{S}}}(t - \tau). \qquad (3.27)$$

Equations (3.26, 3.27) are sets of integro-differential equations. To arrive at a set of ordinary differential equations we wish to introduce the Markov approximation. We first transform to the interaction picture with respect to $\underline{\underline{\Omega}}$ by defining $\tilde{\underline{\underline{\mathfrak{S}}}}(t)$:

$$\underline{\underline{\mathfrak{S}}}(t) = e^{-i\underline{\underline{\Omega}}t}\tilde{\underline{\underline{\mathfrak{S}}}}(t), \qquad (3.28)$$

which yields the following equation of motion

$$\dot{\tilde{\underline{\underline{\mathfrak{S}}}}}(t) = -\int_0^t d\tau\, e^{i\underline{\underline{\Omega}}t}\underline{\underline{\Phi}}(\tau)\, e^{-i\underline{\underline{\Omega}}(t-\tau)}\, \tilde{\underline{\underline{\mathfrak{S}}}}(t - \tau)$$

$$= -\int_0^t d\tau\, e^{i\underline{\underline{\Omega}}t}\underline{\underline{\Phi}}^{(\ell)}(\tau)\, e^{-i\underline{\underline{\Omega}}t}\, \tilde{\underline{\underline{\mathfrak{S}}}}(t - \tau), \qquad (3.29)$$

where $\underline{\underline{\Phi}}^{(\ell)}(\tau)$ is defined by the second equality and can be calculated using the disentanglement theorem [1].

In the Markov approximation it is assumed that $\tilde{\underline{\underline{\mathfrak{S}}}}(t - \tau)$ is essentially constant on the time scale of the decay of $\underline{\underline{\Phi}}^{(\ell)}(\tau)$. Therefore, the τ dependence of $\tilde{\underline{\underline{\mathfrak{S}}}}(t - \tau)$ may be neglected and the upper limit of integration may be shifted to infinity. (One can show [1] that this approximation within the Mori formalism is essentially equivalent to the usual Markov approximation in the Liouville formalism for reduced density operators.) After transforming back to the Schrödinger picture we obtain the following equations for the correlation functions

$$\dot{\underline{\underline{\mathfrak{S}}}}(t) = -i\underline{\underline{\mathfrak{M}}}\,\underline{\underline{\mathfrak{S}}}(t), \qquad (3.30)$$

where

$$\mathfrak{M} = \underline{\underline{\Omega}} - i \int_0^\infty d\tau \, \underline{\underline{\Phi}}^{(\ell)}(\tau). \tag{3.31}$$

Using the definition (3.7) of the system operators O_j, we calculate the matrix Ω_{jk} as

$$\Omega_{jk} = \varepsilon_j \delta_{jk}, \qquad \varepsilon_j = \varepsilon_l - \varepsilon_m. \tag{3.32}$$

If we additionally apply the Born approximation, i.e. replace the exponential operator in (3.23) by $e^{i(1-\tilde{P})(L_S + L_B)\tau}$, the matrix elements of $\Phi^{(\ell)}$ read

$$\Phi_{jk}^{(\ell)}(\tau) = (\tilde{L}_{SB}(-\tau)(1 - \tilde{P})L_{SB} O_j \mid O_k). \tag{3.33}$$

In arriving at this result we have used further relations, namely $[\tilde{P}, L_O] = [\tilde{Q}, L_O] = 0$, where $\tilde{Q} = 1 - \tilde{P}$ and $L_O = L_S + L_B$. Moreover, the operator $\tilde{P}$ can be dropped from (3.33) if we take into account that L_{SB} is linear in the phonon operators.

A different form of the equations of motions for the correlation functions, which will prove useful in the line shape calculation of the next but one section, is obtained immediately from the formal solution of (3.30). Using the definition (3.10) of $S_{jk}(\tau)$ we have $S_{jk}(0) = \delta_{jk}$, i.e. $\mathfrak{S}(0) = 1$, and therefore

$$\mathfrak{S}(\tau) = e^{-i\mathfrak{M}t} \mathfrak{S}(0) = e^{-i\mathfrak{M}t} \tag{3.34}$$

also satisfies

$$\dot{\mathfrak{S}}(t) = -i\mathfrak{S}(t)\mathfrak{M} \tag{3.35}$$

which in index notation reads

$$\dot{S}_{jk}(t) = -\sum_l iS_{jl}(t) M_{lk}. \tag{3.36}$$

Multiplying (3.36) by v_j^* and summing over j we obtain

$$\dot{S}_k(t) = -\sum_l S_l(t) M_{lk}. \tag{3.37}$$

Or else in vector notation, interpreting the S_k as components of a vector

$$\dot{\mathfrak{S}}(t) = -i\mathfrak{M}^T \mathfrak{S}(t). \tag{3.38}$$

For the definition of the $S_k(t)$ we have used (3.8a, 3.9, 3.10), i.e.

$$S_k(t) = \sum_j v_j^* S_{jk}(t) = \text{Tr}(p(\tau) O_k \rho_B). \tag{3.39}$$

For completeness we give the double index notation for Ω_{jk} (3.32) and $\Phi_{jk}^{(\ell)}(\tau)$ (3.33):

$$\Omega_{nrlm} = \text{Tr}(\rho_B(\mathbf{L_S}|n\rangle\langle r|)^+ |l\rangle\langle m|), \tag{3.40}$$

$$\Phi_{nrlm}^{(\ell)}(\tau) = \text{Tr}(\rho_B(\hat{\mathbf{L}}_{SB}(-\tau)\mathbf{L}_{SB}|n\rangle\langle r|)^+ |l\rangle\langle m|). \tag{3.41}$$

4. Guest Molecule Coupled to a Single TLS

In this and the two sections to follow we consider a guest molecule coupled to a single TLS. The motivation for this procedure is two-fold: firstly, one can imagine physical situations where one TLS is much closer to the guest molecule — and therefore much stronger coupled to it — than all the others; secondly, one should learn about the essential line broadening mechanisms even from this simple situation.

The first subsection of this section gives the dipole moment operator whose two-time correlation function determines the optical line shape. The following two subsections present explicitly the equations of motion for the correlation functions, and their solution is given and discussed in the final subsection.

4.1. DIPOLE MOMENT OPERATOR

The dipole moment operator of the guest molecule connected with the optical transition is given by

$$p = \nu(|\alpha\rangle\langle\beta| + |\beta\rangle\langle\alpha|), \tag{4.1}$$

where, without loss of generality, ν is assumed to be real. This operator is defined on the Hilbert space of the guest molecule. To obtain an operator acting on the product space of guest molecule and TLS we could multiply (4.1) by the completeness relation (2.44) and transform to the states $|\rho n\rangle$ defined in (2.50). Instead we can use a slightly different method which allows for direct generalization to the case of a guest molecule interacting with several TLSs. In terms of the states (2.48) the completeness relation is represented as

$$\sum_{n=1}^{2} |\psi_n^\rho\rangle\langle\psi_n^\rho| = 1 \tag{4.2}$$

for fixed value of ρ, either α or β. Multiplication of the operator $|\sigma\rangle\langle\zeta|$ from the left-hand side with (4.2) for $\rho = \sigma$ and from the right-hand side with $\rho = \zeta$

yields

$$|\sigma\rangle\langle\zeta| = \sum_{n,m=1}^{2} |\psi_n^\sigma\rangle\langle\psi_n^\sigma| |\sigma\rangle\langle\zeta| |\psi_m^\zeta\rangle\langle\psi_m^\zeta|$$
$$= \sum_{n,m} |\sigma n\rangle\langle\zeta m| \langle\psi_n^\sigma|\psi_m^\zeta\rangle. \quad (4.3)$$

The matrix element in (4.3) is calculated using (2.48)

$$\langle\psi_n^\sigma|\psi_m^\zeta\rangle = \delta_{nm}\cos(\sigma_s - \zeta_s) + (1 - \delta_{nm})(-1)^n \sin(\sigma_s - \zeta_s). \quad (4.4)$$

The desired representation of p then reads

$$p = v\{\cos(\alpha_s - \beta_s)(|\alpha 1\rangle\langle\beta 1| + |\alpha 2\rangle\langle\beta 2| + |\beta 1\rangle\langle\alpha 1| + |\beta 2\rangle\langle\alpha 2|) +$$
$$+ \sin(\alpha_s - \beta_s)(-|\alpha 1\rangle\langle\beta 2| + |\alpha 2\rangle\langle\beta 1| + |\beta 1\rangle\langle\alpha 2| + |\beta 2\rangle\langle\alpha 1|)\}, \quad (4.5)$$

and using the more compact notation of (2.51) we have

$$p = v\{\cos(\alpha_s - \beta_s)(|1\rangle\langle 3| + |2\rangle\langle 4| + |3\rangle\langle 1| + |4\rangle\langle 2|) +$$
$$+ \sin(\alpha_s - \beta_s)(-|1\rangle\langle 4| + |2\rangle\langle 3| + |3\rangle\langle 2| - |4\rangle\langle 1|)\}. \quad (4.6)$$

For the following calculations it is convenient to mark each transition operator $|\rho n\rangle\langle\sigma m|$ by an index **j**, which is chosen in the following way in order be consistent with recent work [73]:

$$O_1 = O_{\alpha 1}^{\beta 1} = |3\rangle\langle 1|, \qquad O_2 = O_{\alpha 2}^{\beta 2} = |4\rangle\langle 2|,$$
$$O_3 = O_{\alpha 2}^{\beta 1} = |3\rangle\langle 2|, \qquad O_4 = O_{\alpha 1}^{\beta 2} = |4\rangle\langle 1|,$$
$$O_5 = O_1^+ = O_{\beta 1}^{\alpha 1} = |1\rangle\langle 3|, \qquad O_6 = O_2^+ = O_{\beta 2}^{\alpha 2} = |2\rangle\langle 4|,$$
$$O_7 = O_3^+ = O_{\beta 1}^{\alpha 2} = |2\rangle\langle 3|, \qquad O_8 = O_4^+ = O_{\beta 2}^{\alpha 1} = |1\rangle\langle 4|,$$
$$\{O_9, O_{10}, \ldots, O_{16}\} = \{|\rho n\rangle\langle\rho m| \mid \rho = \alpha, \beta; n, m = 1, 2\}. \quad (4.7)$$

As we shall see, there is no need to further specify the indices for operators which are diagonal in the wave functions of the guest molecule. However, it is useful to define an index $\bar{\mathbf{j}}$ which is called conjugate to **j** and is uniquely determined by

$$O_{\bar{j}} = O_j^+. \quad (4.8)$$

With the notations of (4.7) we obviously have

$$\bar{\mathbf{j}} \equiv \mathbf{j} + 4 \bmod 8 \quad (4.9)$$

for $\mathbf{j} \leq 8$. Furthermore, p is represented compactly in the form (3.8a) as

$$p = \sum_j v_j^* O_j^+, \quad (4.10)$$

where

$$v_j = v_j^* = \begin{cases} v \cos(\alpha_s - \beta_s) & \text{for } j = 1, 2 \bmod 4 \\ (-1)^{j+1} v \sin(\alpha_s - \beta_s) & \text{for } j = 0, 3 \bmod 4 \end{cases} \quad (4.11)$$

and $j \leq 8$, and

$$v_j = 0 \text{ for } j > 8. \quad (4.12)$$

A representation of the operator $[p, \rho_S]$ (3.8b) is obtained from

$$[p, \rho_S] = \sum_j v_j [O_j, \rho_S] \quad (4.13)$$

and (2.51, 4.7)

$$[O_{\sigma m}^{\rho n}, \rho_S] = \frac{1}{Z_S} [|\rho n\rangle\langle\sigma m|, e^{-\beta H_S}] = \frac{1}{Z_S} O_{\sigma m}^{\rho n} (e^{-\beta \varepsilon_{\sigma m}} - e^{-\beta \varepsilon_{\rho n}}). \quad (4.14)$$

Using these expressions we immediately have

$$[p, \rho_S] = \sum_j \gamma_j O_j \quad (4.15)$$

with

$$\gamma_j = v_j \xi_j, \quad (4.16)$$

and

$$\xi_j = \frac{1}{Z_S} (e^{-\beta \varepsilon_{\sigma m}} - e^{-\beta \varepsilon_{\rho n}}). \quad (4.17)$$

4.2. Equations for correlation functions

To obtain the equations of motion for the correlation functions we have to determine the explicit form of the matrix $\mathfrak{M}$ which is given by (3.31)

$$M_{jk} = \Omega_{jk} - i \int_0^\infty d\tau \, \Phi_{jk}^{(\ell)}(\tau). \quad (4.18)$$

Ω_{jk} has been determined in (3.32, 3.40), where

$$\varepsilon_j = \varepsilon_{\rho n} - \varepsilon_{\sigma m} \equiv \varepsilon_{\sigma m}^{\rho n}. \quad (4.19)$$

The evaluation of

$$\Phi_{jk}^{(\ell)}(\tau) = \text{Tr}(\rho_B O_k^+ \tilde{L}_{SB}(-\tau) L_{SB} O_j)^*$$
$$= \{\text{Tr}_S(O_k^+ \text{Tr}_B \rho_B \, e^{-iL_0\tau} L_{SB} \, e^{iL_0\tau} L_{SB} \, O_j)\}^* \quad (4.20)$$

(see (3.33, 3.41)) is straightforward but somewhat tedious. After the evaluation of the time integral in (4.18) we arrive at

$$\int_0^\infty d\tau \, \Phi_{jk}^{(\ell)}(\tau) = \delta_{\zeta\rho}\delta_{\xi\sigma}(\delta_{l\tilde{n}}\delta_{rm}\{-iP_1(A_\rho(-1)^n - A_\sigma(-1)^m) +$$

$$+ (D - iP_2)(A_\rho(-1)^n - A_\sigma(-1)^m)A_\rho(-1)^n + (D_{\rho n\tilde{n}} - iP_{\rho n\tilde{n}})B_\rho^2 -$$

$$- (D + iP_2)(A_\rho(-1)^n - A_\sigma(-1)^m)A_\sigma(-1)^m + (D_{\sigma m\tilde{m}} + iP_{\sigma m\tilde{m}})B_\sigma^2\} +$$

$$+ \delta_{l\tilde{n}}\delta_{r\tilde{m}}\{-(D_{\rho\tilde{n}n} - iP_{\rho\tilde{n}n})B_\rho B_\sigma - (D_{\sigma\tilde{m}m} + iP_{\sigma\tilde{m}m})B_\rho B_\sigma\} +$$

$$+ \delta_{l\tilde{n}}\delta_{rm}\{-iP_1 B_\rho - (D + iP_2)B_\rho A_\sigma(-1)^m +$$

$$+ (D - iP_2)B_\rho A_\rho(-1)^{\tilde{n}} +$$

$$+ (D_{\rho\tilde{n}n} - iP_{\rho\tilde{n}n})(A_\rho(-1)^n - A_\sigma(-1)^m)B_\rho\} +$$

$$+ \delta_{ln}\delta_{r\tilde{m}}\{iP_1 B_\sigma - (D - iP_2)B_\sigma A_\rho(-1)^n +$$

$$+ (D + iP_2)B_\sigma A_\sigma(-1)^{\tilde{m}} -$$

$$- (D_{\sigma\tilde{m}m} + iP_{\sigma\tilde{m}m})(A_\rho(-1)^n - A_\sigma(-1)^m)B_\sigma\}). \qquad (4.21)$$

In this expression we have used the following abbreviations

$$P_1 = \sum_q 2D_q S_{-q} \frac{P}{\omega_q}, \qquad (4.22a)$$

$$P_2 = \sum_q |D_q|^2 \frac{P}{\omega_q}, \qquad (4.22b)$$

$$P_{\rho n\tilde{n}} = \sum_q |D_q|^2 \left\{ (n_q + 1) \frac{P}{\omega_q + \varepsilon_{\rho n}^{\rho\tilde{n}}} - n_q \frac{P}{\omega_q - \varepsilon_{\rho n}^{\rho\tilde{n}}} \right\}$$

$$= \sum_q |D_q|^2 \left\{ (n_q + 1) \frac{P}{\omega_q - \varepsilon_{\rho\tilde{n}}^{on}} - n_q \frac{P}{\omega_q + \varepsilon_{\rho\tilde{n}}^{on}} \right\} \qquad (4.22c)$$

$$D = \sum_q |D_q|^2 (2n_q + 1)\pi\delta(\omega_q) \qquad (4.22d)$$

$$D_{\rho n\tilde{n}} = \sum_q |D_q|^2 (n_q + \delta_{n2})\pi\delta(\omega_q - |\varepsilon_{\rho\tilde{n}}^{on}|). \qquad (4.22e)$$

P reminds that in evaluating integrals only the principal value has to be taken. $D_\mathbf{q}$ and $S_\mathbf{q}$ are defined in (2.28), and

$$n_\mathbf{q} = 1/(e^{\beta\omega_\mathbf{q}} - 1) \tag{4.23}$$

is the mean phonon occupation number of mode $\mathbf{q} = (\mathfrak{q}, \sigma)$ in thermal equilibrium and stems from the evaluation of bath averages.

Equation (4.21) together with (4.19) and (3.32) determines the tensor $\mathfrak{M}$. These expressions show that $M\binom{\rho n}{\sigma m}\binom{\zeta l}{\xi r}$ (the notation is obvious from (4.7)) is diagonal in the indices which characterize the states of the guest molecule. This means that the system of equations for the 16 correlation functions, which can be formed by the various index combinations, factorizes into four 4×4 problems. For the description of optical transitions we only need the cases $(\rho, \sigma) = (\alpha, \beta)$ and $(\rho, \sigma) = (\beta, \alpha)$ as is seen from the representation (4.5) of the dipole moment. Furthermore, one can show that the correlation functions in the two cases can be derived from each other. For the investigation of the optical line shape we therefore have to solve one 4×4 eigenvalue problem.

To obtain equations for the relevant correlation functions we denote the vector of the first four components (see (3.37) and (4.7)) of $\mathfrak{S}(t)$ by $\mathbf{S}(t)$ and the transposed of the corresponding upper left part of $-i\mathfrak{M}$ by $\mathbf{N}$. Thus from (3.38) we arrive at

$$\dot{\mathbf{S}}(t) = \mathbf{N}\mathbf{S}(t). \tag{4.24}$$

If a complete set of eigenvectors $\{\mathbf{x}^k\}$ of the (neither hermitean nor antihermitean) matrix $\mathbf{N}$ exists, $\mathbf{S}(t)$ can be written

$$\mathbf{S}(t) = \sum_{k=1}^{4} s_k e^{R_k t} \mathbf{x}^k. \tag{4.25}$$

The R_k are the generally complex eigenvalues of $\mathbf{N}$, the s_k are determined by the initial condition $\mathbf{S}(0)$. The optical line shape (3.1) is then given by a superposition of 16 (complex) Lorentzian lines. Their central frequencies are determined by the imaginary parts of the eigenvalues, their widths by the real parts. Therefore, the solution of the eigenvalue problem

$$(\mathbf{N} - R\mathbf{1})\mathbf{x} = 0 \tag{4.26}$$

provides a first piece of information on the optical linewidth of the guest molecule.

4.3. EVALUATION OF THE COEFFICIENTS (DEBYE MODEL)

To evaluate the coefficients (4.22) we introduce densities of states (readopting temporarily a notation in which the phonon branch indices are explicitly written):

$$\sum_{q\sigma} \ldots \to \sum_{\sigma} \int_0^\infty d\omega \, g_\sigma(\omega) \ldots \quad (4.27)$$

Starting with the terms containing δ functions, we obtain

$$D = \sum_\sigma \frac{\pi f_\sigma^2}{8\rho V c_\sigma^2 \omega_c^{\kappa-1}} \lim_{\omega \to 0} g_\sigma(\omega) \omega^\kappa (2n(\omega) + 1), \quad (4.28)$$

where

$$f_\sigma^2 \equiv |f_\sigma^l - f_\sigma^u|^2 \quad (4.29)$$

and

$$n(\omega) = \frac{1}{e^{\beta\omega} - 1}. \quad (4.30)$$

As $\omega \to 0$, the density of states $g_\sigma(\omega)$ is proportional to ω^2 (in a three-dimensional system). This enables us to conclude that

$$D = 0 \quad (4.31)$$

for $\kappa > -1$. Abbreviating $\varepsilon_{\rho 1}^{\rho 2}$ by ε, we get for the next expression

$$D_{\rho n \bar{n}}(\varepsilon) = \sum_\sigma \frac{\pi f_\sigma^2}{8\rho V c_\sigma^2} \varepsilon \left(\frac{\varepsilon}{\omega_c}\right)^{\kappa-1} g_\sigma(\varepsilon)(n(\varepsilon) + \delta_{n2}). \quad (4.32)$$

The principal value terms are given by

$$P_1 = \sum_\sigma \frac{f_\sigma^{(d)}}{4\rho V c_\sigma^2} \int_0^\infty d\omega \left(\frac{\omega}{\omega_c}\right)^{\kappa-1} g_\sigma(\omega), \quad (4.33a)$$

where

$$f_\sigma^{(d)} \equiv (f_\sigma^u)^2 - (f_\sigma^l)^2, \quad (4.33b)$$

and

$$P_2 = \sum_\sigma \frac{f_\sigma^2}{8\rho V c_\sigma^2} \int_0^\infty d\omega \left(\frac{\omega}{\omega_c}\right)^{\kappa-1} g_\sigma(\omega), \quad (4.34)$$

$$P_{\rho n \bar{n}}(\varepsilon) = \sum_\sigma \frac{f_\sigma^2}{8\rho V c_\sigma^2} \{I_{\sigma 1}((-1)^n \varepsilon) + (-1)^n I_{\sigma 2}(\varepsilon)\}, \quad (4.35)$$

where

$$I_{\sigma 1}(\varepsilon) = \int_0^\infty d\omega \left(\frac{\omega}{\omega_c}\right)^{\kappa-1} \omega g_\sigma(\omega) \frac{P}{\omega - \varepsilon}, \qquad (4.36)$$

$$I_{\sigma 2}(\varepsilon) = \int_0^\infty d\omega \left(\frac{\omega}{\omega_c}\right)^{\kappa-1} \omega g_\sigma(\omega) n(\omega) \left\{\frac{P}{\omega - \varepsilon} - \frac{P}{\omega + \varepsilon}\right\}. \qquad (4.37)$$

The case $\kappa = 1$ is especially simple: the integrals in (4.33, 4.34) just yield the normalization constant. We now introduce an average velocity of sound as well as phonon coupling constant by the relation

$$\frac{f^2}{c^5} = \frac{1}{3} \sum_{\sigma=1}^3 \frac{f_\sigma^2}{c_\sigma^5} \qquad (4.38)$$

and replace $g_\sigma(\omega)$ by

$$g_{\sigma D}(\omega) \equiv \frac{V\omega^2}{2\pi^2 c_\sigma^3} \Theta(\omega_D - \omega) \qquad (4.39)$$

which results in

$$\sum_\sigma \frac{f_\sigma^2}{c_\sigma^2} g_\sigma(\omega) \to 3 \frac{f^2}{c^2} \frac{V\omega^2}{2\pi^2 c^3} \Theta(\omega_D - \omega) \equiv \frac{f^2}{c^2} g_D(\omega), \qquad (4.40)$$

the last identity defining a Debye density of states. Because of the prefactors f_σ^2/c_σ^2, our Debye velocity of sound is defined in a manner differing from the usual way [106]. However, from experiments [107, 108] one concludes (at least for silica glasses) that $f_L^2 \approx 2f_T^2$ which keeps these prefactors near to 1. Moreover, we assume that $f_\sigma^{(d)}/f_\sigma^2 = f^{(d)}/f^2$ does not depend on the polarization σ and thus an averaging procedure similar to (4.38) leads to a single coupling constant in the calculation of P_1, too.

The Debye frequency is then given by

$$\sum_\sigma \int_0^\infty d\omega \frac{f_\sigma^2}{c_\sigma^2} g_\sigma(\omega) = 3 \frac{f^2}{c^2} \int_0^{\omega_D} d\omega \frac{V\omega^2}{2\pi^2 c^3} = 3N \frac{f^2}{c^2}, \qquad (4.41)$$

where $3N$ is the number of 'acoustic degrees of freedom' of the material (which is usually smaller than three times the number of atoms). The characteristic frequency ω_c is put equal to the Debye frequency.

Introducing reduced quantities by

$$\omega = \omega_D x, \qquad \varepsilon = \omega_D e,$$

$$T = T_r \theta_D \left(\theta_D = \frac{\omega_D}{k} \right) \Rightarrow \beta\omega = \frac{x}{T_r} \equiv y,$$

$$n(\omega) = n_r(y),$$

$$g_D(\omega) = \frac{3V\omega_D^2}{2\pi^2 c^3} g_r(x) \Rightarrow g_r(x) = x^2 \Theta(1-x), \tag{4.42}$$

we end up with

$$D_{\rho n\bar{n}}(\varepsilon) = \omega_D \pi F e^\kappa g_r(e) \left(n_r \left(\frac{e}{T_r} \right) + \delta_{n2} \right), \tag{4.43}$$

$$P_2 = \int_0^1 dx\, x^{\kappa-1} g_r(x) = \omega_D F\, \frac{1}{\kappa+2} \tag{4.44}$$

$$P_1 = \frac{2f^{(d)}}{f^2} P_2, \tag{4.45}$$

$$P_{\rho n\bar{n}}(\varepsilon) = \omega_D F \{ i_1((-1)^n e) + (-1)^n e i_2(e, T_r) \}, \tag{4.46}$$

$$i_1(e) = \int_0^1 dx\, \frac{x^\kappa g_r(x) - e^\kappa g_r(e)}{x-e} + e^\kappa g_r(e) \int_0^1 dx\, \frac{P}{x-e}$$

$$= \int_0^1 dx\, \frac{x^\kappa g_r(x) - e^\kappa g_r(e)}{x-e} + e^\kappa g_r(e) \ln\frac{1-e}{e}, \tag{4.47}$$

$$i_2(e, T_r) = 2 \int_0^1 dx\, \frac{x^\kappa g_r(x) n_r\left(\frac{x}{T_r}\right) - e^\kappa g_r(e) n_r\left(\frac{e}{T_r}\right)}{x^2 - e^2} +$$

$$+ e^{\kappa-1} g_r(e) n_r\left(\frac{e}{T_r}\right) \ln\frac{1-e}{1+e}, \tag{4.48}$$

where F is a dimensionless constant, given by

$$F = \frac{3f^2 \omega_D^2}{16\pi^2 \rho c^5} = 0.0793\, \frac{(f/[\text{eV}])^2 (\theta_D/[\text{K}])^2}{(\rho/[\text{g/cm}^3])(c/[\text{km/s}])^5}. \tag{4.49}$$

From (4.47, 4.48) we gather that the principal value terms (4.46) diverge at $e = 1$. Reconsidering (4.33)–(4.35), however, we recognize that this is an artifact of the Debye model the density of states of which contains a singularity at the Debye frequency which is of a type not to be found in real materials. From the well-known behavior of singularities of densities of states in real systems [106, 109] (e.g. van Hove type) we conclude that the integrals (4.34, 4.35) converge everywhere. Because we are treating a low-temperature model, there is some hope that the weak (logarithmic) divergencies at high frequencies in (4.47, 4.48) will not be too important.

4.4. Solution to the Eigenvalue Problem

The calculation of the eigenvalues of **N** is rather tedious. A detailed description of the procedure is given in [1]. In this place, we restrict ourselves to the starting point of the calculation and the result. Matrix **N** is shown in Table I. Some new shorthand notations have been introduced for the coefficients calculated in the preceding subsection, namely

$$D_{12} \equiv D_{\alpha 12}(\varepsilon_{\alpha 1}^{\alpha 2}) = \pi \omega_D F e_{21}^{\kappa+2} n_r \left(\frac{e_{21}}{T_r} \right), \tag{4.50a}$$

$$D_{21} \equiv D_{\alpha 21}(\varepsilon_{\alpha 1}^{\alpha 2}) = \pi \omega_D F e_{21}^{\kappa+2} \left(n_r \left(\frac{e_{21}}{T_r} \right) + 1 \right), \tag{4.50b}$$

$$D_{34} \equiv D_{\beta 12}(\varepsilon_{\beta 1}^{\beta 2}) = \pi \omega_D F e_{43}^{\kappa+2} n_r \left(\frac{e_{43}}{T_r} \right), \tag{4.50c}$$

$$D_{43} \equiv D_{\beta 21}(\varepsilon_{\beta 1}^{\beta 2}) = \pi \omega_D F e_{43}^{\kappa+2} \left(n_r \left(\frac{e_{43}}{T_r} \right) + 1 \right), \tag{4.50d}$$

$$P_{12} \equiv P_{\alpha 12}(\varepsilon_{\alpha 1}^{\alpha 2}), \tag{4.51a}$$

$$P_{21} \equiv P_{\alpha 21}(\varepsilon_{\alpha 1}^{\alpha 2}), \tag{4.51b}$$

$$P_{34} \equiv P_{\beta 12}(\varepsilon_{\beta 1}^{\beta 2}), \tag{4.51c}$$

$$P_{43} \equiv P_{\beta 21}(\varepsilon_{\beta 1}^{\beta 2}), \tag{4.51d}$$

After a renormalization transformation of the starting Hamiltonian (2.72), which enables us to take into account contributions of P_1 to infinite order of perturbation theory, the eigenvalue problem can be solved exactly. The solution is written

$$R_k = -\tfrac{1}{2}A + \tfrac{1}{2} i s_k \sqrt{B + 2 t_k \sqrt{C}}, \tag{4.52}$$

TABLE I
Matrix **N** $(=-i\boldsymbol{\Omega} - \int_0^\infty d\tau\,\boldsymbol{\Phi}'(\tau))$

$-i\varepsilon_{31} - D_{12}B_\alpha^2 - D_{34}B_\beta^2 - iP_{12}B_\alpha^2 + iP_{34}B_\beta^2 + iP_1(A_\alpha - A_\beta) - iP_2(A_\alpha^2 - A_\beta^2)$	$(D_{12} + D_{34})B_\alpha B_\beta + (iP_{12} - iP_{34})B_\alpha B_\beta$	$-D_{12}(A_\alpha + A_\beta)B_\alpha - iP_{12}(A_\alpha + A_\beta)B_\alpha - iP_1 B_\alpha + iP_2(A_\alpha + A_\beta)B_\alpha$	$-D_{34}(A_\alpha + A_\beta)B_\beta + iP_{34}(A_\alpha + A_\beta)B_\beta + iP_1 B_\beta - iP_2(A_\alpha + A_\beta)B_\beta$
$(D_{21} + D_{43})B_\alpha B_\beta + (iP_{21} - iP_{43})B_\alpha B_\beta$	$-i\varepsilon_{42} - D_{21}B_\alpha^2 - D_{43}B_\beta^2 - iP_{21}B_\alpha^2 + iP_{43}B_\beta^2 - iP_1(A_\alpha - A_\beta) - iP_2(A_\alpha^2 - A_\beta^2)$	$D_{43}(A_\alpha + A_\beta)B_\beta - iP_{43}(A_\alpha + A_\beta)B_\beta + iP_1 B_\beta + iP_2(A_\alpha + A_\beta)B_\beta$	$D_{21}(A_\alpha + A_\beta)B_\alpha + iP_{21}(A_\alpha + A_\beta)B_\alpha - iP_1 B_\alpha + iP_2(A_\alpha + A_\beta)B_\alpha$
$D_{21}(A_\alpha - A_\beta)B_\alpha + iP_{21}(A_\alpha - A_\beta)B_\alpha - iP_1 B_\alpha - iP_2(A_\alpha - A_\beta)B_\alpha$	$D_{34}(A_\alpha - A_\beta)B_\beta - iP_{34}(A_\alpha - A_\beta)B_\beta + iP_1 B_\beta + iP_2(A_\alpha - A_\beta)B_\beta$	$-i\varepsilon_{32} - D_{21}B_\alpha^2 - D_{34}B_\beta^2 - iP_{21}B_\alpha^2 + iP_{34}B_\beta^2 - iP_1(A_\alpha + A_\beta) - iP_2(A_\alpha^2 - A_\beta^2)$	$(D_{21} + D_{34})B_\alpha B_\beta + (iP_{21} - iP_{34})B_\alpha B_\beta$
$-D_{43}(A_\alpha - A_\beta)B_\beta + iP_{43}(A_\alpha - A_\beta)B_\beta + iP_1 B_\beta - iP_2(A_\alpha - A_\beta)B_\beta$	$-D_{12}(A_\alpha - A_\beta)B_\alpha - iP_{12}(A_\alpha - A_\beta)B_\alpha - iP_1 B_\alpha + iP_2(A_\alpha - A_\beta)B_\alpha$	$(D_{12} + D_{43})B_\alpha B_\beta + (iP_{12} - iP_{43})B_\alpha B_\beta$	$-i\varepsilon_{41} - D_{12}B_\alpha^2 - D_{43}B_\beta^2 - iP_{12}B_\alpha^2 + iP_{43}B_\beta^2 + iP_1(A_\alpha + A_\beta) - iP_2(A_\alpha^2 - A_\beta^2)$

where the signs s_k and t_k are chosen according to

$$s_k = \begin{cases} (-1)^k \operatorname{sign}(\varepsilon_{21} - \varepsilon_{43}) & \text{for } k = 1, 2 \\ (-1)^{k+1} & \text{for } k = 3, 4 \end{cases} \quad (\operatorname{sign}(0) \equiv 1), \quad (4.53a)$$

$$t_k = \begin{cases} -1 & \text{for } k = 1, 2 \\ 1 & \text{for } k = 3, 4 \end{cases}, \quad (4.53b)$$

and

$$A = i(\varepsilon_{31} + \varepsilon_{42}) + r_{12} + r_{21} + r_{34} + r_{43}, \quad (4.54a)$$

$$B = -((i\varepsilon_{43} + r_{43} - r_{34})^2 + (i\varepsilon_{21} + r_{12} - r_{21})^2 + \\ + 2(r_{12} + r_{34})(r_{21} + r_{43}) + 2(r_{21} + r_{34})(r_{12} + r_{43})) \quad (4.54b)$$

$$C = (\varepsilon_{21}\varepsilon_{43} - i\varepsilon_{43}(r_{12} - r_{21}) + i\varepsilon_{21}(r_{34} - r_{43}))^2 - \\ - 4(B_\beta^2 - B_\alpha^2)(\varepsilon_{21}^2 R_{34} R_{43} B_\beta^2 - \varepsilon_{43}^2 R_{12} R_{21} B_\alpha^2), \quad (4.54c)$$

where

$$\varepsilon_{21} = \varepsilon_\alpha, \quad \varepsilon_{43} = \varepsilon_\beta, \quad \varepsilon_\rho = \sqrt{W^2 + [\Delta - 2(V_\rho - P_1)]^2}, \quad (4.55)$$

and

$$R_{12} = D_{12} + iP_{12} - iP_2, \quad (4.56a)$$

$$R_{21} = D_{21} + iP_{21} - iP_2, \quad (4.56b)$$

$$R_{34} = D_{34} - iP_{34} + iP_2, \quad (4.56c)$$

$$R_{43} = D_{43} - iP_{43} + iP_2, \quad (4.56d)$$

$$r_{12} = R_{12} B_\alpha^2, \quad r_{21} = R_{21} B_\alpha^2, \quad r_{34} = R_{34} B_\beta^2, \quad r_{43} = R_{43} B_\beta^2. \quad (4.57)$$

From the equation of motion for the reduced density matrix (see [73]), it can be concluded, that the r_{jk} are transition rates corresponding to phonon transitions within the renormalized TLSs. For example, r_{12} is the rate of transitions from state $|1\rangle$ to $|2\rangle$ under phonon absorption, whereas r_{21} describes transitions from $|2\rangle$ to $|1\rangle$, i.e. phonon emission processes. $B_\rho = W/\varepsilon_\rho$ in analogy to (2.70). Because we are dealing with complex numbers, we have to be more specific about square roots. We define

$$\sqrt{z} \equiv \sqrt{|z|}\, e^{i\phi}, \quad \text{where } \phi \in \left(-\frac{\pi}{2}, \frac{\pi}{2}\right]. \quad (4.58)$$

As is shown in [1], the full inclusion of principal value terms in Table I leads to inconsistencies such as exploding correlation functions (which is not, however, a consequence of the divergences at $e = 1$, see (4.47, 4.48)). Therefore, we shall neglect the principal value terms as is usually done in the literature [110, 111]. This means that the R_{jk} in (4.56a–d) are replaced by the D_{jk} (of (4.50)).

For these eigenvalues, the choice of signs given in (4.53a,b) (together with definition (4.58)) guarantees that the association between optical transitions and eigenvalues remains the same as in Figure 5.

4.5. DEPENDENCE OF THE EIGENVALUES ON SYSTEM PARAMETERS

We investigate the behavior of the eigenvalues, especially their real parts, as functions of the system parameters to get some insight into the relative importance of these parameters.

First, we mention that the deviations of the imaginary parts from the unperturbed energies are small which confirms that perturbation theory (i.e. the Born approximation) is applicable. More details regarding this point are given in [1].

Let us now consider the real parts of the eigenvalues. In Figure 6 they are given as functions of the asymmetry parameter Δ for fixed values of λ, ΔV and the temperature. To facilitate the discussion the energy splittings ε_{21} and ε_{43} are shown as well.

Considering the linewidth of transition R_1 we recognize that it has a maximum near $\Delta = 2V_\alpha$ and $\Delta = 2V_\beta$ where the energies ε_{21} and ε_{43} take on their minima. Because the width of R_1 is determined by phonon absorption alone, it becomes large where the phonon occupation number does so. This holds even though the density of phonon states becomes minimal there, because the occupation number varies stronger with frequency than the density of states. Thus the maxima of the linewidth are only slightly shifted as against to the minima of ε_{21}, ε_{43}.

There are jump discontinuities in the linewidth at $\Delta = 2V_{\alpha/\beta} \pm \sqrt{\omega_D^2 - W^2}$. This is a consequence of the Debye density of states having a jump discontinuity. The figure shows that below $\Delta \approx 0.2\omega_D$ there is no dephasing at all, because both ε_{21} and ε_{43} are larger than ω_D and no phonon transitions are possible. At $\Delta \approx 0.2\omega_D$, ε_{21} becomes smaller than ω_D, the levels ε_1 and ε_2 get finite widths, the linewidth of the optical transition R_1 jumps to a finite value. Another jump happens at $\Delta \approx 0.4\omega_D$, because now the lifetime of the upper level ε_3 becomes limited by phonon (absorption) processes.

Comparing the real part of R_4 with that of R_1 we see that below $\Delta \approx 0.4\omega_D$ the two linewidths coincide. This is because the phonon system does not see a difference between the upper levels ε_3 and ε_4 whereas the lower level is equal for both transitions. With increasing Δ, two competing tendencies become effective. The densities of states decrease (ε_{21} and ε_{43} become smaller), the mean phonon numbers increase. In the case of R_1, where only absorption is important, the second effect overcompensates the first, the negative real part increases. For R_2, R_3 (above $\Delta \approx 0.2\omega_D$), and R_4 (above $\Delta \approx 0.4\omega_D$), however, there are emission processes, especially the temperature independent spontaneous emission. As long as the mean phonon occupation number is smaller than 1, the decrease in dephasing by spontaneous emission is larger than the increase by induced emission or absorption. This is because the spontaneous emission works as if it were induced

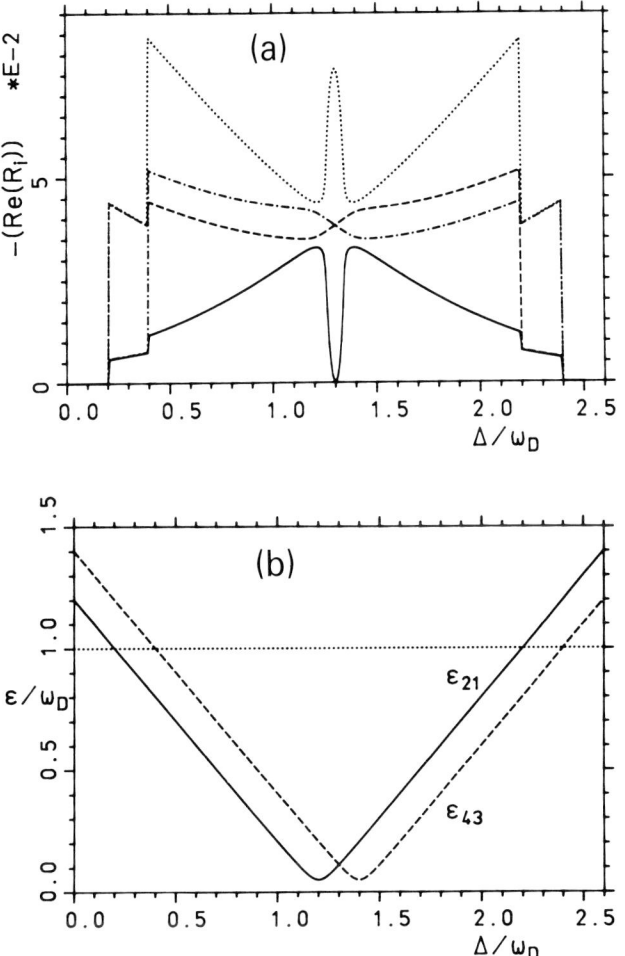

Fig. 6. (a) Negative real parts (in units of ω_D) of the eigenvalues of the single-TLS problem. Full line: R_1; dotted line: R_2; dashed line: R_3; dash-dotted line: R_4. Parameter values: $\lambda = 3$, $T = 0.5\theta_D$, $V_a = 0.6\omega_D$, $\Delta V = 0.1\omega_D$, $\theta_D = 300$ K, $c = 3.75$ km/s, $\rho = 2$ g/cm^3, $f = 1$ eV. (b) Energies ε_{21}, ε_{43} as a function of Δ for the same parameter values.

by one phonon. Therefore, the linewidths of these transitions decrease with increasing Δ until $\Delta \approx 2V_a$ is reached. That the real part of R_1 goes to zero as Δ approaches $\Delta = V_\alpha + V_\beta$ is a motional narrowing effect. For around this point the difference between ε_{21} and ε_{43} becomes small as compared to the rates of scattering between the transitions ε_{31} (R_1) and ε_{42} (R_2).

The symmetry of the picture with respect to the axis $\Delta = V_\alpha + V_\beta$ is a consequence of the symmetry with respect to index conjugation and subsequent complex conjugation. We have chosen a rather large temperature in the figure to obtain a visible linewidth of R_1, which at lower temperatures becomes much smaller than the other widths. Furthermore, with decreasing temperature the dip in the R_1 width (at $\Delta = V_\alpha + V_\beta$) becomes narrower, the other transitions become more and more dominated by spontaneous emission. An increase in λ has a similar effect as regards the dip; moreover, it decreases the distance between the four eigenvalues at $\Delta \approx 2V_\alpha, 2V_\beta$. For details on the dependence of the eigenvalue real parts on λ see [1].

Because the density of states has a strong effect on the real parts of R_2 to R_4, the failure of the Debye model at high frequencies will become important for these eigenvalues. It can be shown [1], however, that the temperature dependent part of the line width of the transition j, i.e. $-\text{Re}(R_j(T) - R_j(T = 0))$ is close to $-\text{Re}(R_1)$ (except in the neighborhood of $\Delta = V_\alpha + V_\beta$), which in turn is rather insensitive to the special shape of the density of states.

From the above reflections we expect our theory to give a better approximation for R_1 than for the other eigenvalues and, therefore, restrict our considerations to the former. This will be justified further in Section 7.

Figure 7 shows the dependence of the R_1 linewidth on ΔV (2.54) for different temperatures. The log–log plot displays some interesting features. First, over several decades, the real part of R_1 is almost independent of ΔV (except for the peaks in Figure 7(b) which stem from the maximum at $\Delta \approx 2V_\beta$). Only for very small ΔV does the linewidth fall off proportional ΔV^2 ($\Delta \neq 2V_\alpha, 2V_\beta$) or ΔV^4 ($\Delta = 2V_\alpha, \Delta = 2V_\beta$). Therefore, the region where the linewidth can be expanded in powers of ΔV is rather small. Second, the ΔV value where the crossover to a stronger ΔV dependence takes place, shifts to lower values as temperature

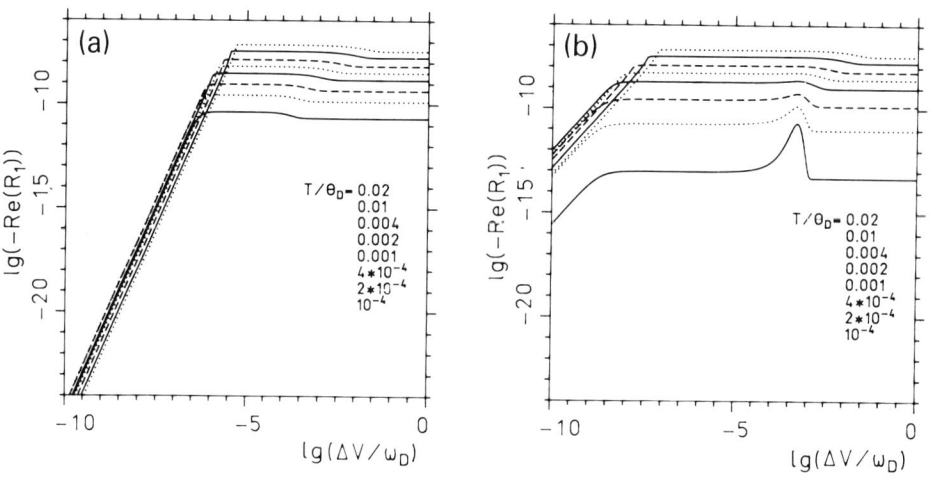

Fig. 7. Linewidth (in units of ω_D) of the optical transition R_1 as a function of ΔV. $\lambda = 5$, same material parameters (θ_D, c, ρ, f) as in Figure 6. (a) $\Delta = 2V_\alpha$; the straight curve segments on the lefthand side of the picture describe a ΔV^4 power law. (b) $\Delta = 2V_\alpha + 10^{-3}\omega_D$, to the left of their breaks the curves are proportional ΔV^2.

decreases. This fact is rather important: it means that if there are many TLSs with small couplings ΔV, more of them contribute significantly to the linewidth at low than at high temperatures. Below the 'critical' ΔV value, the contribution of most TLSs is negligible because of its fast decrease as a function of ΔV; above, however, all TLSs contribute almost equal amounts. In these arguments implicitly the assumption is contained that there are single-TLS contributions which add up, an assumption which still has to be justified.

4.6. ANALYTICAL APPROXIMATIONS

We consider two limiting cases in which analytical approximations yield simpler formulae than (4.52)–(4.54). The first case has been treated in the literature [67, 68, 90], though not yet completely. We start with the supposition

$$V_\alpha, V_\beta \ll E = \sqrt{\Delta^2 + W^2}. \tag{4.59}$$

Introducing the abbreviations

$$G \equiv \frac{3f^2}{16\pi\rho c^5} \left(= \frac{\pi F}{\omega_D^2} \right), \quad n_{jk} = n(\varepsilon_{jk}), \tag{4.60}$$

and putting $\kappa = 1$ we have for the first eigenvalue

$$R_1 = -\tfrac{1}{2}A - i\tfrac{1}{2}\mathrm{sign}(\varepsilon_{21} - \varepsilon_{43})\sqrt{B - 2\sqrt{C}}, \tag{4.61}$$

with

$$A = i(\varepsilon_{31} + \varepsilon_{42}) + GW^2[\varepsilon_{21}(2n_{21} + 1) + \varepsilon_{43}(2n_{43} + 1)], \tag{4.62a}$$

$$B = \varepsilon_{21}^2 + \varepsilon_{43}^2 + 2iGW^2(\varepsilon_{21}^2 - \varepsilon_{43}^2) -$$
$$- G^2 W^4 (\varepsilon_{21}(2n_{21} + 1) + \varepsilon_{43}(2n_{43} + 1))^2, \tag{4.62b}$$

$$C = \varepsilon_{21}^2 \varepsilon_{43}^2 + 4G^2 W^4 (\varepsilon_{21}^2 - \varepsilon_{43}^2)(\varepsilon_{21}^2 n_{21}(n_{21} + 1) - \varepsilon_{43}^2 n_{43}(n_{43} + 1)). \tag{4.62c}$$

We now expand R_1 up to second order in V_α, V_β. For the inner square root we then have:

$$\sqrt{C} = \varepsilon_{21}\varepsilon_{43} + 2G^2 W^4 \frac{\varepsilon_{21}^2 - \varepsilon_{43}^2}{\varepsilon_{21}\varepsilon_{43}} \times$$
$$\times \{\varepsilon_{21}^2 n_{21}(n_{21} + 1) - \varepsilon_{43}^2 n_{43}(n_{43} + 1)\} + O(V^3). \tag{4.63}$$

If we moreover assume

$$V_\alpha, V_\beta \ll GW^2 |\varepsilon_{jk}|, \tag{4.64}$$

the outer square root becomes

$$\sqrt{B - 2\sqrt{C}} = i\,\text{sign}(\varepsilon_{21} - \varepsilon_{43})\,GW^2 x \left(1 + i\,\frac{GW^2(\varepsilon_{21}^2 - \varepsilon_{43}^2)}{-G^2 W^4 x^2} + \right.$$

$$+ \frac{1}{2} \frac{(\varepsilon_{21} - \varepsilon_{43})^2}{-G^2 W^4 x^2} - \frac{2G^2 W^4(\varepsilon_{21}^2 - \varepsilon_{43}^2) y}{-\varepsilon_{21}\varepsilon_{43} G^2 W^4 x^2} -$$

$$\left. - \frac{1}{8} \frac{-4G^2 W^4 (\varepsilon_{21}^2 - \varepsilon_{43}^2)^2}{G^4 W^8 x^4} + O(V^3) \right), \qquad (4.65)$$

where

$$x = \varepsilon_{21}(2n_{21} + 1) + \varepsilon_{43}(2n_{43} + 1), \qquad (4.66a)$$

$$y = \varepsilon_{21}^2 n_{21}(n_{21} + 1) - \varepsilon_{43}^2 n_{43}(n_{43} + 1). \qquad (4.66b)$$

The signum function appears in (4.65) because the imaginary part of $B - 2\sqrt{C}$ changes its sign at $\varepsilon_{21} = \varepsilon_{43}$ while its real part is negative. This signum just compensates the one of (4.61).

Some algebraic manipulations [1] lead to:

$$\text{Re}(R_1) = GW^2 \frac{(\varepsilon_{21}^2 - \varepsilon_{43}^2)}{\varepsilon_{21}\varepsilon_{43}} \frac{\varepsilon_{21}^2 n_{21}(n_{21} + 1) - \varepsilon_{43}^2 n_{43}(n_{43} + 1)}{\varepsilon_{21}(2n_{21} + 1) + \varepsilon_{43}(2n_{43} + 1)} -$$

$$- \frac{(\varepsilon_{21} - \varepsilon_{43})^2}{GW^2} \frac{(\varepsilon_{21} n_{21} + \varepsilon_{43} n_{43})(\varepsilon_{21}(n_{21} + 1) + \varepsilon_{43}(n_{43} + 1))}{(\varepsilon_{21}(2n_{21} + 1) + \varepsilon_{43}(2n_{43} + 1))^2} +$$

$$+ O(V^3). \qquad (4.67)$$

We call the first of these terms I, the second II. If we introduce the further approximation

$$D_{12} = D_{34} = D(E), \quad D_{21} = D_{43} = D'(E) \qquad (4.68a)$$

which is equivalent to

$$\varepsilon_{21}^3(n_{21}[+1]) = \varepsilon_{43}^3(n_{43}[+1]), \qquad (4.68b)$$

the first term becomes

$$\text{I} = -8GW^2 \frac{\Delta^2 (\Delta V)^2}{E^3} \frac{1}{\sinh \beta E} + O(V^3), \qquad (4.69)$$

the Lyo–Orbach result [67], explicitly derived in [73]. It should be noted, however, that approximation (4.69) is not a consistent execution of our expansion, because in assuming (4.68) even terms linear in V_α, V_β are neglected. If (4.67) is evaluated in a consistent manner, one arrives at

$$\mathrm{I} = -4GW^2 \frac{\Delta^2(\Delta V)^2}{E^3} \frac{1}{\sinh \beta E} \left\{ \frac{\beta E}{2} \coth\left(\frac{\beta E}{2}\right) - 1 \right\} + O(V^3), \quad (4.70a)$$

and

$$\mathrm{II} = -\frac{1}{GW^2 E^3} \Delta^2(\Delta V)^2 \frac{\tanh^2\left(\frac{\beta E}{2}\right)}{\sinh \beta E} + O(V^3). \quad (4\text{-}70b)$$

Comparing (4.69) and (4.70a) we find that the first expression is proportional to T for not too low temperatures whereas the second (and (4.70b)) decreases proportional to $1/T$, which may be interpreted as motional narrowing. In a second approximation, we start from the observation that $GW^2\omega_D = \pi F\omega_D e^{-2\lambda}$ usually is a small quantity (F is on the order of 10). We therefore expand $\mathrm{Re}(R_1)$ with respect to GW^2 and obtain (taking advantage of the fact that the r_{jk} are first order in GW^2)

$$\sqrt{C} = \varepsilon_{21}\varepsilon_{43} - i\varepsilon_{43}(r_{12} - r_{21}) + i\varepsilon_{21}(r_{34} - r_{43}) + O(G^2W^4), \quad (4.71)$$

$$B \pm 2\sqrt{C} = \{\varepsilon_{21} \pm \varepsilon_{43} - i((r_{12} - r_{21}) \pm (r_{43} - r_{34}))\}^2 +$$
$$+ \begin{cases} 4(r_{12} + r_{43})(r_{21} + r_{34}) \\ 4(r_{12} + r_{34})(r_{21} + r_{43}) \end{cases} + O(G^2W^4), \quad (4.72)$$

which immediately leads to

$$R_1 = -\tfrac{1}{2}A - i\tfrac{1}{2}\mathrm{sign}(\varepsilon_{21} - \varepsilon_{43})\sqrt{B - 2\sqrt{C}}$$
$$\approx -i\varepsilon_{31} - r_{12} - r_{34}, \quad (4.73)$$

$$R_2 = -\tfrac{1}{2}A + i\tfrac{1}{2}\mathrm{sign}(\varepsilon_{21} - \varepsilon_{43})\sqrt{B - 2\sqrt{C}}$$
$$\approx -i\varepsilon_{42} - r_{21} - r_{43}, \quad (4.74)$$

$$R_3 = -\tfrac{1}{2}A + i\tfrac{1}{2}\sqrt{B + 2\sqrt{C}}$$
$$\approx -i\varepsilon_{32} - r_{21} - r_{34}, \quad (4.75)$$

$$R_4 = -\tfrac{1}{2}A - i\tfrac{1}{2}\sqrt{B + 2\sqrt{C}}$$
$$\approx -i\varepsilon_{41} - r_{12} - r_{43}. \quad (4.76)$$

In detail, our assumptions were

$$|r_{jk}| \ll \varepsilon_{21}, \varepsilon_{43} \quad (4.77)$$

in the cases of R_3 and R_4; to obtain the results for R_1 and R_2 we had to assume additionally

$$|r_{jk}| \ll |\varepsilon_{21} - \varepsilon_{43}| \qquad (4.78a)$$

which for $\varepsilon_{21} \neq \varepsilon_{43}$ and sufficiently large λ is fulfilled if

$$|r_{jk}| \ll 2|\Delta V|. \qquad (4.78b)$$

To compare the approximations (4.73)–(4.76) with the exact eigenvalues we have plotted both in one picture. Figure 8 shows the cases R_1 and R_2. Except in the region around $\Delta = V_\alpha + V_\beta$, i.e. $\varepsilon_{21} = \varepsilon_{43}$, the approximation is fairly good. In the cases of R_3 and R_4 the curves corresponding to the exact result and to the approximation are indistinguishable for the parameter values of Figure 8, and thus we omit this picture.

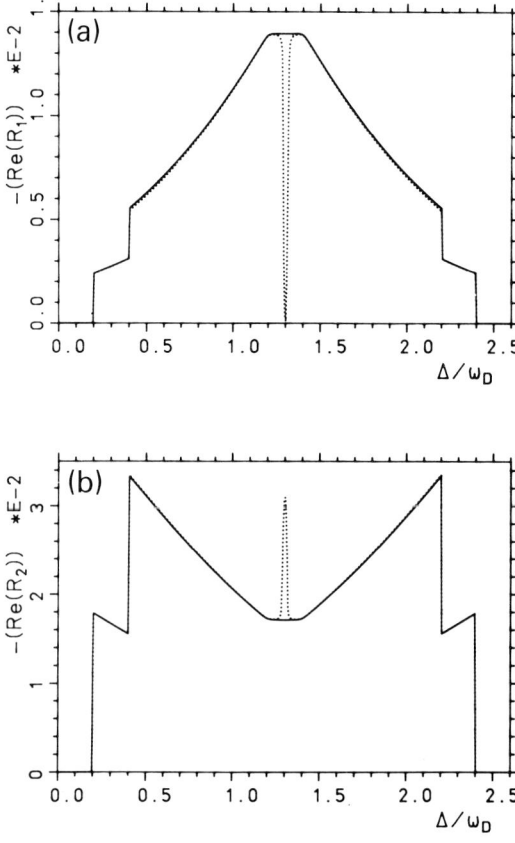

Fig. 8. Comparison of the real parts of the exact eigenvalues (in units of ω_D) (dotted lines) with those of approximation (4.73, 4.74). Parameter values: $\lambda = 4$, $T = 0.5\Theta_D$, $\Delta V = 0.1\omega_D$, $\Theta_D = 300$ K, $c = 3$ km/s, $\rho = 2$ g/cm^3, $f = 1$ eV.

We find that above the critical ΔV value of Figure 7 approximation (4.73)–(4.76) describes the situation very well whereas below that approximation (4.70) holds, if $\Delta \neq 2V_a, 2V_\beta$. Furthermore, we compare (4.73)–(4.76) with a formula given in the literature for relaxation rates of TLSs. The simplest form with a single deformation potential parameter is taken from [93]

$$\tau^{-1} = \left(\frac{1}{c_L^5} + \frac{2}{c_T^5}\right) \frac{B^2 EW^2}{2\pi\rho\hbar^4} \coth\left(\frac{\beta E}{2}\right). \quad (4.79)$$

B is the deformation potential, c_L and c_T are the velocities of sound. With the r_{jk} explicitly written, our relaxation rates read

$$-\text{Re}(R_k) = \frac{3}{8c^5} f^2 \frac{W^2}{2\pi\rho} \left\{ \varepsilon_{21} \left(\frac{\varepsilon_{21}}{\omega_D}\right)^{\kappa-1} (n(\varepsilon_{21})[+1]) \right.$$

$$\left. + \varepsilon_{43} \left(\frac{\varepsilon_{43}}{\omega_D}\right)^{\kappa-1} (n(\varepsilon_{43})[+1]) \right\}. \quad (4.80)$$

(The [ ] describes an option, and $\hbar = 1$ in (4.80).) The factors $3/c^5$ and $1/c_L^5 + 2/c_T^5$ correspond to each other as well as $f^2/8$ and B^2. If we take into consideration that

$$2n(E) + 1 = \coth\left(\frac{\beta E}{2}\right), \quad (4.81)$$

It is clear that for $\kappa = 1$ and $\varepsilon_{21} \approx \varepsilon_{43} \approx E$ formulae (4.79) and (4.80) become equivalent if the eigenvalues R_3 or R_4 are chosen. Whereas τ^{-1} remains finite at all temperatures, there is one eigenvalue (R_1) the real part of which goes to zero as $T \to 0$. As we shall see in Section 7, it is just the optical transitions corresponding to this eigenvalue which give a dominant contribution to the line shape. This shows that there is quite a difference between the optical system and the acoustic one (without guest molecules). The linewidths of TLSs measured in the latter with the help of phonon techniques cannot be compared in a direct way to optical ones. This is because in the optical case not the full TLS linewidths enter the description but the separate widths of the lower and upper TLS levels, respectively.

5. Line Shape

The imaginary part of the dielectric susceptibility is given by (3.5, 3.9, 3.39):

$$\chi''(\omega) = \frac{1}{2} \int_0^\infty d\tau \, (e^{i\omega\tau} - e^{-i\omega\tau}) \sum_k \gamma_k S_k(\tau). \quad (5.1)$$

Because γ_k is different from zero only for $k \leq 8$ and taking advantage of the symmetry under index and complex conjugation, we can write

$$\chi''(\omega) = \frac{1}{2} \int_0^\infty d\tau \, (e^{i\omega\tau} - e^{-i\omega\tau}) \sum_{k=1}^{4} \gamma_k (S_k(\tau) - S_k^*(\tau)), \tag{5.2}$$

a formula which explicitly exhibits that $\chi''(\omega)$ is real. Introducing the scalar product

$$(\mathbf{a}, \mathbf{b}) \equiv \sum_{k=1}^{4} a_k^* b_k \tag{5.3}$$

we obtain

$$\chi''(\omega) = \frac{1}{2} \int_0^\infty d\tau \, (e^{i\omega\tau} - e^{-i\omega\tau})(\boldsymbol{\gamma}, \{\exp(\mathbf{N}\tau) - \exp(\mathbf{N}^*\tau)\} \mathbf{S}(0)), \tag{5.4}$$

where

$$S_l(0) = \nu_l \quad (l = 1, \ldots, 4). \tag{5.5}$$

Formal Fourier transformation yields

$$\chi''(\omega) = -\tfrac{1}{2}(\boldsymbol{\gamma}, \hat{\mathbf{S}}(\omega) - \hat{\mathbf{S}}(-\omega) + \hat{\mathbf{S}}^*(\omega) - \hat{\mathbf{S}}^*(-\omega)) \tag{5.6}$$

with

$$\hat{\mathbf{S}}(\omega) = (\mathbf{N} + i\omega\mathbf{1})^{-1} \mathbf{S}(0). \tag{5.7}$$

This inverse matrix exists for all (real) ω if the real part of the eigenvalues of $\mathbf{N}$ is different from zero. It is easy to give an explicit formula for this matrix.

We denote by $N_{lm}(z)$ the matrix elements of $\mathbf{N} - z\mathbf{1}$ and by $W_{lm}(z)$ the cofactors of these elements [112]. The cofactors are the coefficients in an expansion of the determinant $N(z) \equiv |\mathbf{N} - z\mathbf{1}|$ with respect to a given row or column. If z is not an eigenvalue, we have

$$(\mathbf{N} - z)^{-1} = \frac{1}{N(z)} \mathbf{W}^T(z) \tag{5.8}$$

where $\mathbf{W}^T(z)$ is the transposed of the matrix the elements of which are the cofactors. With some algebra we obtain a compact representation of the line shape using determinants

$$\chi''(\omega) = \text{Re} \left\{ \frac{1}{|\mathbf{N} + i\omega|} \begin{vmatrix} 0 & \boldsymbol{\gamma}^T \\ \boldsymbol{\nu} & \mathbf{N} + i\omega \end{vmatrix} - \frac{1}{|\mathbf{N} - i\omega|} \begin{vmatrix} 0 & \boldsymbol{\gamma}^T \\ \boldsymbol{\nu} & \mathbf{N} - i\omega \end{vmatrix} \right\}. \tag{5.9}$$

The numerators of this expression are 5×5 determinants ($\boldsymbol{\gamma}^T$ and $\boldsymbol{\nu}$ are row and column vectors with 4 elements, $\mathbf{N}$ is a 4×4 matrix), the denominators 4×4 ones. Figure 9 shows two examples of line shapes calculated this way. In Figure

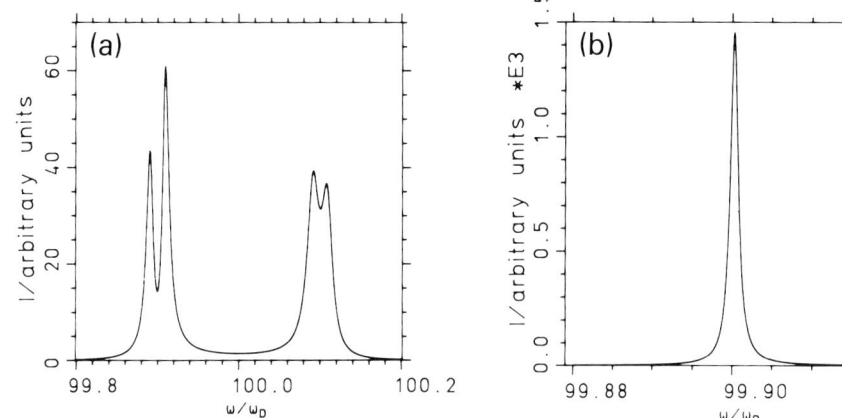

Fig. 9. (a) $\chi''(\omega)$ at $T = 0.2\theta_D$. Parameter values: $\lambda = 4$, $\Delta V = 0.1\omega_D$, $\Delta = 2V_a$, $E_0 = 100\omega_D$ (see 2.53), $\theta_D = 300$ K, $c = 3$ km/s, $\rho = 2$ g/cm^3, $f = 1$ eV. From the left to the right the lines correspond to the transitions: R_3, R_1, R_2, R_4. (b) $T = 0.1\Theta_D$, $\Delta = 2V_a - 0.2\omega_D$, remaining parameters as above.

9(a) the temperature and further parameters have been chosen such that rather broad lines emerge and all four lines can be shown in one graph. Figure 9(b) illustrates how small linewidths can become, at lower temperatures and closer to the range of applicability ($T \leq 0.05\Theta_D$) of the theory.

6. Averaging over Two-Level Systems

6.1. AVERAGING PROCEDURES

Until now we have calculated the linewidth and shape of a guest molecule coupled to a single TLS under the influence of the phonon bath. Because of the disordered structure of a glass, the parameters appearing in this calculation will have a distribution. As to the parameters λ, Δ, we will assume that they are equidistributed within a certain range, following Anderson, Halperin, and Varma [9] and Phillips [10]:

$$P(\lambda, \Delta) = \begin{cases} \bar{P} = 1/\{(\Delta_{max} - \Delta_{min})(\lambda_{max} - \lambda_{min})\} & \text{for } \begin{array}{c} \Delta_{min} \leq \Delta \leq \Delta_{max} \\ \lambda_{min} \leq \lambda \leq \lambda_{max} \end{array} \\ 0 & \text{otherwise} \end{cases} \quad (6.1)$$

In principle, all further parameters are distributed as well. As we have considered different cases we will specify our assumptions concerning these distributions in the course of the discussion. Macroscopic experimental results collect contributions of many statistically distributed microevents, e.g. the absorption of single photons in absorption spectroscopy. The statistical properties are a consequence on the one hand of quantum mechanics, on the other hand — in our system — of the distribution of the relevant parameters. Because macroscopic measurements are reproducible, the sum of micro events does not behave like a 'real' random variable and can be replaced by an ensemble average. The usual procedure [51, 67, 68, 71, 90, 91] is to average the linewidth. Generally this procedure is carried out without further justification. At first view, one would conjecture that an experimentally measured quantity like the spectral intensity, i.e. the line shape, had to be averaged. The linewidth is only a derived quantity in frequency domain experiments (hole-burning). The same holds, although less obvious, for time domain experiments (photon echoes), where the time decay function is measured rather than the decay rate itself.

We give a tentative justification of the averaging of linewidths (procedure 1) which will be substantiated further in Section 7. Let us assume that the TLSs interact with the guest sufficiently independent of each other. This will result in a total relaxation rate which is the sum of relaxation rates stemming from the different relaxation channels provided by the individual TLSs. If the number of interacting TLSs is large enough, statistical fluctuations of this sum will be negligible, it can be replaced by an average. The description should, however, start from a Hamiltonian containing many TLSs, not just one, if procedure 1 is applied. If, on the other hand, the homogeneous optical linewidth is mainly determined by the interaction with a single TLS, the line shape is the quantity to be averaged (procedure 2). The model conceptions implicit in these ideas are visualized in Figure 10.

6.2. NUMERICAL AVERAGING OF THE LINEWIDTH

6.2.1. *Averaging over λ, Δ*

In a first approximation, we average the linewidth over Δ and λ assuming for the

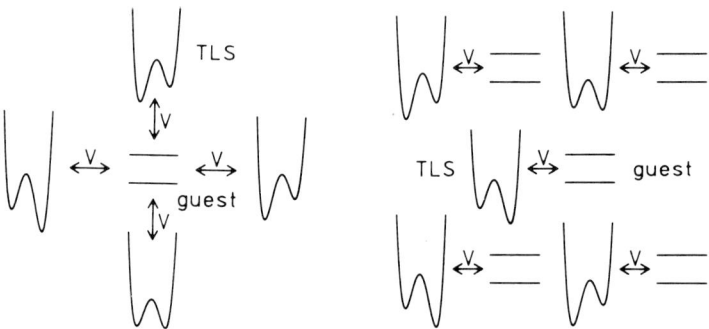

Fig. 10. Physical interpretation of the averaging of the linewidth (l.h.s.) and line shape (r.h.s.), respectively.

present that the distribution of the remaining parameters can be taken into account by using effective values for them. The average is performed by integrating the real part of the eigenvalue over the regime where $P(\lambda, \Delta) \neq 0$ (and dividing the result by the appropriate normalization constant).

In the first averaging, ΔV_{eff} was assumed to be of the order of $0.1\omega_D$. Different values of $\lambda_{\min}$ were considered, $\lambda_{\max}$ was taken to be $\lambda_{\min} + 5$. All Δ values for which the real part of the eigenvalues was different from zero (see Figure 6) were taken into account in the integral. This corresponds to a width of the constant Δ distribution of more than $2\omega_D$. Although this is much more than is until now confirmed by experiments, one may treat the average this way in the case of the temperature dependent part of the linewidth, because the result is not very sensitive to the width of the Δ distribution at low temperatures.

Figures 11 to 13 show the temperature dependence of averaged linewidths obtained in the manner described. Figure 11 shows the dependence on the parameter $\lambda_{\min}$; the values of Debye temperature, velocity of sound and mass density are typical for inorganic materials. Between $T \approx \Theta_D$ and $T \approx \exp(-\lambda_{\min})\Theta_D$ we obtain a quadratic temperature dependence ($\kappa = 1$), for lower temperatures the

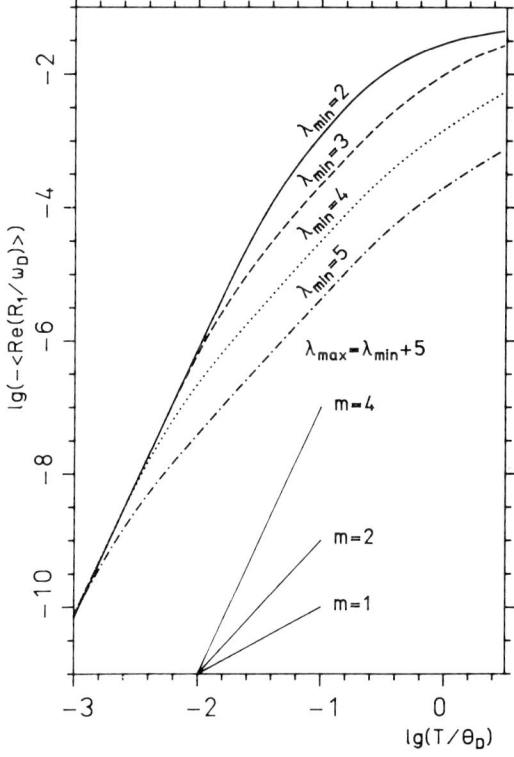

Fig. 11. Linewidth averaged with respect to Δ and λ. The λ ranges are indicated in the picture, the Δ interval is chosen maximal (see text). Remaining parameters: $\Delta V = 0.3\omega_D$, $\Theta_D = 300$ K, $c = 3$ km/s, $\rho = 2$ g/cm^3, $f = 1$ eV.

linewidth is proportional to T^4. For reasons of comparison, straight lines corresponding to the slopes 1, 2, and 4 are plotted in the picture, too.

Figure 12 gives a general impression of the difference in order of magnitude of the linewidths to be expected in inorganic and organic materials, respectively. Mainly this difference is due to the different Debye temperatures and velocities of sound. In reduced units, the linewidth of the organic material surpasses that of the inorganic one by a factor of 10 approximately. Therefore, the absolute linewidths differ by a factor of 30 in the T^2 domain and of 270 in the T^4 domain.

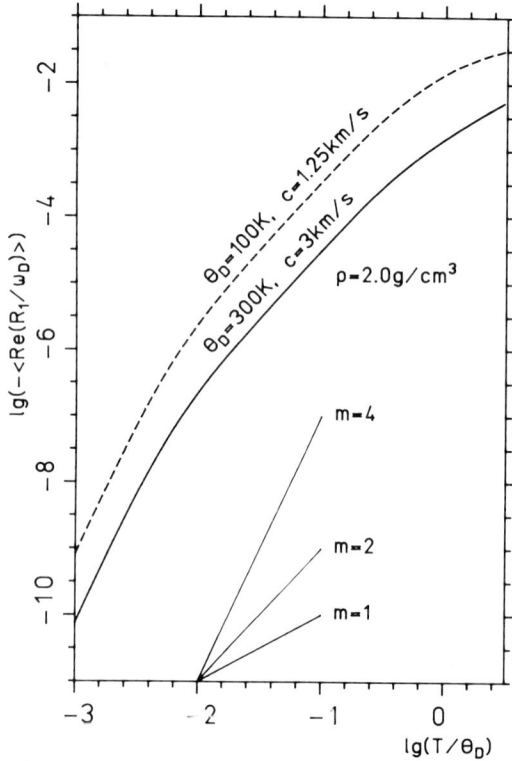

Fig. 12. Dependence on material parameters of the linewidth averaged over λ and Δ; the lower curve is identical to the case $\lambda_{min} = 4$ of Figure 11, the upper differs with respect to Debye temperature and average velocity of sound.

The case of extremely small ΔV is investigated in Figure 13 ($\lambda_{min} = 5$, $\lambda_{max} = 10$). Appreciable changes of the temperature dependence between $10^{-3}\Theta_D$ and $10^{-2}\Theta_D$ do not show up until ΔV becomes smaller than $5 \times 10^{-5}\omega_D$. This result agrees with Figure 7 which displays that for $\lambda = 5$ (at $T = 10^{-3}\Theta_D$) the real part of the eigenvalue is almost independent of ΔV for $\Delta V > 10^{-5}\omega_D$ and Δ values near to the maximum of the linewidth. **Below** the 'critical' ΔV value the contribution of a TLS to the linewidth becomes negligible. Because at higher temperatures this critical value is larger than at lower ones (see Figure 7) the

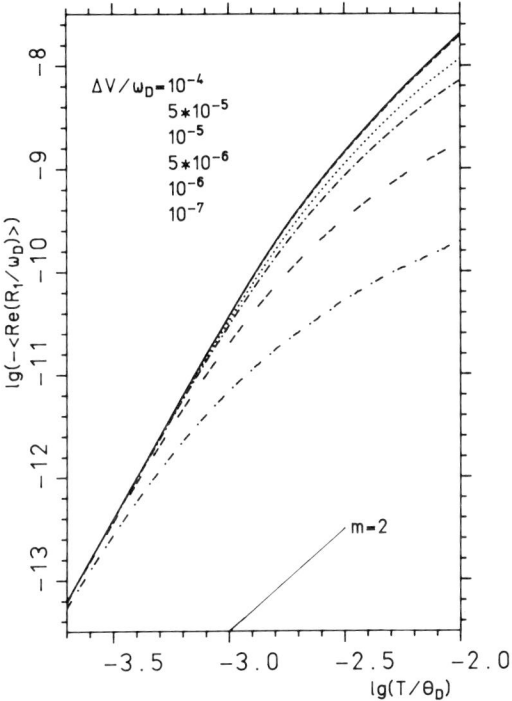

Fig. 13. Linewidth averaged over λ and Δ for small ΔV values. $\lambda_{min} = 5$, maximum Δ range of integration. $\Theta_D = 300$ K, $c = 3.75$ km/s, $\rho = 2$ g/cm^3, $f = 1$ eV.

averaged linewidth decreases faster with decreasing ΔV at higher temperatures and in consequence the curves describing the temperature dependence become less steep.

This result may give rise to misinterpretation. Because within a large ΔV range there is only a weak ΔV dependence of the linewidth one might believe that the averaging over λ and Δ already yields a sufficient description of the experimental linewidth. This conclusion would be correct if the guest molecule were interacting only with TLSs the ΔV values of which are above the critical value at all temperatures considered. Such a situation may arise, e.g., if the coupling to more distant TLSs is screened either because of the high density of less distant TLSs or because of the chemical composition of the system. In these cases, it can be conjectured that the quadratic temperature dependence of Figures 11 and 12 will persist even if an additional averaging over ΔV is performed. If, however, a long-range interaction between TLS and optical guest molecule is assumed, small ΔV values stemming from distant TLSs will become important. These TLSs have to be taken into account in the average and will modify its temperature dependence. Because the number of TLSs with small ΔV will become rather large (being determined by the volume of spherical shells around the guest molecule with large radii) this may have quite important consequences.

6.2.2. Averaging over λ, Δ and ΔV

Because ΔV is the third relevant parameter of the model, the averaging should be extended over this variable, especially if long-range interaction is present. The distributions of λ, Δ and that of ΔV are assumed to be independent. As to the latter distribution, two cases have been considered. The first is dipolar coupling (between electric or elastic dipoles), i.e. $\Delta V \propto 1/r^3$, the second dipole–quadrupole coupling, i.e. $\Delta V \propto 1/r^4$ (conjecturing that the guest molecule behaves like a quadrupole). With the assumption that the TLSs are spatially homogeneously distributed the distribution of ΔV values is then determined.

In these calculations the averaging has been done by numerically generating random values for the parameters, summing up the corresponding linewidths and normalizing the result. The random number generator was based on an analytically solvable special case of the logistic equation which has already been used for similar purposes by Ulam and von Neumann [113, 114]. We have assumed an interaction volume containing 10^6 TLSs in the case of the dipolar coupling and $10^{4.5}$ for the dipole–quadrupole one (taking the same limiting values for the ΔV distribution in both cases). The actual averaging was performed over 4×10^4 to 5×10^5 TLSs which yielded an overall accuracy of 10% (estimated from a comparison of the final curves with those including 10^4 TLSs less). The local accuracy was better at low temperatures.

The results of this calculation, multiplied by the number of TLSs in the interaction volume, are shown in Figure 19 and discussed in Section 7.6 in connection with experimental values. In this place we merely mention that the temperature dependence of the linewidth becomes considerably weaker than the T^2 law obtained from the average over λ and Δ only.

6.3. ANALYTICAL APPROXIMATIONS FOR THE AVERAGED LINEWIDTH

We start from approximation (4.73)–(4.76) and restrict ourselves to the temperature dependent parts of the eigenvalues all real parts of which are equal to $\mathrm{Re}(R_1)$. We have ($w = W/\omega_D$)

$$\mathrm{Re}(R_{1/2/3/4})_{\mathrm{th}} = -\omega_D \pi F w^2 \left(e_{21}^{\kappa-2} g_r(e_{21}) n_r \left(\frac{e_{21}}{T_r} \right) + \right.$$

$$\left. + e_{43}^{\kappa-2} g_r(e_{43}) n_r \left(\frac{e_{43}}{T_r} \right) \right), \quad (6.2)$$

hence ($\delta = \Delta/\omega_D$, $g_r(e_{jk}) = e_{jk}^2 \Theta(1 - e_{jk})$)

$$\frac{\langle \Delta \omega \rangle}{\omega_D} = \frac{\pi F}{N} \int_{\lambda_{\min}}^{\lambda_{\max}} d\lambda \int_{\delta_{\min}}^{\delta_{\max}} d\delta \, e^{-2\lambda} \left\{ e_{21}^{\kappa} \Theta(1 - e_{21}) n_r \left(\frac{e_{21}}{T_r} \right) + \right.$$

$$\left. + e_{43}^{\kappa} \Theta(1 - e_{43}) n_r \left(\frac{e_{43}}{T_r} \right) \right\}, \quad (6.3a)$$

where
$$N = (\lambda_{max} - \lambda_{min})(\delta_{max} - \delta_{min}). \tag{6.3b}$$

If the δ range is limited by the Heavyside functions (maximum range)

$$\Theta(1 - e_{jk}) = \Theta(1 - \sqrt{e^{-2\lambda} + (\delta - 2v_\rho)^2})$$
$$= \begin{cases} 1 & \text{for } 2v_\rho - \sqrt{1 - e^{-2\lambda}} \leq \delta \leq 2v_\rho + \sqrt{1 - e^{-2\lambda}} \\ 0 & \text{otherwise} \end{cases}, \tag{6.4}$$

where $v_\rho \equiv V_\rho/\omega_D$, the linewidth expression becomes

$$\frac{\langle \Delta\omega \rangle}{\omega_D} = \pi F/N \left\{ \int_{\lambda_{min}}^{\lambda_{max}} d\lambda \int_{2v_\alpha - \sqrt{1-e^{-2\lambda}}}^{2v_\alpha + \sqrt{1-e^{-2\lambda}}} d\delta \, e^{-2\lambda} (\sqrt{e^{-2\lambda} + (\delta - 2v_\alpha)^2})^\kappa \times \right.$$
$$\times n_r \left(\frac{\sqrt{e^{-2\lambda} + (\delta - 2v_\alpha)^2}}{T_r} \right) +$$
$$+ \int_{\lambda_{min}}^{\lambda_{max}} d\lambda \int_{2v_\beta - \sqrt{1-e^{-2\lambda}}}^{2v_\beta + \sqrt{1-e^{-2\lambda}}} d\delta \, e^{-2\lambda} (\sqrt{e^{-2\lambda} + (\delta - 2v_\beta)^2})^\kappa \times$$
$$\left. \times n_r \left(\frac{\sqrt{e^{-2\lambda} + (\delta - 2v_\beta)^2}}{T_r} \right) \right\}. \tag{6.5}$$

This can be transformed into

$$\frac{\langle \Delta\omega \rangle}{\omega_D} = 4\pi F/N \int_{\lambda_{min}}^{\lambda_{max}} d\lambda \int_0^{\sqrt{1-e^{-2\lambda}}} d\delta \, e^{-2\lambda} (\sqrt{e^{-2\lambda} + \delta^2})^\kappa n_r \left(\frac{\sqrt{e^{-2\lambda} + \delta^2}}{T_r} \right). \tag{6.6}$$

Commuting the integrals we obtain

$$\frac{\langle \Delta\omega \rangle}{\omega_D} = 4\pi F/N \int_0^{\sqrt{1-\exp(-2\lambda_{max})}} d\delta \int_{\lambda_0(\delta)}^{\lambda_{max}} d\lambda \, e^{-2\lambda} (\sqrt{e^{-2\lambda} + \delta^2})^\kappa \times$$
$$\times \frac{1}{\exp\left(\frac{\sqrt{e^{-2\lambda} + \delta^2}}{T_r}\right) - 1}, \tag{6.7}$$

where
$$\lambda_0(\delta) \equiv \max\left(\lambda_{min}, \frac{1}{2} \ln\left\{\frac{1}{1-\delta^2}\right\}\right) \tag{6.8}$$

as can be seen from Figure 14 showing the integration domain.

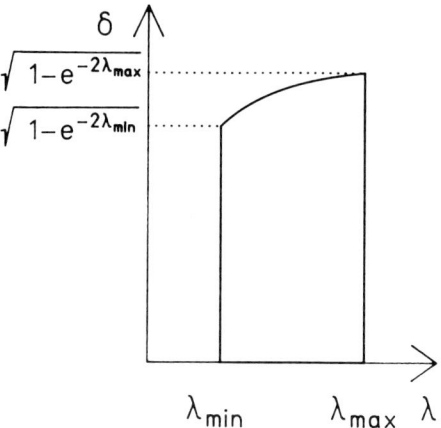

Fig. 14. Integration domain of integrals (6.6, 6.7).

Substituting

$$x = \frac{1}{T_r}\sqrt{e^{-2\lambda} + \delta^2} \quad \Rightarrow \quad dx = -\frac{e^{-2\lambda}\,d\lambda}{T_r^2 x}, \tag{6.9a}$$

we arrive at

$$\frac{\langle \Delta\omega \rangle}{\omega_D} = -\frac{4\pi F}{N}\, T_r^{\kappa+2} \int_0^{\sqrt{1-\exp(-2\lambda_{\max})}} d\delta \int_{x(\lambda_0(\delta))}^{x(\lambda_{\max})} dx\, x^{\kappa+1} \frac{1}{e^x - 1}, \tag{6.9b}$$

where

$$x(\lambda_{\max}) = \frac{1}{T_r}\sqrt{\exp(-2\lambda_{\max}) + \delta^2}, \tag{6.10a}$$

$$x(\lambda_0(\delta)) = \frac{1}{T_r}\min(\sqrt{\exp(-2\lambda_{\min}) + \delta^2},\, 1). \tag{6.10b}$$

Because the inner integrand does not depend on δ, we commute the integrals once more. The integration domain is transformed according to

$$0 \leq \delta \leq \sqrt{1 - \exp(-2\lambda_{\max})},$$

$$\frac{1}{T_r}\sqrt{\exp(-2\lambda_{\max}) + \delta^2} \leq x \leq \frac{1}{T_r}\min(1, \sqrt{\exp(-2\lambda_{\min}) + \delta^2})$$

$$\Leftrightarrow \delta_{\min}(x) \equiv \sqrt{\max(0,\, T_r^2 x^2 - \exp(-2\lambda_{\min}))} \leq \delta$$

$$\leq \sqrt{T_r^2 x^2 - \exp(-2\lambda_{\max})} \equiv \delta_{\max}(x),$$

$$\frac{e^{-\lambda_{\max}}}{T_r} \leq x \leq \frac{1}{T_r}, \tag{6.11}$$

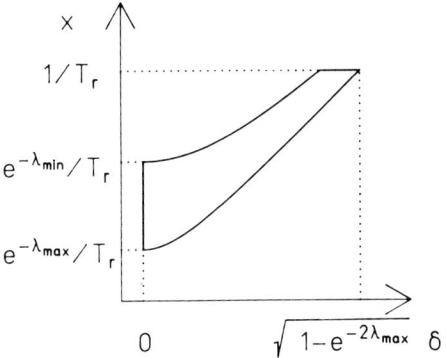

Fig. 15. Integration domain of integral (6.9b).

(compare Figure 15).
Performing now the inner integration we can write

$$\frac{\langle \Delta \omega \rangle}{\omega_D} = \frac{4\pi F}{N} T_r^{\kappa+2} \int_{e^{-\lambda_{max}}/T_r}^{1/T_r} dx\, (\delta_{max}(x) - \delta_{min}(x)) \frac{x^{\kappa+1}}{e^x - 1} \quad (6.12)$$

which more explicitly reads

$$\frac{\langle \Delta \omega \rangle}{\omega_D} = \frac{4\pi F}{N} T_r^{\kappa+3} \left\{ \int_{e^{-\lambda_{max}}/T_r}^{1/T_r} dx\, \frac{x^{\kappa+1}}{e^x - 1} \sqrt{x^2 - (e^{-\lambda_{max}}/T_r)^2} - \right.$$

$$\left. - \int_{e^{-\lambda_{min}}/T_r}^{1/T_r} dx\, \frac{x^{\kappa+1}}{e^x - 1} \sqrt{x^2 - (e^{-\lambda_{min}}/T_r)^2} \right\}. \quad (6.13)$$

Therefore,

$$\frac{\langle \Delta \omega \rangle}{\omega_D} = \frac{4\pi F}{N} T_r^{\kappa+3} \{ I_\kappa(e^{-\lambda_{max}}/T_r) - I_\kappa(e^{-\lambda_{min}}/T_r) \} \quad (6.14)$$

with

$$I_\kappa(a) \equiv \int_a^{1/T_r} dx\, \frac{x^{\kappa+1}}{e^x - 1} \sqrt{x^2 - a^2}. \quad (6.15)$$

We are going to investigate the behavior of the linewidth expression (6.14) in three temperature domains:

(a) $T_r \ll e^{-\lambda_{max}}$ ($\ll 1$),
(b) $e^{-\lambda_{max}} \ll T_r \ll e^{-\lambda_{min}}$,
(c) $e^{-\lambda_{min}} \ll T_r \ll 1$,

(assuming $e^{-\lambda_{max}} \ll e^{-\lambda_{min}} \ll 1$). Obviously, the upper integration limit of $I_\kappa(a)$ can be replaced by ∞ in all three cases.

Consider region (a). Because of $\exp(-\lambda_{max})/T_r \gg 1$ we may replace $1/(e^x - 1)$ by e^{-x} in both integrals of (6.14). This yields

$$I_\kappa(a) \approx \int_a^\infty dx\, x^{\kappa+1} \sqrt{x^2 - a^2}\, e^{-x}$$

$$= a^{\kappa+3} \int_0^\infty du\, (\cosh(u))^{\kappa+1} \sinh^2(u)\, e^{-a\cosh(u)}. \tag{6.16}$$

For integer κ the integral can be expressed by the Bessel functions $K_\nu(z)$ with imaginary argument [115] (no. 3.574,2)

$$I_\kappa(a) = a^{\kappa+3} \left(\frac{d}{d(-a)}\right)^{\kappa+1} \int_0^\infty du\, \sinh^2(u)\, e^{-a\cosh(u)}$$

$$= a^{\kappa+3} \left(\frac{d}{d(-a)}\right)^{\kappa+1} \frac{1}{a} K_1(a). \tag{6.17}$$

Using recurrence relations for the $K_\nu(z)$ (8.486,15 in [115]) we obtain

$$I_1(a) = a^3 \left(K_3(a) - \frac{1}{a} K_2(a)\right). \tag{6.18}$$

With an asymptotic expansion for large arguments (no. 8.451,6 in [115]) we get

$$I_1(a) = a^{5/2} \sqrt{\frac{\pi}{2}}\, e^{-a}(1 + O(1/a)), \tag{6.19}$$

which explicitly yields the approximation ($\kappa = 1$)

$$\frac{\langle \Delta\omega \rangle}{\omega_D} = \frac{4\pi F}{N} \sqrt{\frac{\pi}{2}}\, T_r^{3/2} \exp\{-e^{-\lambda_{max}}/T_r\} \exp\{-(5/2)\lambda_{max}\} \times$$

$$\times \{1 - \exp\{(5/2)(\lambda_{max} - \lambda_{min})\} \exp\{e^{-\lambda_{max}}/T_r - e^{-\lambda_{min}}/T_r\}\} \tag{6.20}$$

At extremely low temperatures, the linewidth is described by the product of a power of the temperature and an Arrhenius factor. Probably until now no experiments exist in this temperature range.

In domain (b) $I_1(\exp(-\lambda_{min})/T_r)$ may again be approximated by expression (6.18), whereas a different approximation has to be found for $I_\kappa(\exp(-\lambda_{max})/T_r)$, i.e. $I_\kappa(a)$ for small a.

We start from the inequalities (valid for $x \geqq a$)

$$x - \frac{a^2}{2x} - \frac{a^4}{2x^3} \leqq \sqrt{x^2 - a^2} \leqq x - \frac{a^2}{2x} - \frac{a^4}{8x^3}, \qquad (6.21)$$

which yield immediately

$$-\int_a^\infty dx \frac{x^{\kappa+1}}{e^x - 1} \frac{a^4}{2x^3} - \int_0^a dx \frac{x^{\kappa+1}}{e^x - 1} \left(x - \frac{a^2}{2x}\right)$$

$$\leqq \int_a^\infty dx \frac{x^{\kappa+1}}{e^x - 1} \sqrt{x^2 - a^2} - \int_0^\infty dx \frac{x^{\kappa+1}}{e^x - 1} \left(x - \frac{a^2}{2x}\right)$$

$$\leqq -\int_a^\infty dx \frac{x^{\kappa+1}}{e^x - 1} \frac{a^4}{8x^3} - \int_0^a dx \frac{x^{\kappa+1}}{e^x - 1} \left(x - \frac{a^2}{2x}\right). \qquad (6.22)$$

The integral with limits 0 and ∞ can be evaluated by use of

$$\int_0^\infty dx \frac{x^{\nu-1}}{e^x - 1} = \Gamma(\nu)\zeta(\nu) \qquad (6.23)$$

(no. 3.411,1 in [115]), where $\zeta(\nu)$ is Riemann's Zeta function. For sufficiently small a we may approximate $e^x - 1$ by x in the integral on the interval $[0, a]$:

$$\int_0^a dx \frac{x^{\kappa+1}}{e^x - 1} \left(x - \frac{a^2}{2x}\right) = \int_0^a dx \frac{x^{\kappa+1}}{x + O(x^2)} \left(x - \frac{a^2}{2x}\right)$$

$$= \left(\frac{1}{\kappa + 2} - \frac{1}{2\kappa}\right) a^{\kappa+2} + O(a^{\kappa+3}). \qquad (6.24a)$$

Hence we get an estimate of the form

$$\tilde{c}_2 a^{\kappa+2} < \int_0^a dx \frac{x^{\kappa+1}}{e^x - 1} \left(x - \frac{a^2}{2x}\right) < \tilde{c}_1 a^{\kappa+2}. \qquad (6.24b)$$

The remaining integral is expressed with the help of a mean value theorem

$$\int_a^\infty dx \frac{x^\kappa}{e^x - 1} \frac{a^4}{x^2} = \frac{\xi^\kappa}{e^\xi - 1} \int_a^\infty dx \frac{a^4}{x^2} = \frac{\xi^\kappa}{e^\xi - 1} a^3, \qquad (6.25)$$

where $\xi \in (a, \infty)$. Inserting the maximum of the ξ dependent function one obtains an upper limit for the integral; for $\kappa \leqq 1$ the maximum is reached at a, for $\kappa > 1$ it is given by the solution of the transcendent equation $x = \kappa(1 - e^{-x})$

and independent of a. Therefore,

$$\int_a^\infty dx \, \frac{x^\kappa}{e^x - 1} \frac{a^4}{x^2} \leq \begin{cases} \tilde{\tilde{c}}_1 a^{\kappa+2} & \text{for } \kappa \leq 1 \\ \tilde{\tilde{c}}_1 a^3 & \text{for } \kappa > 1 \end{cases}. \tag{6.26}$$

For $\kappa \leq 1$ we arrive at

$$-c_1 a^{\kappa+2} \leq I_\kappa(a) - \Gamma(\kappa+3)\zeta(\kappa+3) + \frac{a^2}{2} \Gamma(\kappa+1)\zeta(\kappa+1)$$

$$\leq -c_2 a^{\kappa+2} \tag{6.27a}$$

or

$$I_\kappa(a) = \Gamma(\kappa+3)\zeta(\kappa+3) - \frac{a^2}{2} \Gamma(\kappa+1)\zeta(\kappa+1) + O(a^{\kappa+2}). \tag{6.27b}$$

The case $\kappa = 1$ is somewhat simpler

$$I_1(a) = \frac{\pi^4}{15} - \frac{\pi^2}{12} a^2 + O(a^3). \tag{6.28}$$

A similar estimate can be carried through for $I_\kappa(a)$, when a is large and κ a fraction. This yields

$$2 e^{-a} a^{\kappa+1} \left(1 + O\left(\frac{1}{a}\right)\right) \leq I_\kappa(a) \leq \frac{1}{2} e^{-a} a^{\kappa+2} \left(1 + O\left(\frac{1}{a}\right)\right). \tag{6.29}$$

It is concluded that in region (b) $I_\kappa(\exp(-\lambda_{\min})/T_r)$ may be neglected as compared to $I_\kappa(\exp(-\lambda_{\max})/T_r)$. We then obtain for the linewidth

$$\frac{\langle \Delta \omega \rangle}{\omega_D} \approx \frac{4\pi F}{N} \left\{ T_r^{\kappa+3} \Gamma(\kappa+3)\zeta(\kappa+3) - \right.$$

$$\left. - \frac{1}{2} \exp(-2\lambda_{\max}) T_r^{\kappa+1} \Gamma(\kappa+1)\zeta(\kappa+1) \right\}. \tag{6.30}$$

In the temperature range considered the second term is much smaller than the first and may be omitted. The linewidth is therefore proportional to $T^{\kappa+3}$.

Let us consider region (c). No further estimates are necessary, we just can use (6.27b) for both arguments of I_κ. Because the first term in this relation does not depend on a, it cancels in the difference (6.14). We get (for $\kappa \leq 1$)

$$\frac{\langle \Delta \omega \rangle}{\omega_D} = \frac{2\pi F}{N} \left\{ (e^{-2\lambda_{\max}} - e^{-2\lambda_{\min}}) T_r^{\kappa+1} \Gamma(\kappa+1)\zeta(\kappa+1) + \right.$$

$$\left. + O(e^{-\lambda_{\min}(\kappa+2)} T_r) \right\} \tag{6.31}$$

The averaged linewidth is essentially proportional to $T^{\kappa+1}$. The additional term proportional to T is negligible if κ is not too close to zero.

Looking once more at Figures 11 and 12 we note that our analytical results (6.30) and (6.31) describe the actual situation pretty well, at least for $\kappa = 1$. From Figure 11 we gather that these approximations get worse for small $\lambda_{\min}$ which is because the '$\ll$' in the definition of region (c) becomes invalid.

First measurements [49—52] of the homogeneous optical linewidths seemed to suggest a universal T^2 dependence. It might be argued that this corresponds to the case shown in Figure 11. However, these experiments were carried out at rather high temperatures ($T > 10$ K), where our theory including one-phonon processes only is not expected to be applicable (the validity of this conclusion may be limited somewhat, see Section 7.5). Later experiments below 1 K [60, 61] resulted in $T^{1.3}$ laws which seems to favor the exponent $\kappa = 0.3$. This temperature dependence, however, can be obtained with different assumptions as well. Our starting formula (6.3) did not explicitly refer to the **distribution** of TLSs which showed up only in the normalization constant N. If we express distribution (6.1) in the variables W, ε_ρ, we get

$$P(W, \varepsilon_\rho) = \begin{cases} \varepsilon_\rho/(W\sqrt{\varepsilon_\rho^2 - W^2}) & \text{for } \begin{array}{l} W_{\min} \leq W \leq W_{\max} \\ \sqrt{W^2 + \Delta_{\rho\min}^2} \leq \varepsilon_\rho \leq \sqrt{W^2 + \Delta_{\rho\max}^2} \end{array} \\ 0 & \text{otherwise.} \end{cases} \quad (6.32a)$$

If the actual distribution differs from (6.32a) in that the numerator is replaced by ε_ρ^μ we get

$$P(\lambda, \Delta_\rho) = \begin{cases} \dfrac{1}{N'} (\sqrt{\omega_0^2 e^{-2\lambda} + \Delta_\rho^2})^{\mu-1} & \text{for } \begin{array}{l} \lambda_{\min} \leq \lambda \leq \lambda_{\max} \\ \Delta_{\rho\min} \leq \Delta_\rho \leq \Delta_{\rho\max} \end{array} \\ 0 & \text{otherwise} \end{cases} \quad (6.32b)$$

instead of (6.1). That is we have to replace the exponent κ in (6.3) by $\kappa + \mu - 1$. Every combination of exponents κ, μ with a fixed sum $\kappa + \mu$ yields the same result for the linewidth. A $T^{1.3}$ law may be explained by $\mu = 1$, $\kappa = 0.3$ as well as by $\mu = 0.3$, $\kappa = 1$. At the present stage of the theory a distinction between these possibilities is impossible.

It should be remarked that the modified distributions have been introduced for the TLSs already coupled to the guest molecule. A similar assumption for the TLSs before renormalization would result in changes, however small, if V_α and V_β are sufficiently small.

6.4. AVERAGING OF THE LINE SHAPE

Attempting to describe the homogeneous optical line by an average of line shapes, we assume that $\chi''(\omega)$ contains 'dynamical' effects mainly and does not include 'structural' terms leading to the inhomogeneous line. That is why we have considered a Hamiltonian of a single guest with fixed, i.e. non-statistical energy difference $E_\beta - E_\alpha$. Hence, if we assume that $\chi''(\omega)$ is the contribution of a single TLS-guest system to the homogeneous line, the question arises whether it is possible to measure this line in experiments. Strictly speaking the answer is no. To avoid measuring the inhomogeneous line, in all experiments a narrow-band laser is used to excite a selected group of ions from their ground state into the first excited state. Afterwards the evolution of this excitation is investigated by different means, e.g. measuring an emission spectrum (FLN) or scanning an optical hole with weak laser power. However, because of the finite linewidth not only lines centered at the laser frequency ω_L are excited but also non-resonant ones. Let $g(\omega - \omega_0)$ be the linewidth of the 'true' homogeneous line at central frequency ω_0. If the total absorption probability can be regarded as constant over the width of a homogeneous line (i.e. the inhomogeneous width is much larger than the homogeneous one) the contribution of one homogeneous line to the total absorption is proportional to $g(\omega_L - \omega_0)$. The observed line shape $G(\omega - \omega_L)$ is then

$$G(\omega - \omega_L) = \int_{-\infty}^{\infty} d\omega_0 \, g(\omega - \omega_0) g(\omega_L - \omega_0). \tag{6.33}$$

For a Lorentzian

$$g(\omega - \omega_0) = \frac{\Gamma}{\pi} \frac{1}{(\omega - \omega_0)^2 + \Gamma^2} \tag{6.34a}$$

we obtain

$$G(\omega - \omega_L) = \frac{2\Gamma}{\pi} \frac{1}{(\omega - \omega_L)^2 + 4\Gamma^2}, \tag{6.34b}$$

for a Gaussian

$$g(\omega - \omega_0) = \frac{1}{\sqrt{2\pi}\sigma} \exp\left(-\frac{(\omega - \omega_0)^2}{2\sigma^2}\right), \tag{6.35a}$$

$$G(\omega - \omega_L) = \frac{1}{\sqrt{2\pi}\sqrt{2}\sigma} \exp\left(-\frac{(\omega - \omega_L)^2}{4\sigma^2}\right). \tag{6.35b}$$

In these cases the line shape is conserved by the convolution (6.33), the linewidth grows by a factor 2 or $\sqrt{2}$, respectively. Therefore, the experimental line differs from the theoretical homogeneous line by a factor on the order of 1. It can be shown [1] that this factor is temperature independent, if the line shape function g remains unaltered by temperature changes.

A calculation of $\langle \chi''(\omega - \omega_0) \rangle$ is therefore not expected to yield a complete description of the homogeneous line shape, but the linewidth obtained from it should be at least comparable to experimental ones. Actually, the quantity which ought to be calculated is

$$\int_{-\infty}^{\infty} d\omega_0 \langle \chi''(\omega - \omega_0) \chi''(\omega_L - \omega_0) \rangle$$

$$= \frac{1}{N} \int_{-\infty}^{\infty} d\omega_0 \int_{\lambda_{\min}}^{\lambda_{\max}} d\lambda \int_{\Delta_{\min}}^{\Delta_{\max}} d\Delta \, \chi''(\omega - \omega_0) \chi''(\omega_L - \omega_0). \quad (6.36)$$

Because of the triple integral this calculation would be rather time consuming. To obtain a survey, we will calculate $\langle \chi''(\omega - \omega_0) \rangle$ at first. Figure 16 shows the result of such a line shape calculation. The temperatures investigated in the picture are rather high, because for low temperatures the integration routine converges extremely slowly. In spite of the high temperatures the linewidths are very small, on the order of $10^{-8} \omega_D$, by far too small to describe the experiment. This shows that the situation in a glass is not adequately described by a four level system interacting with the phonon bath. However, such a system may be interesting on its own. To make progress in our problem, we have to consider the case of many TLSs coupled to the guest molecule. From this procedure we shall obtain a justification of the linewidth averaging carried out in the preceding subsections.

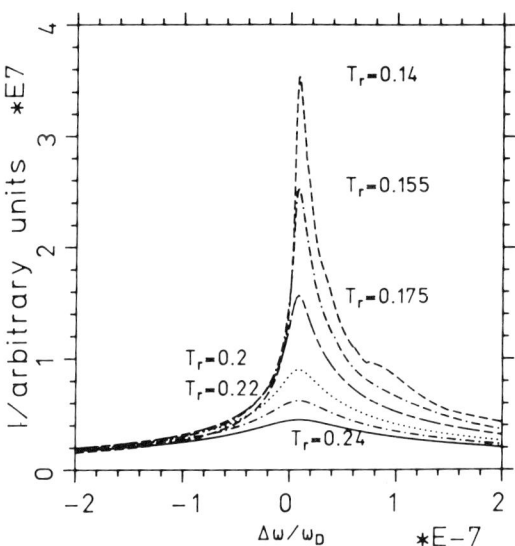

Fig. 16. Average of $\chi''(\omega)$ over Δ and λ. Parameter values: $\lambda_{\min} = 3$, $\lambda_{\max} = 8$, maximum Δ range, $\Delta V = 0.1 \omega_D$, $\Theta_D = 3.75$ km/s, $\rho = 2$ g/cm^3, $f = 1$ eV.

7. Coupling of the Impurity to Several Two-Level Systems

7.1. THE SYSTEM WITHOUT PHONONS: EIGENVALUES AND EIGENSTATES

The system Hamiltonian

$$H_S = \sum_{\rho k n} \varepsilon_{\rho k n} |\rho k n\rangle\langle \rho k n| \tag{7.1}$$

is a sum of N single-particle operators, each 'particle' representing a four-level system indexed by k. These systems do not correspond to physical observables, because the $\varepsilon_{\rho k n}$ (given in (2.61)) are arbitrary to some extent. It is nevertheless useful to visualize the system described by (7.1) as consisting of such entities, because these four-level systems are non-interacting. Therefore, the energy eigenvalues of the many-body system will be sums of the $\varepsilon_{\rho k n}$, whereas the eigenfunctions will be given by appropriate product states.

Multiplying the operator $|\rho k n\rangle\langle \rho k n|$ by the completeness relations

$$\sum_{n_j = 1}^{2} |\psi_{n_j}^{\rho j}\rangle\langle \psi_{n_j}^{\rho j}| = 1, \tag{7.2}$$

for $j \neq k$, $j = 1, 2, \ldots, N$, and using the factorization property

$$|\rho k n_k\rangle = |\rho\rangle |\psi_{n_k}^{\rho k}\rangle \tag{7.3}$$

we obtain a many-particle representation of H_S:

$$H_S = \sum_{k} \sum_{\rho, n_1, n_2, \ldots, n_N} \varepsilon_{\rho k n_k} |\rho\rangle\langle\rho| \prod_{j=1}^{N} |\psi_{n_j}^{\rho j}\rangle\langle \psi_{n_j}^{\rho j}|. \tag{7.4}$$

The product states

$$|\rho, n_1, n_2, \ldots, n_N\rangle \equiv |\rho\rangle |\psi_{n_1}^{\rho 1}\rangle |\psi_{n_2}^{\rho 2}\rangle \ldots |\psi_{n_N}^{\rho N}\rangle, \tag{7.5}$$

form a complete orthonormal set of the system Hilbert space. (Note, however, that $|\psi_n^{\alpha k}\rangle$ and $|\psi_n^{\beta k}\rangle$ are **not** orthogonal in general.) Insertion of (7.5) in (7.4)

$$H_S = \sum_{\rho, \{n_k\}} \left\{ \sum_{k} \varepsilon_{\rho k n_k} \right\} |\rho, n_1, n_2, \ldots, n_N\rangle\langle \rho, n_1, n_2, \ldots, n_N|, \tag{7.6}$$

shows that the states (7.5) are eigenfunctions of H_S, and the eigenvalues are given by

$$\varepsilon_\rho\{n_k\} \equiv \sum_{k} \varepsilon_{\rho k n_k}. \tag{7.7}$$

The symbol $\{n_k\}$ stands for a full sequence of TLS indices $(n_1, \ldots, n_N)$; in a sum it means summation over all possible sequences of this kind.

7.2. REPRESENTATION OF THE DIPOLE MOMENT OPERATOR

To obtain the eigenstate representation of p we use (7.2) in (4.1) which yields

$$p = \nu \sum_{\rho,\sigma} \sum_{\{n_j\},\{m_j\}} (1 - \delta_{\rho\sigma}) \left(\prod_{j=1}^{N} \langle \psi_{n_j}^{oj} | \psi_{m_j}^{oj} \rangle | \psi_{n_j}^{oj} \rangle \langle \psi_{m_j}^{oj} | \right) |\rho\rangle\langle\sigma|$$

$$= \nu \sum_{\rho,\sigma} \sum_{\{n_j\},\{m_j\}} (1 - \delta_{\rho\sigma}) \left(\prod_{j=1}^{N} \langle \psi_{n_j}^{oj} | \psi_{m_j}^{oj} \rangle \right) \times$$

$$\times |\rho, n_1, n_2, \ldots, n_N\rangle\langle\sigma, m_1, m_2, \ldots, m_N|$$

$$\equiv \sum_{\{j\}} \nu_{\{j\}} O_{\{j\}}. \tag{7.8}$$

The last equality defines some new quantities which facilitate the writing of the following formulae. $\{j\}$ is a multiple index combining the sequences $(\rho, n_1, \ldots, n_N)$ and $(\sigma, m_1, \ldots, m_N)$. $\nu_{\{j\}}$ is then given by

$$\nu_{\{j\}} \equiv \nu \begin{pmatrix} \rho & n_1 & n_2 & \ldots & n_N \\ \sigma & m_1 & m_2 & \ldots & m_N \end{pmatrix} = \nu \times (1 - \delta_{\rho\sigma}) \prod \{n_j, m_j\} = \nu_{\{j\}}^*, \tag{7.9a}$$

where

$$\prod \{n_j, m_j\} \equiv \prod_{j=1}^{N} (\delta_{n_j m_j} \cos(\rho_j - \sigma_j) + (1 - \delta_{n_j m_j})(-1)^{n_j} \sin(\rho_j - \sigma_j)), \tag{7.9b}$$

and ρ_j, σ_j are the rotation angles of the transformation leading from $\{|\psi_n^j\rangle\}$ to $\{|\psi_n^{oj}\rangle\}$ (see (2.63, 2.64)). The operator $O_{\{j\}}$ describes a transition of the system starting from a state $|\sigma, m_1, \ldots, m_N\rangle$ and ending at $|\rho, n_1, n_2, \ldots, n_N\rangle$:

$$O_{\{j\}} \equiv O \begin{pmatrix} \rho & n_1 & n_2 & \ldots & n_N \\ \sigma & m_1 & m_2 & \ldots & m_N \end{pmatrix}$$

$$= |\rho, n_1, n_2, \ldots, n_N\rangle\langle\sigma, m_1, m_2, \ldots, m_N|. \tag{7.10}$$

To derive the representation of $[p, \rho_S]$ we remark that

$$\rho_S = \frac{1}{Z_S} e^{-\beta H_S}$$

$$= \frac{1}{Z_S} \sum_{\rho, \{n_k\}} e^{-\beta \varepsilon_\rho \{n_k\}} |\rho, n_1, n_2, \ldots, n_N\rangle\langle\rho, n_1, n_2, \ldots, n_N| \tag{7.11}$$

with the partition function

$$Z_S = \sum_{\rho, \{n_k\}} e^{-\beta \varepsilon_\rho \{n_k\}} = \sum_{\rho, \{n_k\}} \prod_{k=1}^{N} e^{-\beta \varepsilon_{\rho k} n_k}. \tag{7.12}$$

Using

$$[O_{\{j\}}, |\zeta, r_1, \ldots, r_N\rangle\langle\zeta, r_1, \ldots, r_N|]$$
$$= (\delta_{\sigma\zeta}\delta_{m_1, r_1} \cdots \delta_{m_N, r_N} - \delta_{\rho\zeta}\delta_{n_1, r_1} \cdots \delta_{n_N, r_N})O_{\{j\}} \tag{7.13}$$

we obtain

$$[p, \rho_S] = \sum_{\{j\}} \gamma_{\{j\}} O_{\{j\}}, \tag{7.14}$$

where

$$\gamma_{\{j\}} = \frac{1}{Z_s} (e^{-\beta\varepsilon_\sigma\{m_k\}} - e^{-\beta\varepsilon_\rho\{n_k\}}) v_{\{j\}}. \tag{7.15}$$

7.3. CALCULATION OF THE CORRELATION FUNCTION EVOLUTION MATRIX

We want to calculate the optical line shape within the approximation (3.5). Thus we need the dipole moment correlation function of that equation. From Section 7.2 we conclude that the form of expansion (3.9) remains unchanged, i.e.

$$\text{Tr}_S(\langle p(\tau)[p, \rho_S]\rangle) = \sum_{\{j\}, \{k\}} v^*_{\{j\}} \gamma_{\{k\}} \text{Tr}_S(\langle O^+_{\{j\}}(\tau) O_{\{k\}}\rangle) \tag{7.16}$$

and that we should derive equations of motion for the correlation functions

$$S_{\{j\}\{k\}}(t) = \text{Tr}(O^+_{\{j\}}(t) O_{\{k\}} \rho_B) \tag{7.17a}$$

or even better for

$$S_{\{k\}}(t) = \sum_{\{j\}} v^*_{\{j\}} S_{\{j\}\{k\}}(t) = \text{Tr}(p(t) O_{\{k\}} \rho_B). \tag{7.17b}$$

The general proceeding within Mori's formalism is identical to that of Section 3.2 — we just have to replace each double index **j, k, l** by the corresponding multiple index $\{j\}, \{k\}, \{l\}$.

Using

$$(O_{\{j\}} | O_{\{k\}}) = \delta(\{j\}, \{k\}) \tag{7.18}$$

(as in the one-TLS case) we obtain the equations of motion

$$\dot{S}_{\{k\}}(t) = -i \sum_{\{l\}} S_{\{l\}}(t) M_{\{l\}\{k\}}. \tag{7.19}$$

The Kronecker symbol is one if all the subindices of $\{j\}$ are equal to the corresponding indices in $\{k\}$, and is zero otherwise.

The correlation function evolution matrix $\underline{\underline{\mathfrak{M}}}$ may be written

$$M_{\{l\}\{k\}} = \Omega_{\{l\}\{k\}} - i \int_0^\infty d\tau\, \Phi^{(\ell)}_{\{l\}\{k\}}(\tau) \tag{7.20a}$$

with

$$\Omega_{\{l\}\{k\}} = \mathrm{Tr}(\rho_B (\mathbf{L_S} O_{\{l\}})^+ O_{\{k\}}) \tag{7.20b}$$

and

$$\Phi^{(\ell)}_{\{l\}\{k\}}(\tau) = \mathrm{Tr}(\rho_B (\tilde{\mathbf{L}}_{SB}(-\tau)\mathbf{L}_{SB} O_{\{l\}})^+ O_{\{k\}}). \tag{7.20c}$$

These relations are formally the same as in the single-TLS case. Differences will appear only in the course of the explicit evaluation of (7.20c).

$\underline{\underline{\Omega}}$ is easily calculated. Because of

$$\mathbf{L_S} O_{\{j\}} = [H_S, O_{\{j\}}] = (\varepsilon_\rho\{n_k\} - \varepsilon_\sigma\{m_k\}) O_{\{j\}} \equiv \varepsilon\{\mathbf{j}\} O_{\{j\}} \tag{7.21}$$

— the last identity defining $\varepsilon\{\mathbf{j}\}$ — we have immediately

$$\Omega_{\{j\}\{r\}} = \varepsilon\{\mathbf{j}\}\delta(\{\mathbf{j}\},\{\mathbf{r}\}). \tag{7.22}$$

In comparison with the single-TLS case some major changes have to be noted during the calculation of $\Phi^{(\ell)}_{\{j\}\{r\}}(\tau)$. As we know (see e.g. (2.71)) part of the action of $\mathbf{L}_{SB}$ is to flip TLSs from their upper state to the lower one and vice versa. In the single-TLS case there was just one two-level system which could flip, now there are many. Therefore, the operator $O\{\mathbf{j}\}$ is coupled not only to itself and two further operators via $\mathbf{L}_{SB}$ but to $2N + 1$ operators.

For reasons of brevity we introduce some new symbols. If

$$\{\mathbf{j}\} = \begin{pmatrix} \rho & n_1 & \dots & n_N \\ \sigma & m_1 & \dots & m_N \end{pmatrix}, \tag{7.23a}$$

the index with the upper TLS at position k flipped is written

$$\{\mathbf{j}|^k\} \equiv \begin{pmatrix} \rho & n_1 & \dots & \tilde{n}_k & \dots & n_N \\ \sigma & m_1 & \dots & m_k & \dots & m_N \end{pmatrix}. \tag{7.23b}$$

Analogously,

$$\{\mathbf{j}|_l\} = \begin{pmatrix} \rho & n_1 & \dots & n_l & \dots & n_N \\ \sigma & m_1 & \dots & \tilde{m}_l & \dots & m_N \end{pmatrix}. \tag{7.23c}$$

Furthermore, multiple indices differing at two positions from $\{\mathbf{j}\}$ are written $\{\mathbf{j}|^{kl}\}$, $\{\mathbf{j}|^k_l\}$ or $\{\mathbf{j}|_{kl}\}$. To avoid too many index levels we will write these multiple indices on the same level as the corresponding operator. The calculation of $\underline{\underline{\Phi}}^{(\ell)}(\tau)$ is now carried through in the same way as in the single-TLS case. Therefore, we merely give the result (for more details see [1]) which reads

$$-iM(\{j\},\{r\}) = -i\Omega(\{j\},\{r\}) - \int_0^\infty d\tau \, \Phi^{(\ell)}(\{j\},\{r\})(\tau)$$

$$= \delta(\{r\},\{j\})\left\{-i\varepsilon\{j\} - \sum_k [D_{\rho n\tilde{n}}^{kk}(B_\rho^k)^2 + D_{\sigma m\tilde{m}}^{kk}(B_\sigma^k)^2]\right\} +$$

$$+ \sum_{kl} \delta(\{r\},\{j|_l^k\}) B_\rho^k B_\sigma^l (D_{\rho \tilde{n}n}^{lk} + D_{\sigma \tilde{m}m}^{lk}) -$$

$$- \sum_k \delta(\{r\},\{j|^k\}) B_\rho^k \sum_l (A_\rho^l(-1)^{n_l} - A_\sigma^l(-1)^{m_l}) D_{\rho \tilde{n}n}^{lk} +$$

$$+ \sum_k \delta(\{r\},\{j|_k\}) B_\sigma^k \sum_l (A_\rho^l(-1)^{n_l} - A_\sigma^l(-1)^{m_l}) D_{\sigma \tilde{m}m}^{lk} -$$

$$- \sum_{kl} \delta(\{r\},\{j|^{kl}\}) B_\rho^k B_\rho^l D_{\rho \tilde{n}n}^{kl}(1 - \delta_{kl})$$

$$- \sum_{kl} \delta(\{r\},\{j|_{kl}\}) B_\sigma^k B_\sigma^l D_{\sigma \tilde{m}m}^{kl}(1 - \delta_{kl})\bigg\}, \quad (7.24)$$

where

$$D_{\rho n\tilde{n}}^{kl} \equiv \sum_q D_q^k D_{-q}^l \pi (n_q + \delta_{n_l,2}) \delta(\omega_q - \varepsilon_\rho^l). \quad (7.25)$$

In (7.24) the principal value terms have already been neglected.

7.4. APPROXIMATE CALCULATION OF THE EIGENVALUES

Because (7.24) describes a $4^{N+1} \times 4^{N+1}$ matrix, it seems rather hopeless to attempt an exact diagonalization. Let us consider the approximation of retaining only the diagonal elements of $\mathfrak{M}$.

First, with increasing number of two-level systems, the absolute value of the real part of the diagonal elements increases as well, because it is a sum of elements of the same sign (the $D_{\rho n\tilde{n}}^{kk}$ are positive). The terms of the sums appearing with the (non-diagonal) elements $\delta(\{r\}, \{j|^k\})$ and $\delta(\{r\}, \{j|_k\})$ contain factors $D_{\rho n\tilde{n}}^{kl}$ which need not be positive; furthermore, except for some singular configurations (in which all the n_j, m_j are even or all of them are odd) the sign of the terms A_ρ^l, A_σ^l changes. Therefore, in these sums terms of the same order of magnitude but alternating signs are added up which in general will make these sums small in comparison with the sums of the diagonal elements. All further matrix elements do not contain any sums at all, but only a single element of the order of magnitude of one term in the sums already mentioned; they should therefore be about a factor $1/N$ smaller than the real parts of the diagonal elements. Finally, it should be mentioned that the percentage of vanishing matrix elements increases with N; the number of such elements is $4^N(4^N - 2N^2 - 2N - 1)$. These arguments suggest that the approximation

$$-iM(\{\mathbf{j}\}, \{\mathbf{r}\})$$
$$= \delta(\{\mathbf{r}\}, \{\mathbf{j}\}) \left\{ -i\varepsilon\{\mathbf{j}\} - \sum_k [D^{kk}_{\rho n\tilde{n}}(B^k_\rho)^2 + D^{kk}_{\sigma m\tilde{m}}(B^k_\sigma)^2] \right\} \quad (7.26)$$

is probably not too bad. The eigenvalues are then given by

$$R\{\mathbf{j}\} = -i\varepsilon\{\mathbf{j}\} - \sum_k \{(B^k_\rho)^2 D^{kk}_{\rho n\tilde{n}} + (B^k_\sigma)^2 D^{kk}_{\sigma m\tilde{m}}\}. \quad (7.27)$$

Comparing this result to approximation (4.73)–(4.76) within the single-TLS problem (and recalling that r_{12}, r_{21}, etc., correspond to $(B^k_\rho)^2 D^{kk}_{\rho n\tilde{n}}$, etc.) we see that the real parts of the eigenvalues within the present approximation are just sums (over all TLSs) of real parts within the former approximation. However, this approximation could be shown to describe the single-TLS case rather well in most circumstances. Because the approximation should improve with increasing N, this fact supports the arguments given before (7.26).

More important, (7.26) shows that the negative real part, i.e. the relaxation rate, of the many-TLS problem is a sum of relaxation rates of the single-TLS case — at least approximately. This result justifies the averaging of the linewidth in Sections 6.2 and 6.3. However, the linewidth obtained in this way corresponds to a single optical transition and the line shape is a superposition of many single-transition lines (with widths $|\mathrm{Re}(R\{\mathbf{j}\})|$). We have to investigate now whether a similar statement holds for this superposition. This leads to our next step, the calculation of the optical line shape.

7.5. LINE SHAPE FORMULA

Similar to the single-TLS case we can take advantage of symmetry relations (introducing a conjugate index as in (4.8)) to reduce the number of indices over which to sum up. We obtain for the line shape (in a vector and matrix notation)

$$\chi''(\omega) = \frac{1}{2} \int_0^\infty d\tau \, (\gamma, [e^{(i\omega + \mathbf{N})\tau} - e^{(-i\omega + \mathbf{N})\tau} -$$
$$- e^{(i\omega + \mathbf{N}^*)\tau} + e^{(-i\omega + \mathbf{N}^*)\tau}]\mathbf{S}(0)), \quad (7.28)$$

where

$$N(\{\mathbf{r}\}, \{\mathbf{j}\}) = -iM(\{\mathbf{j}\}, \{\mathbf{r}\}), \quad (7.29a)$$

and

$$\{j\} = \begin{pmatrix} \beta & n_1 & n_2 & \ldots & n_N \\ \alpha & m_1 & m_2 & \ldots & m_N \end{pmatrix},$$

$$\{r\} = \begin{pmatrix} \beta & l_1 & l_2 & \ldots & l_N \\ \alpha & r_1 & r_2 & \ldots & r_N \end{pmatrix} \tag{7.29b}$$

are multiple indices describing a **fixed** transition $|\alpha\rangle \rightarrow |\beta\rangle$ of the optical guest. (Consequently, **N** is a $4^N \times 4^N$ matrix, whereas $\mathfrak{M}$ has $4^{N+1} \times 4^{N+1}$ elements.)

Because **N** is diagonal we can easily invert this matrix to obtain the line shape formula

$$\chi''(\omega) = i\omega \sum_{\{j\}} \gamma\{j\} \nu\{j\} \times$$

$$\times \left(\frac{1}{\omega^2 - (\varepsilon\{j\})^2 + (\phi\{j\})^2 + 2i\varepsilon\{j\}\phi\{j\}} - \text{c.c.} \right)$$

$$= \sum_{\{j\}} \gamma\{j\} \nu\{j\} \frac{4\omega\varepsilon\{j\}\phi\{j\}}{[\omega^2 - (\varepsilon\{j\})^2 + (\phi\{j\})^2]^2 + 4(\varepsilon\{j\}\phi\{j\})^2}, \tag{7.30}$$

where

$$\phi\{j\} = \sum_k ((B_\beta^k)^2 D_{\beta n\tilde{n}}^{kk} + (B_\alpha^k)^2 D_{\alpha m\tilde{m}}^{kk}). \tag{7.31}$$

To discuss this formula we first give a more detailed expression for the prefactor of the fraction. From (7.15), (7.7) and (7.21) we obtain

$$\gamma\{j\}\nu\{j\} = (\nu\{j\})^2 \frac{1}{Z_S} (\exp(-\beta\varepsilon_\alpha\{m_k\}) - \exp(-\beta\varepsilon_\beta\{n_k\}))$$

$$= (\nu\{j\})^2 \frac{1}{Z_S} \exp(-\beta\varepsilon_\alpha\{m_k\})(1 - \exp(-\beta\varepsilon\{j\})), \tag{7.32}$$

where (see (7.9))

$$(\nu\{j\})^2 = \nu^2 \prod_{k=1}^{N} (\delta_{n_k m_k} \cos^2(\beta_k - \alpha_k) + (1 - \delta_{n_k m_k}) \sin^2(\beta_k - \alpha_k)). \tag{7.33}$$

With the definition

$$E_0 = E_\beta - E_\alpha + \sum_{k=1}^{N} (S_\beta^k - S_\alpha^k) \tag{7.34}$$

and after a few algebraic transformations of (7.12) the line shape formula may be written

$$\chi''(\omega) = \frac{4\nu^2}{\prod_{k=1}^{N}(1+e^{-\beta\varepsilon_a^k}) + \exp\left\{-\beta\left(E_0 + \frac{1}{2}\sum_{k=1}^{N}(\varepsilon_\beta^k - \varepsilon_a^k)\right)\right\}\prod_{k=1}^{N}(1+e^{-\beta\varepsilon_\beta^k})} \times$$

$$\times \sum_{n_1, m_1, n_2, \ldots, m_N} \exp\left\{-\frac{1}{2}\sum_{k=1}^{N}((-1)^{m_k}+1)\varepsilon_a^k\right\}(1-e^{-\beta\varepsilon\{\mathbf{j}\}}) \times$$

$$\times \prod_{k=1}^{N}\{\delta_{n_k,m_k}\cos^2(\beta_k - \alpha_k) + (1-\delta_{n_k,m_k})\sin^2(\beta_k - \alpha_k)\} \times$$

$$\times \frac{\omega\varepsilon\{\mathbf{j}\}\phi\{\mathbf{j}\}}{[\omega^2 - (\varepsilon\{\mathbf{j}\})^2 + (\phi\{\mathbf{j}\})^2]^2 + 4(\varepsilon\{\mathbf{j}\}\phi\{\mathbf{j}\})^2}. \quad (7.35)$$

Within the sum of (7.35) we can distinguish the frequency dependent **shape** term describing the line shape of an optical transition indicated by $\{\mathbf{j}\}$ and its **trigonometric** (the product over TLSs) and **thermal** prefactors (the exponential expressions). Because $\phi\{\mathbf{j}\} \ll |\varepsilon\{\mathbf{j}\}|$ the **shape** term takes on its maximum value near $\omega = \varepsilon\{\mathbf{j}\}$, where it is proportional to $1/\phi\{\mathbf{j}\}$. It may be easily checked that $\phi\{\mathbf{j}\}$ is the linewidth of the one shape term, i.e. the width of a single optical transition. Therefore, the narrowest optical lines contribute the most important shape terms.

The calculation of $\phi\{\mathbf{j}\}$ within the Debye model yields

$$\phi\{\mathbf{j}\} = \phi_{\text{th}} + \phi_{\text{ath}}\{\mathbf{j}\}, \quad (7.36a)$$

where

$$\phi_{\text{th}} = \sum_{k=1}^{N}\frac{\pi f^2}{8\rho V c^2}\sum_{\rho=a}^{\beta}\left(\frac{\varepsilon_\rho^k}{\omega_D}\right)^{\kappa-1}\varepsilon_\rho^k g_D(\varepsilon_\rho^k)n(\varepsilon_\rho^k)(B_\rho^k)^2 \quad (7.36b)$$

and

$$\phi_{\text{ath}}\{\mathbf{j}\} = \sum_{k=1}^{N}\frac{\pi f^2}{8\rho V c^2}\left\{\left(\frac{\varepsilon_\beta^k}{\omega_D}\right)^{\kappa-1}\varepsilon_\beta^k g_D(\varepsilon_\beta^k)(B_\beta^k)^2\delta_{n_k,2} + \right.$$

$$\left. + \left(\frac{\varepsilon_a^k}{\omega_D}\right)^{\kappa-1}\varepsilon_a^k g_D(\varepsilon_a^k)(B_a^k)^2\delta_{m_k,2}\right\}. \quad (7.36c)$$

Only the linewidth at $T=0$, $\phi_{\text{ath}}\{\mathbf{j}\}$ depends explicitly on the transition $\{\mathbf{j}\}$ in our approximation. The **thermal** contribution ϕ_{th} to the linewidth is equal for all transitions.

Now there is exactly one eigenvalue $R\{\mathbf{j}\}$ the real part of which goes to zero as $T \to 0$. This eigenvalue is characterized by $n_k = m_k = 1$ for all k. Furthermore, the thermal prefactor $\exp(-\frac{1}{2}\beta\sum_{k=1}^{N}((-1)^{m_k}+1)\varepsilon_a^k)$ takes on its maximum value (namely 1) for this eigenvalue. (The factor $(1 - \exp(-\beta\varepsilon\{\mathbf{j}\}))$ may be replaced by 1 because $\varepsilon\{\mathbf{j}\}$ is an optical energy and therefore $\beta\varepsilon\{\mathbf{j}\} \gg 1$ for temperatures below the Debye temperature.)

Investigating the trigonometric prefactor one finds that the sine factors are small if Δ^k is not in the interval $[2V_\alpha^k, 2V_\beta^k]$ whereas the cosine factors are almost one everywhere except in this interval. This behavior is shown in Figure 17.

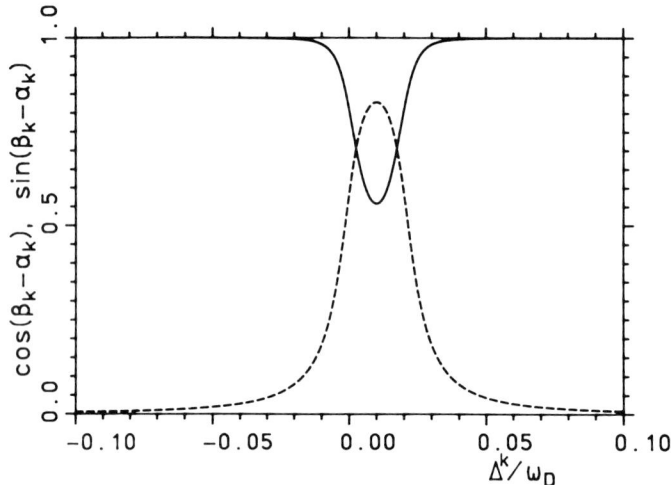

Fig. 17. Dependence of the trigonometric factors on system parameters: $\lambda = 5$, $\Delta V = 0.01\omega_D$, $V_a = 0$. Full line: $\cos(\beta_k - \alpha_k)$, dashed line: $\sin(\beta_k - \alpha_k)$.

Now the trigonometric prefactor consists of N factors each of which is a cosine or sine term. For a given configuration of Δ^k values it takes on its maximum, if a cosine term is taken for all of the TLSs for which Δ^k lies outside the interval mentioned and a sine term otherwise. Usually, ΔV is a small quantity ($\Delta V \ll \omega_D$), i.e. the cosine term is larger than the sine one in most cases. Therefore, the trigonometric prefactor of the eigenvalue $R\binom{\beta\ 1\ \cdots\ 1}{\alpha\ 1\ \cdots\ 1}$ will be larger than most of the other eigenvalues. This leads to the conclusion that at sufficiently low temperatures the sum in (7.34) will be dominated by one term corresponding to $\{\mathbf{j}\} = \binom{\beta\ 1\ \cdots\ 1}{\alpha\ 1\ \cdots\ 1}$ which we will call **principal** term. The contributions by the thermal prefactor, the trigonometric prefactor (most probably) and the shape term are maximal for this index. For $T = 0$, the principal term becomes a δ function whereas the heights of the other terms remain finite — they are $\propto 1/\phi_{ath}\{\mathbf{j}\}$. One expects that at these temperatures the whole linewidth will be well approximated by the width of the principal term. Therefore, the linewidth is given by a sum or average over linewidths $(-\mathrm{Re}(R_1))$ of the single-TLS problem.

As temperature increases the occupation numbers $n(\varepsilon_o^k)$ increase as well and ϕ_{th} becomes comparable to the smallest values of $\phi_{\text{ath}}\{\mathbf{j}\}$ (i.e. those for which only a few of the indices n_k, m_k are different from 1). Because some of the corresponding lines (e.g. that of $R\bigl(\begin{smallmatrix}\beta\,2\,1\,\cdots\,1\\ \alpha\,1\,1\,\cdots\,1\end{smallmatrix}\bigr)$) have the same thermal, others (e.g. $R\bigl(\begin{smallmatrix}\beta\,2\,1\,\cdots\,1\\ \alpha\,2\,1\,\cdots\,1\end{smallmatrix}\bigr)$) the same trigonometric prefactors as the principal term they will contribute a comparable amount to the total line shape as soon as the thermal part of their linewidth becomes comparable with the athermal one. Therefore, the number of single transitions contributing significantly to the line shape goes up with temperature, i.e. the linewidth increases faster than that of the principal (or any other) term, because an increasing number of lines with athermal contributions to the linewidth are added. One concludes that a crossover from the temperature dependence describing the width of ϕ_{th} at low temperatures to a stronger temperature law at higher temperatures may occur. This crossover would not come about through a dynamical effect but simply through the increase in number of single transitions contributing relevantly to the total line shape. If they cannot overcompensate this effect, many-phonon processes may be quite unimportant at higher temperatures.

One should be aware, however, that up to now the argument is qualitative only, because the crossover temperature has not yet been estimated. In the following section we shall assume that this temperature is sufficiently high in order to allow a description of the total optical line shape by the principal term.

7.6. LINEWIDTH CALCULATION: COMPARISON WITH EXPERIMENT

Within our approximation, the total linewidth is a sum of single transition widths. To evaluate the sum, we have to know the parameter values Δ^k, λ^k, ΔV^k of the TLSs coupled to the guest molecule. These values can be drawn from a probability distribution. As to Δ, λ, we take the distribution given in (6.1). The distribution of ΔV values depends on various circumstances, especially on the functional relationship assumed between ΔV and the distance of the TLS from the optical guest. We consider two limiting cases.

First, we suppose that all TLSs within a certain (spherical) volume around the guest molecule are interacting with it and none outside this volume, which contains the tacit assumption that the interaction abruptly falls to zero as the boundary of the volume is crossed (case 1). If the interaction for the most distant TLSs taken into account is still larger than the critical ΔV value discussed in Figure 7, approximation (4.73) holds for each of these TLSs and the sum ϕ_{th} will be a good description of the linewidth.

In the second case, the interaction is assumed to go smoothly to zero with increasing distance (case 2). Of course, major contributions to the linewidth will still arise from TLSs satisfying (4.73), because as ΔV becomes so small that the quadratic ΔV dependence of the single-TLS linewidth holds, TLSs which are even more distant may be neglected (if ΔV goes to zero faster than $1/r^{1+\varepsilon}$). But the number of TLSs which fulfill (4.73) in the single-TLS approximation increases when the temperature goes down. Therefore, we may accept the approximation of adding up single-TLS contributions and take ϕ_{th} as a measure for the linewidth, but we should increase the radius of the 'interaction sphere' with decreasing temperature. To avoid a tedious calculation of this radius we can, however, just add up the real parts of the exact solution of the single-TLS problem, if we take a sufficiently large radius of influence for the interaction to include all relevant TLSs at the lowest temperature considered. (For higher temperatures this has the consequence that many TLSs are taken into account that could in principle be neglected — the other way round it would be worse.) Figure 18 shows a calculation according to case 1 in comparison with experimental data obtained below 1 K [60—62]. We have assumed that $\Delta V \propto 1/r^3$, but we may gather that the weak ΔV dependence of the single-TLS width results in a temperature dependence of the linewidth which is independent of the specific relationship between ΔV and r. From Section 6 we expect a $T^{1+\kappa}$ power law in a certain temperature range.

Assuming $\kappa = 0.3$ the experimental data can be described quite satisfactorily as the figure shows. The linewidth magnitudes result from the assumption that 1000 TLSs were interacting with the guest molecule. This means that the most

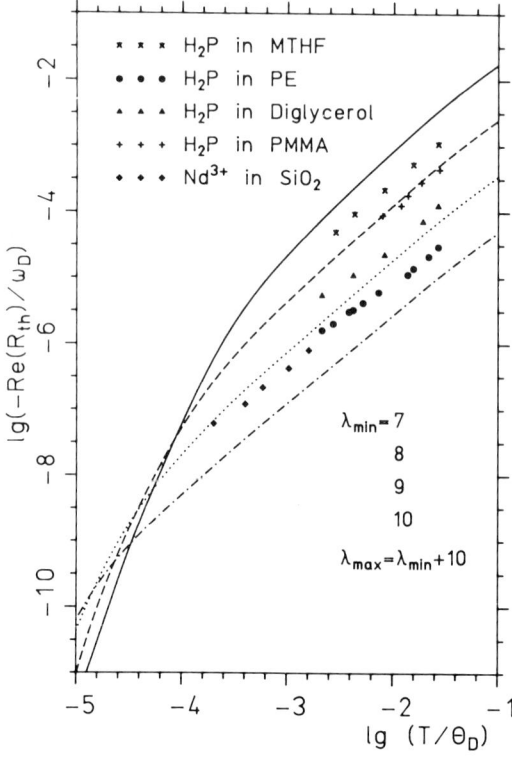

Fig. 18. Width of the most intense single transition line contributing to the line shape (7.35) in comparison with experimental data taken from [60—62]. Short-range interaction is assumed. Parameter values: $-0.1 < \Delta_a/\omega_D < 0.1$, $\Delta V_{\max} = 0.1446\omega_D$, $(\langle \Delta V \rangle = 10^{-3}\omega_D)$, $\Delta V \propto 1/r^3$, $r_{\max} = 10 r_{\min}$, $\Theta_D = 150$ K, $c = 2$ km/s, $\rho = 1$ g/cm^3, $f = 1$ eV, $\kappa = 0.3$.

distant TLSs interacting are about ten times farther away from the guest molecule than the nearest ones. There is, however, no independent justification for the value $\kappa = 0.3$.

Let us therefore consider case 2. We add up the exact real parts of the single-TLS case and choose the maximum distance so large that all TLSs with significant contributions to the linewidth are taken into account. Figure 19 shows the results of such a calculation compared to the same experimental data as in Figure 18. The upper curves correspond to dipolar coupling ($\Delta V \propto 1/r^3$) the lower ones to quadrupole–dipole coupling ($\Delta V \propto 1/r^4$). The TLSs are assumed to be spatially homogeneously distributed. The minimum value of ΔV considered is 10^{-6} of its maximum value which means that for dipolar coupling the most distant TLS is a hundred times as far from the guest molecule as the nearest, i.e. the number of TLSs taken into account is 10^6. In the $1/r^4$ case it is $10^{4.5}$.

The figure shows that the data points are not as well described as in case 1, but this is outweighed by the fact that we did not fit any parameters. (κ was put equal

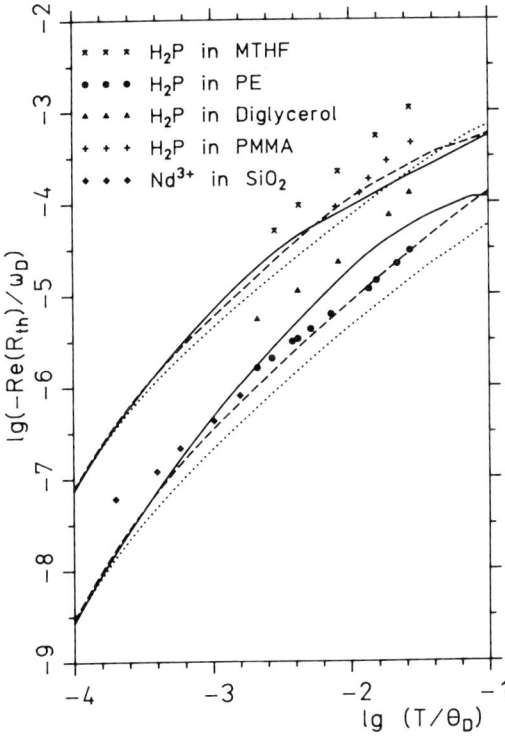

Fig. 19. Sum of (exact) real parts of the single-TLS problem describing the linewidth. Distant TLSs are taken into account. Parameter values: $-0.1 < \Delta_a/\omega_D < 0.1$, $\Delta V_{max} = 0.72382 \times 10^{-3}\omega_D = 10^6 \Delta V_{min}$ ($\Rightarrow \langle \Delta V \rangle = 10^{-8} \omega_D$ for the upper curves), $\Theta_D = 200$ K, $c = 2$ km/s, $\rho = 2$ g/cm^3, $f = 1$ eV, $\kappa = 1.0$. Full lines: $\lambda_{min} = 6$; dashed lines: $\lambda_{min} = 7$, dotted lines: $\lambda_{min} = 8$. $\lambda_{max} = \lambda_{min} + 7$. Upper curves: $\Delta V \propto 1/r^3$, lower curves: $\Delta V \propto 1/r^4$.

to one, which is the standard value.) A further advantage of case 2 is its capability to yield a possible explanation for the crossover from one power law of the temperature dependence of the linewidth to a different one seen in some experiments [61]. Whereas in case 1 the functional dependence of ΔV on r does not play any role, in case 2 the curves corresponding to a steeper r dependence show a stronger temperature dependence, too. Now it is well known that there are two mechanisms of interaction between the guest molecule and the TLSs. The first is **electrostatic** coupling [67] stemming from the electric dipole moments of both TLSs and the guest molecule (which may instead have a quadrupole moment only) the second **elastic** coupling [69] through the orbit-lattice interaction (the TLSs are regarded as elastic dipoles, their deformation fields influencing the guest molecule). The second interaction usually is stronger than the first. If one thinks in terms of screening, one would therefore expect that these interactions have different ranges, the elastic coupling being stronger for nearby TLSs but also being screened stronger, therefore falling off rapidly at some distance where the electrostatic coupling, which is screened more weakly, still shows a smooth r dependence. Qualitatively, we obtain the picture given in Figure 20. At low temperatures the TLSs far away from the guest molecule and outnumbering the nearer ones will each give a contribution to the linewidth of the guest which is comparable to one of the latter. With increasing temperature, according to (4.73) the contribution of a single TLS increases, but the number of TLSs for which this approximation holds decreases and the rest are contributing only negligible amounts. Therefore, the increase in total linewidth is less than to be expected from that of the single-TLS width. Now when a temperature is reached, where only TLSs with r values smaller than or equal to that of the jump discontinuity in the picture contribute significantly, a further increase in temperature will no longer diminish the number of TLSs which are not negligible — the linewidth increases faster. Whether the following smooth r dependence at high ΔV's is to

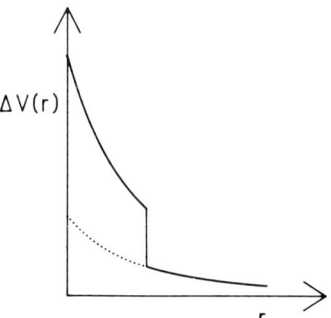

Fig. 20. General behavior of the TLS–ion interaction if screening is important. The full line represents the total interaction which is the sum of elastic and electrostatic couplings on the left-hand side of the jump discontinuity (the dotted line shows the electrostatic part of the interaction) and on the right-hand side consists of electrostatic interaction only, if one assumes that the elastic coupling is screened totally.

be seen in experiment or not, depends on the temperature above which only these ΔV values would give significant contributions. If it is too high, the effect will be covered either by many-phonon and activated processes or by the non-dynamical effects discussed at the end of the last section.

In numerical calculations we have been able to see different power laws resulting from the effect described. But up to now we have not found a crossover from a $T^{1.0}$ to a $T^{1.3}$ dependence which is observed experimentally.

7.7. NUMERICAL LINE SHAPES

It is not possible to evaluate (7.35) numerically exactly for a sufficiently large number of TLSs ($N > 100$), because the CPU time needed increases $\propto 4^N$. A calculation taking into account 8 TLSs with representative TLS parameters yields Figure 21. Each curve consists of $4^8 = 65\,536$ lines, but only a single line is seen.

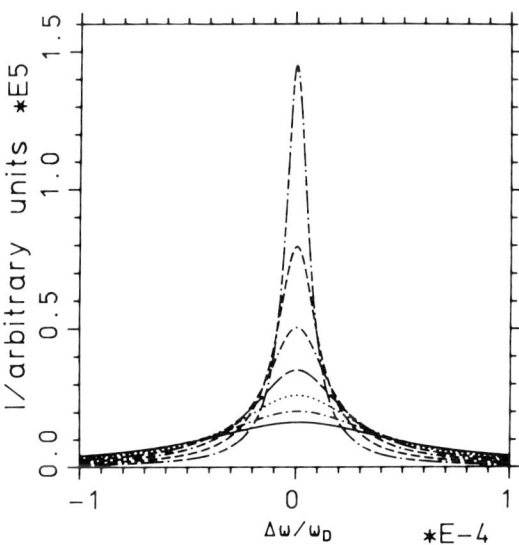

Fig. 21. Line shape according to (7.35). 8 TLSs. Different curves correspond to different temperatures, ranging from $0.002\Theta_D$ (most intense line shape) to $0.005\Theta_D$ (least intense line shape) in steps of $0.005\Theta_D$. Parameter values of λ, Δ_a/ω_D, $\Delta V/\omega_D$ (given in this order): $5, 0, 1.1 \times 10^{-5}$; $5, 2 \times 10^{-2}, 1.2 \times 10^{-5}$; $6, -2 \times 10^{-2}, 1.1 \times 10^{-5}$; $6, 0, 1.0 \times 10^{-5}$; $7, 0, 1.1 \times 10^{-5}$; $7, 1.0 \times 10^{-2}$, 1.1×10^{-5}; $8, 0, 1.0 \times 10^{-5}$; $8, 10^{-2}, 1.2 \times 10^{-5}$. Remaining parameters: $\kappa = 1$, $\Theta_D = 200$ K, $c = 2$ km/s, $\rho = 2$ g/cm^3, $f = 1$ eV, ($E_0 = 100\omega_D$).

Therefore, one is inclined to believe that the spectrum is dominated by this line. This may be tested in a direct manner by calculating the single line belonging to the eigenvalue $R\begin{pmatrix}\beta & 1 & \cdots & 1 \\ \alpha & 1 & \cdots & 1\end{pmatrix}$. The result is shown in Figure 22. The similarity between the figures is striking. But regarding the intensities of the lines which are given in the same (arbitrary) units for both figures, we recognize that although the lines of Figure 22 probably are the most intense single transition lines they only

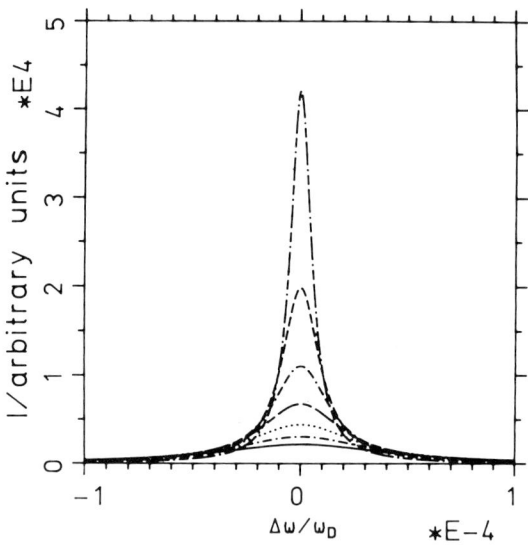

Fig. 22. Line shape of the $R\binom{\beta\;|\;\cdots\;|}{a\;|\;\cdots\;|}$ line. Same parameter values as in Figure 21.

contribute a third or less of the total line intensity. This situation may change somewhat in favor of the arguments given in Section 7.5, if the number of TLS considered is large, a question which therefore should be investigated further.

8. Concluding Remarks

To derive the optical line shape of guest molecules or atoms in glasses we have used a model which couples the guest to the TLSs, the characteristic degrees of freedom of glasses, and the TLSs to the virbrations of the glass. The direct coupling of the guest to the vibrations was assumed to contribute only negligibly to the optical linewidth. The vibrations were treated as a heat bath and eliminated from the equations of motion through Mori's formalism. Principal value terms, occurring in this procedure, have been neglected as usual. Furthermore, the linewidth of the guest molecule, i.e. its dephasing rate, was calculated as the sum of the contributions of the TLSs coupled to it, a procedure which was justified in the framework of an approximate many-particle treatment.

It was shown that the temperature dependence of the observed linewidth can be explained in one of the following four ways. The first three possibilities are mathematically equivalent, but allow for three different physical interpretations; the fourth possibility requires a different mathematical treatment.

(1) The first interpretation assumes that the matrix elements h_q describing the coupling of the TLSs to the vibrations are proportional to $\sqrt{\omega_q^\kappa}$ with the exponent $\kappa \approx 0.3$. This assumption means that the 'effective deformation

potential' $f\sqrt{(\omega/\omega_c)^{\kappa-1}}$ increases with decreasing frequency. As long as $\kappa > 0$ the interaction matrix elements h_q disappears for $\omega \to 0$, i.e. there are no divergences. But from a theoretical point of view the phenomenological exponent κ is somewhat disturbing, because in crystals $h_q \propto \sqrt{\omega_q}$ for acoustic and $\propto 1/\sqrt{\omega_q}$ for optical phonons. However, the TLSs themselves represent a phenomenological microscopic component of the model and therefore the assumed frequency dependence for h_q cannot be excluded at present. In one of the earlier papers [116] in connection with the optical line shape it is stressed that in glasses one cannot assume $h_q \propto \sqrt{\omega_q}$ except for extremely low frequencies.

(2) Instead of assuming a phenomenological frequency dependence of the deformation potential one can mathematically equivalently assume a modified distribution of TLSs. This interpretation results in a higher density of TLSs at lower excitation energies. The advantage of this interpretation is that it merely represents a modification of the distribution of the TLSs [117, 118], a quantity which is unknown *a priori*.

(3) Finally, also a modification of the density of the vibrational degrees of freedom results in the same expression for the linewidth, namely leaving h_q and the TLS density unchanged and replacing the Debye density of states $\propto \omega^2$ by one $\propto \omega^{\kappa+1}$. An interpretation of this kind has recently been given by Lyo and Orbach [72] in the framework of a fracton model. Whether this interpretation is applicable remains an open question, because fractons are excitations of short wavelength whereas at low temperatures long-wavelength excitations are important.

(4) Another possibility to interpret the experimentally observed temperature dependence of the linewidth leaves the quantities unchanged considered so far in this section and takes into account a weak decrease with increasing distance of the coupling between the guest and the TLSs implying a long range of the interaction. For that reason a large number of TLSs contribute to the linewidth of the guest. A problem of this interpretation is the fact that very many TLSs have to be coupled to the guest molecule in order to describe the experimentally observed linewidth numerically. An advantage is definitely that no further assumptions are needed. In this interpretation it is possible to give a rough estimate of the temperature dependence of the linewidth, which represents a lower bound. The analytical calculations in Section 6.3 show that within a certain temperature range the linewidth, averaged over Δ and λ, is proportional to T^2. From Figure 7 we can estimate that for Δ values close to the maximum of the linewidth the critical value of ΔV, above which TLSs contribute considerably to the linewidth, is proportional to T. Assume that the functional relationship between the distance r of the TLS from the guest and the coupling ΔV is given by $\Delta V \propto r^{-s}$. Then the radius of the sphere around the guest molecule, within which TLSs contribute considerably to the linewidth, is $\propto T^{-1/s}$ and their number $\propto T^{-3/s}$. If one assumes that all TLSs give the same contribution, the linewidth is $\propto T^{2-3/s}$. In the case of

dipolar interaction ($s = 3$) the linewidth is $\propto T$ and for quadrupolar interaction ($s = 4$) it becomes $\propto T^{1.25}$. The numerically determined linewidth is somewhat steeper because the real situation is not as simple as assumed in this estimate.

As regards the second of the foregoing alternatives, Hunklinger [119] shows that in silica glasses the assumption of a constant distribution $P(\lambda, \Delta)$ is very well satisfied indeed, whereas in polymers a larger part of more slowly relaxing TLSs seem to exist (i.e. TLSs with larger values of λ and thus smaller energy splittings). Thus for polymers the tendency seems to be correct, however it is unclear whether the portion of these TLSs is large enough. Most acoustic experiments in glasses are interpreted successfully with the assumption of a constant deformation potential which would outrule the first alternative. From these experiments, however, the product of the density of states of the TLSs and of the square of the deformation potential is extracted. Thus an increase in this product can be attributed to the deformation potential or to the density of states, and occasionally modified distributions for the TLSs are used [22, 117, 118].

The $T^{1.3}$ dependence of the linewidth is also obtained in a model by Jackson and Silbey [76] who use a combination of two relaxation processes, one on the basis of TLSs, the other being induced by the direct coupling of the guest to librational modes. A different treatment in the framework of spectral diffusion also yields a $T^{1.3}$ dependence of the dephasing rate [75, 120]. As in our model, the guest molecule is coupled to several TLSs. An interesting question to be investigated in the future is whether there exists a connection between both treatments.

References

1. K. Kassner, Thesis, Univ. of Ulm (1985).
2. R. C. Zeller and R. O. Pohl, *Phys. Rev.* **B4**, 2029 (1971).
3. R. Berman, *Proc. Roy. Soc. London* **A208**, 90 (1951).
4. A. C. Anderson, W. Reese, and J. C. Wheatley, *Rev. Sci. Instrum.* **34**, 1386 (1963).
5. R. A. Fisher, G. E. Brodale, E. W. Hornung, and W. F. Giauque, *Rev. Sci. Instr.* **40**, 365 (1969).
6. D. Redfield, *Phys. Rev. Lett.* **27**, 730 (1971).
7. P. Fulde and H. Wagner, *Phys. Rev. Lett.* **27**, 1280 (1971).
8. S. Takeno and M. Goda, *Prog. Theor. Phys.* **48**, 1468 (1972).
9. P. W. Anderson, B. I. Halperin, and C. M. Varma, *Phil. Mag.* **25**, 1 (1972).
10. W. A. Phillips, *J. Low Temp. Phys.* **7**, 351 (1972).
11. H. B. Rosenstock, *J. Non-Cryst. Solids* **7**, 123 (1972).
12. H. P. Baltes, *Solid State Commun.* **13**, 225 (1973).
13. L. S. Kothari, Usha, *J. Non-Cryst. Solids* **15**, 347 (1974).
14. G. J. Morgan and D. Smith, *J. Phys. C* **7**, 649 (1974),
15. D. Walton, *Solid State Commun.* **14**, 335 (1974).
16. P. R. Couchman, R. L. Reynolds, and R. M. Cotteril, *Nature* **264**, 534 (1976).
17. W. H. Tantilla, *Phys. Rev. Lett.* **39**, 554 (1977).
18. S. Hunklinger, W. Arnold, S. Stein, R. Nava, and K. Dransfeld, *Phys. Lett.* **A42**, 253 (1972).
19. B. Golding, J. E. Graebner, B. I. Halperin, and R. J. Schutz, *Phys. Rev. Lett.* **30**, 223 (1973).

20. A. C. Anderson, 'Thermal Conductivity' in *Amorphous Solids, Low-Temperature Properties (Topics in Current Physics)*, Vol. 24 (ed. W. A. Phillips), Springer-Verlag, Berlin, Heidelberg, New York (1981), p. 65.
21. R. B. Stephens, *Phys. Rev.* **B8**, 2896 (1973).
22. J. C. Lasjaunias, A. Ravex, M. Vandorpe, and S. Hunklinger, *Solid State Commun.* **17**, 1045 (1975).
23. M. T. Loponen, R. C. Dynes, V. Narayanmurti, and J. P. Garno, *Phys. Rev. Lett.* **45**, 457 (1980).
24. M. Meissner and K. Spitzmann, *Phys. Rev. Lett.* **46**, 265 (1981).
25. L. Piché, R. Maynard, S. Hunklinger, and J. Jäckle, *Phys. Rev. Lett.* **32**, 1426 (1974).
26. S. Hunklinger and M. v. Schickfus, 'Acoustic and Dielectric Properties of Glasses at Low Temperatures' in *Amorphous Solids, Low-Temperature Properties (Topics in Current Physics)*, Vol. 24 (ed. W. A. Phillips), Springer-Verlag, Berlin, Heidelberg, New York (1981), p. 81.
27. M. Schmidt, R. Vacher, J. Pelous, and S. Hunklinger, *J. Phys. (Paris)* **43**, C9-501 (1983).
28. W. Arnold and S. Hunklinger, *Solid State Commun.* **17**, 883 (1975).
29. M. W. Klein, B. Fischer, A. C. Anderson, and P. J. Anthony, *Phys. Rev.* **B18**, 5887 (1978).
30. J. E. Graebner and B. Golding, *Phys. Rev.* **B19**, 964 (1979).
31. J. L. Black and B. I. Halperin, *Phys. Rev.* **B16**, 2879 (1977).
32. M. v. Schickfus, S. Hunklinger, and L. Piche, *Phys. Rev. Lett.* **35**, 876 (1975).
33. W. A. Phillips, 'The Thermal Expansion of Glasses' in *Amorphous Solids, Low-Temperature Properties (Topics in Current Physics)*, Vol. 24 (ed. W. A. Phillips), Springer-Verlag, Berlin, Heidelberg, New York (1981), p. 53.
34. M. v. Haumeder, U. Strom, and S. Hunklinger, *Phys. Rev. Lett.* **44**, 84 (1982).
35. H. Tokumoto, K. Kajimura, S. Yamasuki, and K. Tanaka, 'Observation of Two-Level Tunneling States in Amorphous Silicon by Surface Acoustic Waves' in *Proceedings of the 17th International Conference on Low Temperature Physics*, Vol. 1 (eds U. Eckern, A. Schmid, W. Weber, H. Wühl) North-Holland, Amsterdam, Oxford, New York, Tokyo (1984), p. 381.
36. B. M. Kharlamov, R. I. Personov, and L. A. Bykovskaya, *Optics Commun.* **12**, 191 (1974).
37. A. A. Gorokhovskii, R. K. Kaarli, and L. A. Rebane, *Zh. Eksp. Teor. Fiz. Pis'ma Red* **20**, 474 (1974) [*JETP Lett.* **20**, 216 (1974)].
38. H. P. H. Thijssen, R. van den Berg, and S. Völker, 'Hole-Burning in Organic Glasses at Temperatures Down to 0.3 K' in *Conf. Proc. of the VW Symposium, Mittelberg 16.—21.9.84, on Photoreaktive Festkörper* (ed. H. Sixl), Wahl-Verlag, Karlsruhe (1984), p. 763.
39. S. Völker and J. H. van der Waals, *Mol. Phys.* **32**, 1703 (1976).
40. J. M. Hayes, R. P. Stout, and G. J. Small, *J. Chem. Phys.* **74**, 4266 (1981).
41. R. M. Shelby and R. M. Macfarlane, *Chem. Phys. Lett.* **64**, 545 (1979).
42. A. M. Dicker, L. W. Johnson, S. Völker and J. H. van der Waals, *Chem. Phys. Lett.* **100**, 8 (1983).
43. K. K. Rebane and R. A. Avarmaa, *J. Photochemistry* **17**, 311 (1981).
44. J. M. Hayes and G. J. Small, *Chem. Phys. Lett.* **54**, 435 (1978).
45. J. M. Hayes and G. J. Small, *Chem. Phys.* **27**, 151 (1978).
46. F. G. Patterson, W. H. Lee, R. W. Olson, and M. D. Fayer, *Chem. Phys. Lett.* **84**, 59 (1981).
47. L. A. Rebane, A. A. Gorokhovskii, and J. V. Kikas, *Appl. Phys.* **B29**, 235 (1982).
48. J. Friedrich and D. Haarer, *J. Chem. Phys.* **76**, 61 (1982); *Angew. Chemie (Int. Ed. Engl.)* **23**, 113 (1984).
49. A. A. Gorokhovskii and L. A. Rebane, *Optics Commun.* **20**, 144 (1982).
50. A. A. Gorokhovskii, Y. V. Kikas, V. V. Pal'm, and L. A. Rebane, *Sov. Phys. Sol. State* **23**, 602 (1981).
51. P. M. Selzer, D. L. Huber, D. S. Hamilton, W. M. Yen, and M. J. Weber, *Phys. Rev. Lett.* **36**, 817 (1976).
52. J. Hegarty and W. M. Yen, *Phys. Rev. Lett.* **43**, 1126 (1979).
53. P. Avouris, A. Campion, and M. A. El-Sayed, *J. Chem. Phys.* **67**, 3397 (1977).
54. J. R. Morgan and M. A. El-Sayed, *Chem. Phys. Lett.* **84**, 213 (1981).

55. R. M. Macfarlane and R. M. Shelby, *Optics Commun.* **45**, 46 (1983).
56. R. M. Shelby, *Optics Lett.* **8**, 88 (1983).
57. A. I. M. Dicker and S. Völker, *Chem. Phys. Lett.* **87**, 481 (1982).
58. H. P. H. Thijssen, A. I. M. Dicker, and S. Völker, *Chem. Phys. Lett.* **92**, 7 (1982).
59. H. P. H. Thijssen, S. Völker, M. Schmidt, and H. Port, *Chem. Phys. Lett.* **94**, 537 (1983).
60. H. P. H. Thijssen, R. E. van den Berg, and S. Völker, *Chem. Phys. Lett.* **97**, 295 (1983).
61. H. P. H. Thijssen, R. E. van den Berg, and S. Völker, *Chem. Phys. Lett.* **103**, 23 (1983).
62. J. Hegarty, M. M. Broer, B. Golding, J. R. Simpson, and J. B. MacChesney, *Phys. Rev. Lett.* **51**, 2033 (1983).
63. W. Breinl, J. Friedrich, and D. Haarer, *J. Chem. Phys.* **80**, 3496 (1984).
64. W. Breinl, J. Friedrich, and D. Haarer, *Chem. Phys. Lett.* **106**, 487 (1984).
65. W. Breinl, J. Friedrich, and D. Haarer, *J. Chem. Phys.* **81**, 3915 (1984).
66. W. Richter, G. Schulte, and D. Haarer, *Opt. Commun.* **51**, 412 (1984).
67. S. K. Lyo and R. Orbach, *Phys. Rev.* **B22**, 4223 (1980).
68. P. Reineker and H. Morawitz, *Chem. Phys. Lett.* **86**, 359 (1982).
69. T. L. Reinecke, *Solid State Commun.* **32**, 1103 (1979).
70. J. R. Klauder and P. W. Anderson, *Phys. Rev.* **125**, 912 (1962).
71. S. K. Lyo, *Phys. Rev. Lett.* **48**, 688 (1982).
72. S. K. Lyo and R. Orbach, *Phys. Rev.* **B29**, 2300 (1984).
73. P. Reineker, H. Morawitz, and K. Kassner, *Phys. Rev.* **B29**, 4546 (1984).
74. S. Hunklinger and M. Schmidt, *Z. Phys. B — Condensed Matter* **54**, 93 (1984).
75. D. L. Huber, M. M. Broer, and B. Golding, *Phys. Rev. Lett.* **52**, 2281 (1984).
76. B. Jackson and R. Silbey, *Chem. Phys. Lett.* **99**, 331 (1983).
77. I. S. Osad'ko, *Chem. Phys. Lett.* **115**, 411 (1985).
78. W. Zachariasen, *J. Amer. Chem. Soc.* **54**, 3841 (1932).
79. J. M. Ziman, *Models of Disorder*, Cambridge Univ. Press, Cambridge, London, New York, Melbourne (1979), S. IX.
80. J. D. Bernal, *Nature* **183**, 141 (1959); *Nature* **185**, 68 (1960); *Nature* **188**, 910 (1960).
81. D. Weaire, *Contemp. Phys.* **17**, 173 (1976).
82. N. Rivier, *Phil. Mag. A* **40**, 859 (1979); 'Topological Structure of Glasses' in *Amorphous Materials: Modeling of Structure and Properties* (ed. V. Vitek) Conf. Proc. of the Metallurgical Soc. of AIME (1982), p. 83.
83. N. Rivier, *Rev. Bras. Fis.* (submitted for publication).
84. J. Jäckle, L. Piché, W. Arnold, and S. Hunklinger, *J. Non-Cryst. Solids* **20**, 365 (1976).
85. J. A. Sussmann, *Phys. kond. Materie* **2**, 146 (1964); *J. Phys. Chem. Solids* **28**, 1643 (1967).
86. H. Haken, *Quantenfeldtheorie des Festkörpers*, Teubner, Stuttgart 1973, p. 78.
87. C. Kittel, *Quantentheorie der Festkörper*, Oldenbourg, München (1970), p. 32.
88. W. A. Phillips, *J. Low Temp. Phys.* **11**, 757 (1973).
89. J. Joffrin and A. Levelut, *J. Phys. (Paris)* **36**, 811 (1975).
90. H. Morawitz and P. Reineker, *Solid State Commun.* **42**, 609 (1982).
91. P. Reineker, K. Kassner, and H. Morawitz, *Materials Science* **10**, 221 (1984).
92. P. Reineker, K. Kassner, and H. Morawitz, 'Dephasing Optischer Anregung von Gastmolekülen in Gläsern' in *Conf. Proc. of the VW Symposium, Mittelberg 16.—21.9.84, Photoreaktive Festkörper* (ed. H. Sixl), Wahl-Verlag, Karlsruhe (1984), p. 65.
93. J. Jäckle, *Z. Physik* **257**, 212 (1972).
94. R. Orbach and M. Tachiki, *Phys. Rev.* **158**, 524 (1967).
95. J. Friedrich, J. D. Swalen, and D. Haarer, *J. Chem. Phys.* **73**, 705 (1980).
96. P. Reineker, 'Exciton Dynamics in Molecular Crystals and Aggregates' in *Springer Tracts in Modern Physics*, Vol. 94 (ed. G. Höhler), Springer-Verlag, Berlin, Heidelberg, New York (1982), p. 111.
97. P. Reineker, Thesis, Univ. of Stuttgart (1971).
98. R. Kubo, *J. Phys. Soc. Japan* **12**, 570 (1957).
99. H. Mori, *Prog. Theor. Phys.* **33**, 423 (1965).
100. R. F. Fox, *Physics Reports* **48**, 179 (1978).

101. D. Forster, *Hydrodynamic Fluctuations, Broken Symmetry, and Correlation Functions*, Benjamin, Reading, Massachusetts (1975).
102. H. Grabert, *Projection Operator Techniques in Nonequilibrium Statistical Mechanics*, Springer-Verlag, Berlin, Heidelberg, New York (1982), p. 36.
103. W. Götze, *Phil. Mag.* **43**, 219 (1981).
104. P. Talkner, Thesis, Univ. of Stuttgart (1979).
105. R. P. Feynman, *Phys. Rev.* **84**, 108 (1951).
106. B. Donovan and J. F. Angress, *Lattice Vibrations*, Chapman and Hall, London (1971), p. 45.
107. S. Hunklinger and L. Piché, *Solid State Commun.* **17**, 1189 (1975).
108. B. Golding, J. E. Graebner, and R. J. Schutz, *Phys. Rev.* **B14**, 1660 (1976).
109. L. Van Hove, *Phys. Rev.* **89**, 1189 (1953).
110. H. Haken, *Handbuch der Physik*, Band XXV/2c, *Licht und Materie*, Springer-Verlag, Berlin, Heidelberg, New York (1970), p. 40.
111. A. Abragam, *The Principles of Nuclear Magnetism*, Clarendon Press, Oxford (1970), p. 405.
112. T. Muir, *A Treatise on the Theory of Determinants*, Dover Publications, New York (1960).
113. S. M. Ulam and J. von Neumann, *Bull. Am. Math. Soc.* (*abstr.*) **53**, 1120 (1947).
114. S. M. Ulam, *Proceedings of the 11th International Congress of Mathematicians, Cambridge, Massachusetts*, Vol. II, Am. Math. Soc., Providence R.I. (1950), p. 264.
115. I. S. Gradshteyn and I. M. Ryzhik, *Table of Integrals, Series and Products*, Academic Press, London, 1965.
116. D. L. Huber, *J. Non-Cryst. Solids* **51**, 241 (1982).
117. B. Golding, J. Graebner, and A. B. Kane, *Phys. Rev. Lett.* **41**, 1487 (1978).
118. P. Doussineau, C. Frenois, R. G. Leisure, A. Levelut, and J.-Y. Prieur, *J. Phys.* (*Paris*) **41**, 1193 (1980).
119. S. Hunklinger, 'Low-Energy Excitations in Disordered Solids: New Aspects' in *Phonon Scattering in Condensed Matter* (eds W. Eisenmenger, K. Laßmann, S. Döttinger), Springer-Verlag, Berlin, Heidelberg, New York, Tokyo (1984), p. 378.
120. M. M. Broer, B. Golding, W. H. Haemmerle, J. R. Simpson and D. L. Huber, Preprint.

STRUCTURAL RELAXATION PROCESSES IN POLYMERS AND GLASSES AS STUDIED BY HIGH RESOLUTION OPTICAL SPECTROSCOPY

J. FRIEDRICH and D. HAARER

Physikalisches Institut der Universität Bayreuth, Postfach 3008, 8580 Bayreuth, F.R.G.

1. Introduction

Optical spectroscopy of glasses can be subdivided into two classes of experiments. One class deals with the intrinsic optical properties of disordered and glassy solids. Experiments of this kind are prevalent in the field of amorphous semiconductors. The optical experiments are mainly directed towards investigating the optical properties of the band-to-band transitions and yield information on the density of states of the disordered materials. One observes that the band gap, which is well defined for crystalline substances, is washed out in the case of amorphous materials and, hence, there are states in energy regimes which are forbidden in the crystalline solid.

Our article deals with a second class of experiment, in which the glassy material is used as host material for color center-like states or molecular states which are due to a well-defined doping of the glassy material. A historic example for those doped glasses are the beautiful stained glass windows of cathedrals, which owe their magnificent colors to metal ions or metal colloid particles, giving rise to optical absorption bands in the visible spectral range. In this and in other examples of doped glasses the dopant materials determine the various optical properties of the amorphous host—guest system.

The topic of this article is to demonstrate, that high-resolution optical spectroscopy can add a completely new dimension to the spectroscopy of glasses, a field, which hitherto, has received no or little attention, because of various serious intrinsic limitations. These limitations can, in part, be overcome by techniques like hole-burning [1, 2] or fluorescence line-narrowing [3, 4] (for reviews see [5—8]).

The main limitation of the spectroscopy of glasses is the very large inhomogeneous optical bandwidth, which is a consequence of the disorder in the host material and, which usually limits the spectral resolution to an extent, that typical features of optical lines, such as phonon-structures or even vibrational structures are 'smeared out'. This situation is schematically depicted in Figure 1. The figure shows schematically the random nature of the solvent cage. Very often a solvent cage, i.e. a local host—guest configuration is called a 'site', although this term is not very clearly defined. In Figure 1, three different 'sites' are shown. The black rectangles with numbers 1, 2 and 3 symbolize molecules (for instance, dye molecules) which are embedded in an organic glass (for instance, solid ethanol or a polymer glass like PMMA (polymethylmethacrylate)). Due to the different

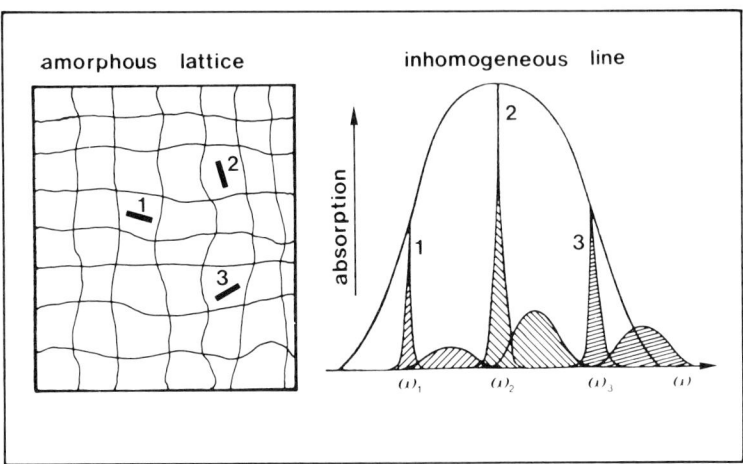

Fig. 1. Three sites in an amorphous host matrix and their respective absorption spectra (see text).

environment of the three absorbes, their spectrum appears at different frequencies in the optical band as is symbolically depicted in the right-hand part of Figure 1. As we can see from the figure, the details of the molecular site spectra are smeared out and one expects for a macroscopic sample a broad line shape, obscuring the various features of the molecular spectra. Some typical spectral features are depicted in Figure 2. They are:

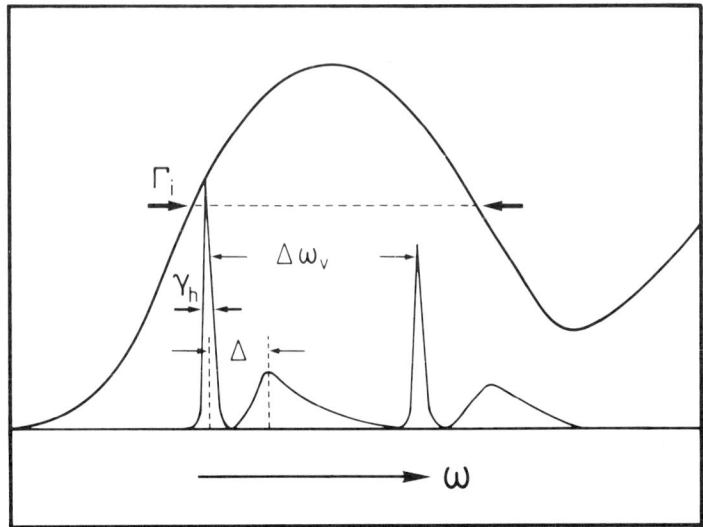

Fig. 2. Spectral parameters of a molecular site such as: the inhomogeneous linewidth Γ_i, the homogeneous linewidth γ_h, the spacing of the phonon sideband Δ (Stokes-shift) and the spacing of a vibrational transition $\Delta\omega_v$.

(a) the shape and width γ_h of the homogeneous zero-phonon absorption line;
(b) its relative intensity (Debye—Waller factor);
(c) the line shape of the phonon sideband;
(d) the Stokes shift parameter Δ;
(e) the spectral positions and intensities of vibrational lines.

The scheme of photochemical hole-burning allows us to 'label' one site spectroscopically by performing 'site-selective' photochemistry with a laser light source which is, ideally, infinitely narrow in spectral width. Usually one performs the photochemistry in the region of the lowest allowed (singlet) absorption band of the guest molecule. Figure 3 shows a typical reaction scheme, in which the photochemistry occurs in the triplet manifold of the molecular absorber (typical examples are, free base porphin [9], free base phthalocyanine [1], quinizarin [10]). The hole-burning spectrum is a 'negative image' of the site absorption spectrum, as is shown in Figure 4 for the case of free base phthalocyanine.

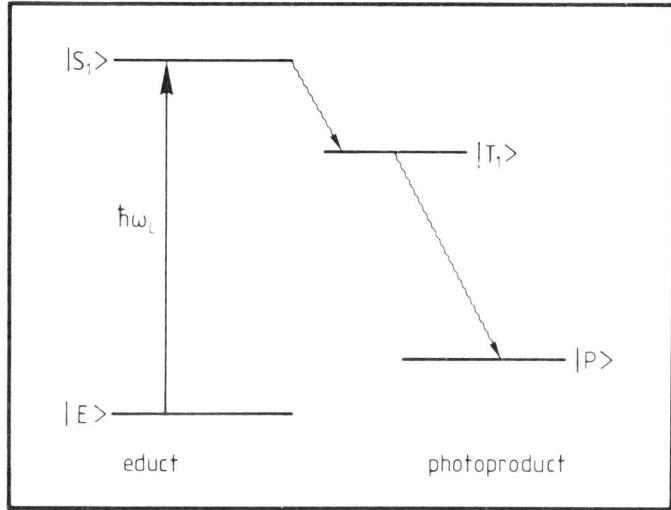

Fig. 3. Reaction scheme for triplet state photochemistry of the hole-burning reaction.

In this article we do not intend to give an exhaustive review of the field of photochemical and photophysical hole-burning. For such a review we refer to [5, 7, 8]. Instead, we would like to focus our attention to the question: What can one learn by hole burning experiments about the nature of glassy host systems and how does a glassy system change, if various external parameters are changed. The most obvious parameter is the time t. Here we have to assume that the glass state is a non-equilibrium state and allows, even at very low temperatures (1 K), relaxation processes to occur. Other parameters, whose influence on the glass system can be observed are external electric fields and strain fields. All of the above

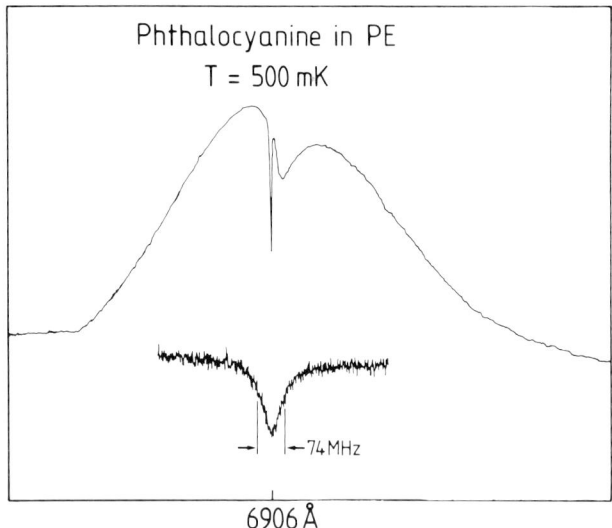

Fig. 4. Hole-burning spectrum of free base phthalocyanine in polyethylene. The zero-phonon line is shown below on an enlarged scale. The broad feature on the right-hand side of the zero-phonon line is the pseudo-phonon sideband. This feature will not be dealt with in this article (for a review see [8, 13]).

questions are based on rather recent experimental work and, hence, the article is bound to reflect the present state of the art.

2. The 'Site-Memory' Function

In the following paragraphs we would like to focus our attention to small spectroscopic changes, which occur in the spectrum of doped amorphous solids. In the introduction it was pointed out, that hole-burning is a powerful spectroscopic method of detecting subtle changes which occur as time, tempertaure or external electric fields or strain fields are changed. In the following we would like to put the above statement into a more quantitative form.

Let us assume that the original absorption spectrum is described by a series of electronic transitions. The lowest allowed transition is assumed to be at ω_0. Due to matrix inhomogeneities, the value of ω_0 is spread out over a frequency range Γ_i and we therefore have to assume, in the simplest case, a Gaussian distribution of sites around the center frequency ω_0. We call this distribution the 'site distribution function' $N^0(\omega' - \omega_0)$. Its normalized form can be written in the following fashion:

$$\frac{N^0(\omega' - \omega_0)}{N} = 2\sqrt{\frac{\ln 2}{\pi}} \cdot \frac{1}{\Gamma_i} \exp\left[\frac{-4(\omega' - \omega_0)^2 \ln 2}{\Gamma_i^2}\right]. \tag{1}$$

Let us now assume that we irradiate the sample for some time τ at the center frequency $\omega_0 = \omega_L$; ω_L is the laser frequency. After laser irradiation for a time τ, the site distribution function will have changed by the amount $\delta^\tau N(\omega' - \omega_L)$:

$$\delta^\tau N(\omega' - \omega_L) = \frac{N^0(\omega' - \omega_0) - N^\tau(\omega' - \omega_L)}{N}. \tag{2}$$

For simplicity we can center our ω-axis at $\omega_L = \omega_0 = 0$ and, hence, have a change in the site distribution function given by $\delta^\tau N(\omega')$. Its detailed spectroscopic shape is of no interest at the moment. However, we assume, that its half width γ is small compared to the width Γ_i of the inhomogeneous line.

So far we have only dealt with the site distribution functions before and after photochemical changes. In order to perform an absorption experiment, however, one has to ascribe to each site an integrated absorption cross section σ and a normalized line shape function $g(\omega - \omega')$. We assume that σ is constant for all sites at the various ω' frequencies and, hence, we get a spectrum of a 'hole' which is the convolution of the site-change function $\delta^\tau N(\omega')$ and a characteristic molecular absorption function $g(\omega - \omega')$:

$$L^\tau(\omega) = \sigma \int_{-\infty}^{+\infty} \delta^\tau N(\omega') g(\omega - \omega') \, d\omega' \tag{3}$$

Usually $g(\omega - \omega')$ is the characteristic line shape function consisting of a zero-phonon part $z(\omega - \omega')$ of width γ_h and a phonon sideband $p(\omega - \omega' - \Delta)$ [11–14]. The width γ_h is given by the molecular relaxation processes at the temperature of the experiment T and is (FWHH; in units of $2\pi/s$)

$$\gamma_h = \frac{1}{T_1} + \frac{2}{T_2^*}, \tag{4}$$

where T_1 is the lifetime of the excited state and T_2^* is the pure dephasing time of the molecular ensemble (see, for instance, (8)).

In the following we do not consider the phonon sideband (for more details see [8]) but restrict our considerations to the zero-phonon line. If we rewrite Equation (3) for this special case (negligible electron–phonon coupling) we get:

$$L^\tau(\omega, T) = \sigma \int_{-\infty}^{+\infty} \delta^\tau N(\omega', T) z(\omega - \omega', T) \, d\omega' \tag{5}$$

having in mind that at each temperature T at which we perform our absorption experiment, the width of the line shape function z, and also the site memory function, is different due to the temperature dependence of the pertinent relaxation times.

Since the hole-burning photochemistry can be performed at 'a burn temperature' T_b which is different from the temperature T at which the absorption spectrum is monitored, Equation (5) has to be written in a way which includes both temperatures:

$$L^\tau(\omega, T, T_b) = \sigma \int_{-\infty}^{+\infty} \delta^\tau N(\omega', T_b, T) \cdot z(\omega - \omega', T) \, d\omega'. \tag{6}$$

One of the most interesting aspects of hole-burning spectroscopy is the fact that the function $\delta^\tau N(\omega', T_b, T)$ is a function which describes the original photochemical change of the sample immediately after the laser chemistry at $t = 0$. Each subsequent matrix change which occurs in an irreversible fashion with time (spectral diffusion) or due to the change of external parameters such as temperature, electric, magnetic or strain fields, etc., will be registered by the $\delta^\tau N(\omega', T_b, T)$-function. If one assumes, for instance, that there is a change in the site-distribution function as the time evolves, one can use the hole-burning experiment to monitor how the glass system, which has been put into in a special non-equilibrium state by laser irradiation, approaches its former state. Therefore, Equation (6) has to be written as

$$L^\tau(\omega, T, T_b, t, q) = \sigma \int_{-\infty}^{+\infty} \delta^\tau N(\omega', T_b, T, t, q) z(\omega - \omega', T) \, d\omega', \tag{7}$$

where q is any one of the above-mentioned external parameters. (We do not consider the possibility that z may also depend on q.)

The premise that the 'site memory function' $\delta^\tau N(\omega')$ monitors the history of the glass has been made rather early and has been documented by cycling the sample between the burn temperature T_b and some higher temperatures T [15, 16]. This led to an irreversible change in the hole spectrum. The measured change, which can be looked upon as a change of the site memory function $\delta^\tau N$, was measured and attributed to ground state redistribution processes. A formal theory describing the expected hole-burning line shape has been developed [17]. A more explicit theory of spectral diffusion processes will be given in the following section in conjunction with the observed spectral changes of photochemical holes as a function of time.

Before we continue with the rather subtle subject of spectral diffusion, we should mention that various theories on the shape of photochemical holes in the absence of spectral diffusion have been presented by different authors (see [8] and references therein). If, for instance, we postulate, that spectral diffusion is zero and that $T = T_b$, then Equations (5), (6) and (7) take the well-known form [8]

$$L^\tau(\omega) = \frac{I\tau\sigma^2\alpha^2\Phi}{\hbar\omega_L} \int_{-\infty}^{+\infty} z(\omega')z(\omega - \omega') \, d\omega' \tag{8}$$

Here $I \cdot \tau/\hbar\omega_L$ is the integrated number of photons during the photochemical irradiation time τ and Φ is the photochemical quantum yield. Note that the integrated absorption cross section appears quadratically in the above expression, because the hole-burning experiment is based on two subsequent events. First, the absorption of a photon and the subsequent photochemistry and, second, the detection of the hole via an absorption or fluorescence experiment.

If one inserts in Equation (8) for $z(\omega')$ a Lorentzian line with half width γ_h, one gets the well-known result of a photochemical hole with the width of $\gamma = 2\gamma_h$.

3. The Non-Equilibrium Nature of Glasses and its Relation to Optical Properties

The liquid-to-glass transition is a very general phenomenon and can be observed practically for most kinds of materials depending on the experimental conditions. (For reviews see [18–20].) Unlike a liquid-to-crystal transition the glass transition is no thermodynamic phase transition, in the sense that there are no diverging thermodynamic quantities connected with it. It is of a kinetic nature. At low temperatures the glass-forming molecules cannot relax into equilibrium on the time scale of the experiment, hence are frozen in a local minimum of their free energy surface. This 'freezing-in' occurs within a finite but rather narrow temperature interval of a few K. Hence, the potential distribution and the molecular packing is random. Connected with this randomness are large voids of free volume within which part of the glass-forming molecules can move quasi-liquid-like. Hence, one can look upon a glass as a solid cluster of infinite volume, which is interrupted by smaller solid- and liquid-like clusters. The glass-to-liquid transition can be viewed as a percolation transition, where a liquid-like cluster of infinite extension is formed at the glass transition temperature. Such a free volume percolation model explains rigorously the temperature dependence of the viscosity in the range of the transition temperature, known as the Vogel–Fulcher law.

A percolation model of the glassy state has some other interesting consequences: since a percolation cluster has a rather complex shape, e.g. like a sponge, it can no longer be regarded as a three-dimensional entity but is characterized by a fractal dimension, i.e. in this case a dimension varying between 2 and 3. Consequently, the density of phonon states of such clusters (the so-called fractons), which is of great influence on the various relaxation processes of the glass, is lower than that of a three-dimensional crystal [21]. However, the pecularity of a fractal dimension gets lost in case the wavelength of the excited phonons gets large enough to surpass the size of the percolution cluster. This occurs at a critical temperature T_c, below which the glass may be considered as a normal three-dimensional solid [22]. (For reviews see the papers by Lyo and by Reinecker and Kassner in this volume.)

Apart from the free volume model of the glass there is another very useful approach to account for the randomness of the potential distribution, the so-called TLS model. (For reviews see [23, 24].) Figure 5 shows a sketch of the random potential along a configuration space coordinate. This sketch clearly shows the non-equilibrium nature of the glass: suppose the system is frozen-in in a certain configuration, say q_1 then it can still relax to another configuration q_2. This

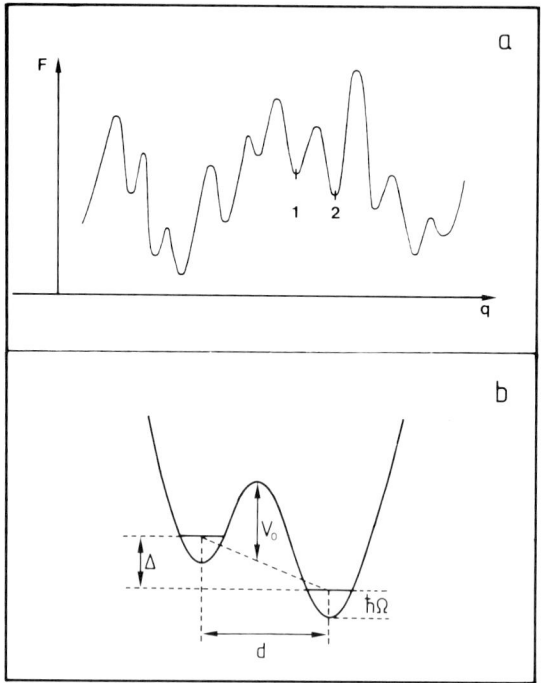

Fig. 5. (a) Potential energy along a configurational space coordinate of the glass. (b) Characteristic parameters of a two-well potential: barrier height V_0, energy asymmetry Δ, tunneling distance d and zero point vibration of energy $\hbar\Omega$.

would be possible even at zero temperature, by tunneling. Such relaxation processes, which change the structure of the glass, are characteristic for non-equilibrium systems. The random potential in Figure 5 can be viewed as an ensemble of double minimum potentials (5b). This is the simplest element of the complicated potential, which allows for a dynamic description. In case higher levels can be neglected the two lowest states of the double well form a two-level system, a so-called TLS. The TLS degrees of freedom are characteristic for amorphous solids. As a rule, they do not exist in a crystal. Within the frame of the TLS concept the thermodynamic properties of amorphous solids at low temperatures can be rather well described. The basic requirement to describe macroscopic properties is the distribution of the density of states involved. In the case of low-lying excitations this density of states can be obtained from experiments on the specific heat as a function of temperature. In glasses it has been found that, below 1 K, where the thermodynamics is determined by the TLS, the specific heat increases linearly with temperature [25]. From a linear specific heat it follows that the corresponding density of states, in this case of the TLS, is constant in energy [26]:

$$p(E)\,dE = p_0\,dE \tag{9}$$

where $p(E)$ is the number of TLS states with energy E per unit volume. The energies $E_\pm$ of the states of a double well depend on two parameters, e.g. the asymmetry parameter Δ and the tunneling parameter λ (Figure 5(b)):

$$E_\pm = \pm \tfrac{1}{2} E_{\Delta\lambda}$$
$$E_{\Delta\lambda} = \sqrt{\Delta^2 + [\hbar\Omega]^2 \, e^{-2\lambda}} \tag{10}$$

$\hbar\Omega$ is an energy of the order of the zero point vibrations in the double well. The quantity $\hbar\Omega \, e^{-\lambda}$ is referred to as the tunneling matrix element Δ_0:

$$\Delta_0 = \hbar\Omega \, e^{-\lambda} \tag{11}$$

λ can be estimated from Gamov's formula (for example, see [27]):

$$\lambda = \left[\frac{mV_0 d^2}{2\hbar^2} \right]^{\frac{1}{2}}. \tag{12}$$

V_0 is the barrier height, d the tunnel distance and m the mass of the tunneling species. At this point it is worthwhile to point out that stationary specific heat experiments do not yield an estimate of the boundaries of the distribution $p(E)$, since at higher temperatures ($T > 1$ K), the Debye phonons dominate the TLS contribution and consequently $p(E)$ cannot be investigated within this energy range. The lower limit is confined by the experimental experimentally achievable temperature. Hence, all the information which the c_p experiments can provide is that $p(E)$ is constant on the scale of kT.

It was assumed by Anderson et al. [26] that if the density of TLS states is uniformly distributed in E, it is also uniformly distirbuted in Δ and λ. That a uniform distribution of Δ and λ is rather consistent with $p(E) = $ const is shown by integrating the densities $p(\Delta) = p_\Delta$ and $p(\lambda) = p_\lambda$ under the constraints of Equation (10):

$$p(E) \, dE = p_\Delta p_\lambda \int_{\Delta_{\min}}^{\Delta_{\max}} d\Delta \int_{\lambda_{\min}}^{\lambda_{\max}} d\lambda \, \delta(E - E_{\Delta\lambda}) \, dE. \tag{13}$$

Transforming the integration over $d\lambda$ into an integration over $dE_{\Delta\lambda}$, one readily gets

$$p(E) = \frac{1}{2} p\Delta p\lambda \ln \left[\frac{1 + (\Delta_{\max}(E)/E)}{1 - (\Delta_{\max}(E)/E)} \cdot \frac{1 - (\Delta_{\min}/E)}{1 + (\Delta_{\min}/E)} \right]. \tag{14}$$

Thus, one has the important result that if Δ and λ are independently uniformly distributed, the corresponding states are also almost uniformly distributed in energy space; that is, they show only a very weak logarithmic dependence on energy. This is rather consistent with the observed linear specific heat behavior.

As can be seen from Equation (14), the density of TLS states diverges as E approaches its minimum value $\Delta_{\max}$. In this case λ would have to be infinity

(Equation (10)). This is, however, unrealistic since $\lambda \to \infty$ implies that the corresponding barrier heights (or tunnel distances) would also have to approach infinity. Thus, in a real situation Δ_{max} has to be smaller than E, but is, in any case, close to E.

As far as Δ_{min} is concerned, it is usually assumed to be zero, thus allowing for symmetric double wells, where the total energy originates from the splitting of the tunneling states. Although this choice may be justified in case nothing is known about the TLS states involved, one can imagine situations where the symmetry of the problem excludes TLS states with $\Delta_{min} = 0$.

In the case $\Delta_{min} = 0$, Equation (14) can be simplified by using Equation (10) and the fact that $E \approx \Delta_{max}$

$$p(E)\, dE \to p_\Delta p_\lambda \ln\left[\frac{2E}{\Delta_{0,min}}\right] dE \qquad (15)$$

A different view of the complex problem of the density of TLS states was given by Jäckle [28]. Based on the assumption of a uniform distribution of Δ and λ, it could be shown that the distribution $p(R)$ of tunneling relaxation rates R (at a constant energy E, Figure 6) obeys the following relation:

$$p(R)\, dR\, dE = \frac{1}{2} p_\Delta p_\lambda \frac{dR\, dE}{R(1 - R/R_{max})^{1/2}} \qquad (16)$$

where R (for a one-phonon process) is given by

$$R = \left[\frac{1}{C_l^5} + \frac{2}{C_t^5}\right]\left[\frac{\partial \Delta}{\partial \xi}\right]^2 \frac{\Delta_0^2}{2\pi \rho \hbar^4} E \,\text{ctgh}\, \frac{E}{2kT} \qquad (17)$$

with the longitudinal and transverse sound velocities c_l and c_t, the deformation

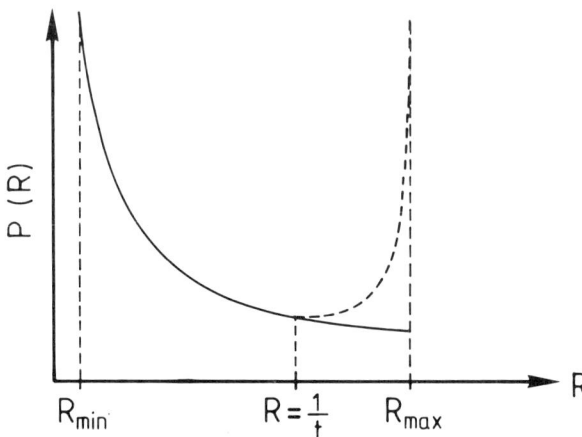

Fig. 6. Distribution of relaxation rates according to [28].

potential parameter $\partial\Delta/\partial\zeta$, the strain coordinate ζ and the mass density ρ. The above distribution quite clearly reflects a wide distribution of relaxation rates in the range

$$0 < R < R_{max}(E) \tag{18}$$

which is characteristic for the amorphous state. $R_{max}(E)$ is the maximum rate which corresponds to the fastest relaxation processes. The slow processes in the vicinity of $R = 0$ prevent the system from equilibrating on a measureable time scale. If

$$R_{max} \geqq t_{exp}^{-1} \tag{19}$$

with t_{exp} reflecting the experimental time scale, then, the above distribution warrants that the non-equilibrium nature of the glass can be observed experimentally. When the tunneling potentials are asymmetric, the related tunneling processes correspond to a structural relaxation, because the mean positions of the glass-forming molecules are changed.

However, $p(R)$ presents the same problems as $p(E)$ (Equation (14)): There are two singularities at $R = R_{max}$ and at $R = 0$, respectively. The singularity at R_{max} is integrable; however, the pole at $R = 0$ is not. It carries most of the population of the TLS states. $R = 0$ is, however, as unphysical as $E = \Delta_{max}$, since again it would mean an infinite λ (Equations (11) and (17)) and, hence, infinite barrier heights or tunneling distances (Equation (12)). Therefore, we have to conclude that there has to be some cut-off limit corresponding to a minimum relaxation rate R_{min}, so that the distribution can be normalized. It is a question of great interest to get information about R_{min} and R_{max}, because, these two parameters determine the complete relaxation behavior of the glass.

It is worthwhile to point out that the singularity at $R = R_{max}$ is a consequence of the fact that the distribution of TLS states contains symmetric double wells ($\Delta = 0$). If these symmetric wells are neglected because of the symmetry of the problem considered, the singularity at R_{max} vanishes.

Up to now only little is known about the nature of the TLS states; that is, about the microscopic tunneling coordinate. It was assumed that the TLS states may be strongly related to the voids of free volume into which adjacent molecules could tunnel [18]. This, however, tells us only that the free volume concept is rather consistent with the TLS model. But there is no definite microscopic information.

4. Dynamic and Adiabatic Optical Relaxation Processes

In the following we discuss the interrelation between the non-equilibrium nature of the glassy state with its wide distribution of structural (tunneling) relaxation rates and the optical relaxation processes, characterized by the time constants T_1 and T_2^*, which determine the homogeneous line shape function.

The homogeneous linewidth is, at finite temperatures, determined by fast phonon scattering processes. When the lattice is characterized by a fractal structure one talks of fracton scattering processes. In some cases also the lifetime of the excited state may be very fast so that radiative or radiationless T_1 processes may play an important role in the dephasing time. Since in these latter processes the lattice is only indirectly involved via the width of the final states, we cannot learn very much about the amorphous solid state and, hence, do not consider these processes in more detail.

We divide the various dephasing processes into two categories depending on their time scale. We call all those processes occurring on a time scale

$$t \lesssim T_1 \tag{20}$$

dynamic, and those processes occurring on a time scale

$$t > T_1 \tag{21}$$

adiabatic scattering processes [29].

The reason for distinguishing between adiabatic and dynamic processes is the fact that there are several types of experiments to measure the dephasing times, e.g. 'dynamic' experiments, like photon-echo or fluorescence line narrowing and adiabatic experiments like persistent (i.e. photochemical, photophysical or long-time scale population) hole-burning [30]. Clearly a dynamic experiment can only provide information on dynamic scattering processes. They are limited in their time range by the lifetime of the molecule. Persistent holes, on the other hand, remain in the sample for long time periods and thus are sensitive also to adiabatic scattering processes. All the very slow structural relaxation events occurring on a wide distribution of time scales (Equation (16)) may show up in a change of the linewidth and the line shape of the hole with time [30—34]. Since the holes are, at low temperatures, very sharp compared to the inhomogeneous width which is a typical measure of the fluctuation of the molecule matrix interaction, very small structural changes can be investigated with remarkable accuracy. It follows from the above arguments, that the width γ of a persistent spectral hole obeys the inequality.

$$\gamma \gtrsim \gamma_h \tag{22}$$

for any time $t > T_1$, with γ_h being the homogeneous width, which according to this definition is solely due to dynamical scattering processes. Due to the relaxation processes given by the distribution (16), the hole width is expected to be time dependent. Hence, if one wants to get information on γ_h from a hole-burning experiment one has to make sure that the time evolution of the measured hole on time scales larger than T_1 can be neglected under the experimental conditions. Recently it was demonstrated by Molenkamp and Wiersma that hole-burning and photon-echo experiments may yield quite different results [35].

5. Reversibility and Irreversibility

Apart from the time dependence of the line shape and linewidth of persistent spectral holes, there are other optical properties of the amorphous state which originate from its non-equilibrium nature, as has been already mentioned in a foregoing section. One of these is the irreversibility of hole width and shape as a function of external parameters, such as temperature, electric fields or pressure.

Let us, for the moment, consider the temperature dependence of spectral holes more closely, which has been investigated extensively. (For reviews see [5, 7, 8].)

The temperature dependence of the homogeneous optical linewidth yields valuable information on the host matrix, since it can provide the scientists with detailed knowledge about the density of states involved in the phonon or fracton scattering, and on the specific mechanism (resonance scattering, one phonon, two phonon); it also provides information about the coupling mechanism between probe molecule and lattice (e.g. dipole–dipole, quadrupole–dipole, etc.). However, in amorphous solids, the situation is being complicated due to the structural relaxation processes, which, as has been discussed above, occur spontaneously; they can, however, also be triggered by external parameters, e.g. temperature, strain fields, electric fields, etc. Suppose a hole is burnt at a burning temperature T_b and its temperature dependence is investigated. Then, the shape $L(T_b, T)$ and the width $\gamma(T_b, T)$ of the hole depend in a very complicated fashion on the temperature of the bath. There are of course the dynamic scattering processes which are assumed to be reversible with temperature, and which are characterized by a Lorentzian line shape. However, if the temperature is raised from T_b to T, then all these TLS states which are characterized by barrier heights $V_0 \leq kT$ can undergo a temperature activated crossing of the TLS barriers or a thermally assisted tunneling irrespective of the size of the energy asymmetry Δ [15, 16]. On the time scale of the experiment, these activated processes can be considered as irreversible. The whole system relaxes into a new equilibrium position q_2 (Figure 7). The relaxing TLSs create random diagonal energy changes, and, consequently, the probe molecules will absorb at different energies and the width and shape

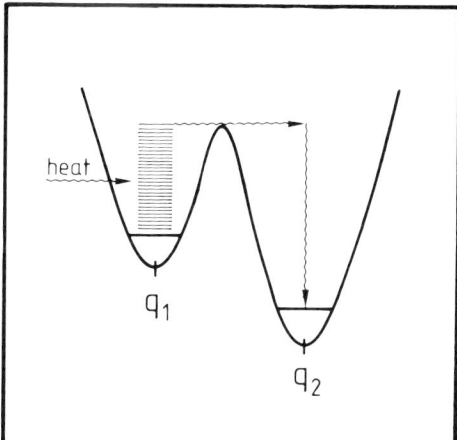

Fig. 7. Temperature induced structural relaxation of a glass (see text).

will be changed. These changes are superimposed on the contribution of the dynamical processes. Their irreversible part can be measured separately by cycling the temperature back from T to T_b.

The quantity

$$\gamma(T_b \to T \to T_b) - \gamma(T_b) = \gamma_{ir} \qquad (23)$$

can be considered as the irreversible contribution to the width due to temperature induced structural relaxation (Figure 8). The corresponding line shape is given by a convolution of the site memory function $\delta N(\omega', T_b, T)$ which contains all the irreversible processes, with the homogeneous line shape function $z(\omega - \omega', T_b)$ which contains only reversible processes (see Section 1, Equation (6)).

Since the spontaneous structural processes, as discussed in the foregoing section, are slow compared to the temperature-induced processes, they can be neglected on the time scale of the experiment. A mathematical treatment of the above model (see also Equation (6)) was presented in [17].

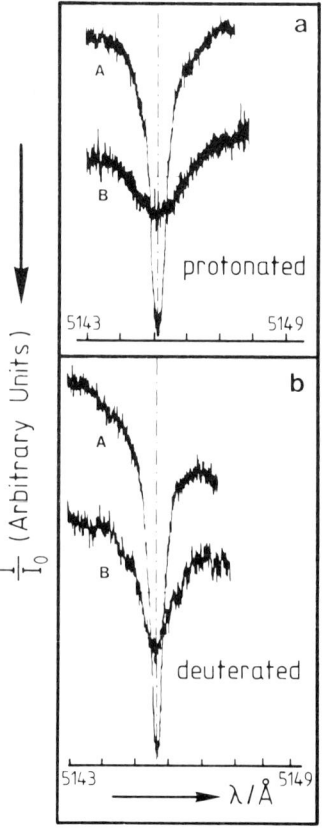

Fig. 8. Irreversible line-broadening effects for (a) the protonated and (b) the deuterated system of quinizarin in alcohol glass [15]. A: hole burnt at 5 K. B: After a closed cycle from 5 to 20 K to 5 K.

To summarize, we have shown that due to the random nature of the amorphous potential, and the concomitant structural relaxation processes, persistent spectral holes can lead to a diffusion in ω-space as time goes on and (or) as the temperature is increased. Under certain conditions the diffusion Kernel can be shown to be Lorentzian. In this case, Equation (23) holds exactly. A basic theoretical outline concerning spectral diffusion is given in Section 7. Experiments on pressure induced spectral diffusion are discussed in Section 10.

6. The Residual Linewidth

A third, very interesting — but experimentally unproven — consequence of the structural relaxation dynamics of the glassy state is the fact that there might be a residual optical width as T goes to zero. In crystalline solids, for example, the optical linewidth γ_0 at very low temperature is determined by Heisenberg's law

$$1/\gamma_0 = \tau_0, \qquad (24)$$

where τ_0 is the lifetime of the excited electronic state of the probe molecule considered ('lifetime limit'). Equation (24) has been verified for a series of organic materials, even glasses [36—39].

In contrast to the situation of defects in crystals, the level scheme of an impurity center in an amorphous solid is characterized by a four-level scheme (Figure 9), due to the coupling of the electronic states to the nearest TLS lattice states. The residual width (if there is any) of the absorption line is caused by the relaxation processes in the electronic excited state, as shown by the wavy arrow. The related processes in the electronic ground state (dotted arrow) would only contribute to the fluorescence — or hot absorption linewidth. The relaxation processes in the electronic excited state can be induced by the two optical transitions: $|1\rangle \to |1^*\rangle$ and $|2\rangle \to |1^*\rangle$. The first transition, however, will disappear as the temperature approaches zero, and, the second transition is only partially allowed because it involves a flip of the TLS state. However, we point out that the above scheme and its selection rules would break down if the system considered is far off thermal equilibrium. Hence, a residual linewidth, different from the lifetime limit is in principle possible. So far, however, the experimental verification is still an open question.*

7. Spectral Diffusion and Structural Relaxation: Model Description

As was shown in the discussions above there is a wide distribution of tunneling relaxation rates. The question is whether one can get information on the time scales on which these processes occur, that is on R_{min} and R_{max}, by using highly

---

* We point out, however, that recent hole burning experiments by Gorokhovskii *et al.* [72] down to temperatures as low as 0.05 K do not give the lifetime limited hole width. The discrepancy is more than a factor of 2.

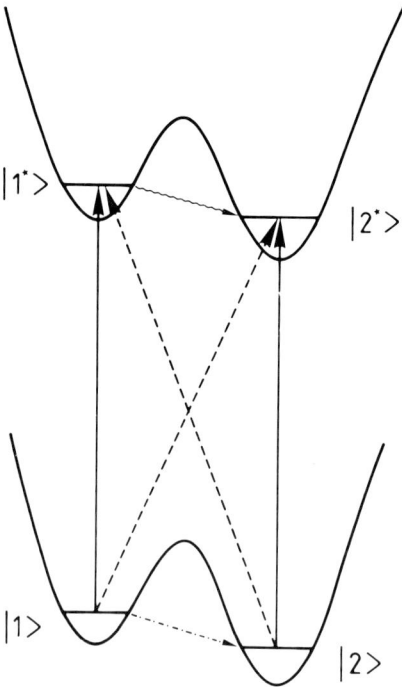

Fig. 9. Four-level scheme of the lowest optical transition of an 'optical center' in a glass. $|1\rangle$ and $|2\rangle$ are the electronic ground states, $|1^*\rangle$ and $|2^*\rangle$ are the electronic excited states.

resolved spectral holes in the optical domain as a-sensitive probe. To get a dynamical informtaion one has to study the time evolution of the linewidth, the lineshape and the area of the holes.

The quantity which is easiest and most straightforward to interpret is the area A of the hole.

7.1. THE DECAY LAW OF PERSISTENT SPECTRAL HOLES IN AMORPHOUS SOLIDS

Suppose a hole is measured at an arbitrary time t_1, then its area A_1 is determined by the number N_1 of molecules which are at t_1 in the photoproduct state $|P\rangle$ (see Figure 10). The population in $|P\rangle$ at a certain time t is determined by the relaxation rates between the product and the educt state. We assume that the photoproduct is stabilized against the educt by the potential barriers involved which reflect the amorphous nature of the matrix. Then $|P\rangle$ and $|E\rangle$ form a special kind of TLS system which is photochemically induced, a so-called TLS_p system [30, 33]. At this point we introduce a concept which is of crucial importance for the remainder of this article. We assume that the distribution function of the photochemically induced TLS system (TLS_p) can be treated in analogy to the normal TLSs which dominate the low-temperature specific heat.

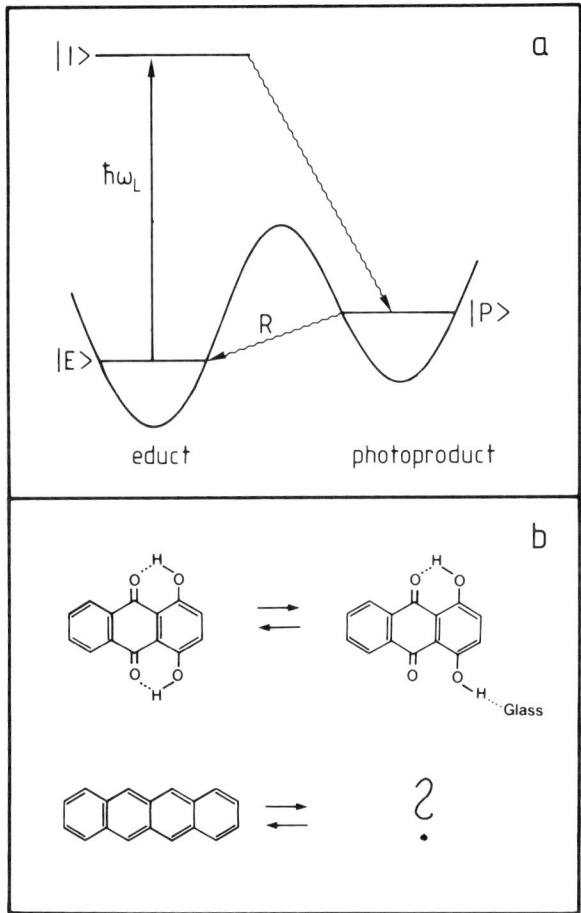

Fig. 10. (a) Photochemical creation of a product state $|P\rangle$ and its 'dark decay' through relaxation R from $|P\rangle$ to $|E\rangle$. (b) Two classes of molecules showing 'dark decay': quinizarin and tetracene.

With this assumption the number N_1 can be calculated using Equation (16). For the calculation we correlate the rates as given by Equation (16) with the experimental time regime by assuming that, in the statistical limit, those centers will relax at time t_1 which are characterized by rates

$$R_1 = 1/t_1. \tag{25}$$

Hence, all centers with $R < R_1$ are, at t_1, in the product state $|P\rangle$ and the number N_1 can be determined by integrating the distribution $p(R)$ form R_1 to a final value R_{min}. The latter characterizes the slowest relaxation processes involved. As mentioned above, the assumption $R_{min} = 0$ is unphysical and, indeed, the distribution cannot be normalized assuming $R_{min} = 0$. The integration can be evaluated exactly. But we assume, for the moment, that R_1 is very small compared to R_{max}.

Hence, the time scale of the experiment is slow compared to the fastest processes involved. Then

$$N_1 = \frac{1}{2} p_\Delta p_\lambda \int_{R_{min}}^{R_1} \frac{dR}{R} = \frac{1}{2} p_\Delta p_\lambda \ln \frac{R_1}{R_{min}}$$

Similarly, one calculates the number $N(t)$ of molecules which are at time t, in the product state $|P\rangle$ by integrating $p(R)$ from R_{min} to $R = 1/t$:

$$N(t) = \tfrac{1}{2}(p_\Delta p_\lambda/2) \ln R/R_{min}. \tag{26}$$

The relative hole area $A(t)/A_1$ as a function of time, which is the quantity that can be measured experimentally, is then given by

$$A(t)/A_1 = N(t)/N_1 = 1 - [\ln R_1/R_{min}]^{-1} \ln R_1 t. \tag{27}$$

We therefore conclude that given the distribution (16), the hole area decays linearly on a logarithmic time scale. This logarithmic law holds within the range $R_{min} \ll t^{-1} \ll R_{max}$ [33].

At this point it is worthwhile to point out that if one can rule out the occurrence of symmetric double wells, the singularity at R_{max} in the distribution (16) is absent and one gets, within the frame of the model, a logarithmic law extending to $t = 1/R_{max}$. Beyond the limits $t_{max} = 1/R_{max}$ and $t_{min} = 1/R_{min}$, the decay approaches an exponential form. Equation (27) was also derived by Fushman [40].

If one succeeds in measuring the logarithmic decay law, one can determine the slope factor of the decay law. It is given by

$$s_a = 1/(\ln R_1/R_{min}) \tag{28}$$

allowing for an estimate of the minimum structural relaxation rate constant R_{min}.

The slope factor contains all the properties of the material, characterized by the rate R_{min}. However, since the slope itself depends only in a logarithmic fashion on the material parameter R_{min} we expect that it is rather insensitive to experimental parameters like temperature, external fields, pressure, etc. Hence, the above logarithmic decay law is assumed to be of a rather uniform nature for all amorphous materials. This was verified experimentally [33, 60] for several systems (Section 9).

In the above paragraph, we drew the conclusion that parameters which enter into the relaxation rate in a linear or algebraic fashion will, generally, give rise to only minor changes in the slope due to the logarithmic dependency. In cases, however, in which molecular parameters change the tunneling rates in an exponential fashion (e.g. isotopic substitution) one would expect considerable changes even in the logarithmic slope factor (see below).

7.2. THE TIME EVOLUTION OF THE OPTICAL WIDTH

To get an idea as to how structural relaxation processes influence the time evolution of the hole width we employ the concept of spectral diffusion based on ideas of Klauder and Anderson in their treatment of the inhomogeneous width of diluted spin systems [41]. The principle of the concept is sketched in Figure 11. We consider a particular dye molecule, which is coupled to an ensemble of TLS systems via electric or strain fields. The TLS systems can be viewed as a pseudo-spin-$\frac{1}{2}$ system [42], with the spin either up or down. These two-spin configurations correspond to either an upper or a lower occupied TLS level, respectively. Now the energy of the probe molecule certainly depends on the configuration of the whole ensemble of interacting pseudo spins. Since there is a very large number of configurations, one can ask for the probability of a molecule absorbing at a certain energy $\varepsilon(\mathbf{r}_1, \mathbf{r}_2, \ldots, \mathbf{r}_N)$ around its unperturbed energy E_0. N is the total number of interacting spins, the vectors $\mathbf{r}$ describe their positions with respect to the position of the probe molecule. This probability distribution $L(\varepsilon)$ is equivalent to the line shape which is, in this case, inhomogeneous. $L(\varepsilon)$ can be calculated by calculating the number of configurations which lead to the same energy shift ε from the unperturbed line center. This problem was solved by Klauder and Anderson for a real spin-$\frac{1}{2}$ system by integrating over the configuration space under the constraint that

$$\varepsilon = \varepsilon(\mathbf{r}_1, \mathbf{r}_2, \ldots, \mathbf{r}_N)$$

$$L(\varepsilon) = V_c^{-1} \int dV_c \delta[\varepsilon - \varepsilon(\mathbf{r}_1, \ldots, \mathbf{r}_N)] \tag{29}$$

V_c is the volume of the total configurational space. The above phase integral can

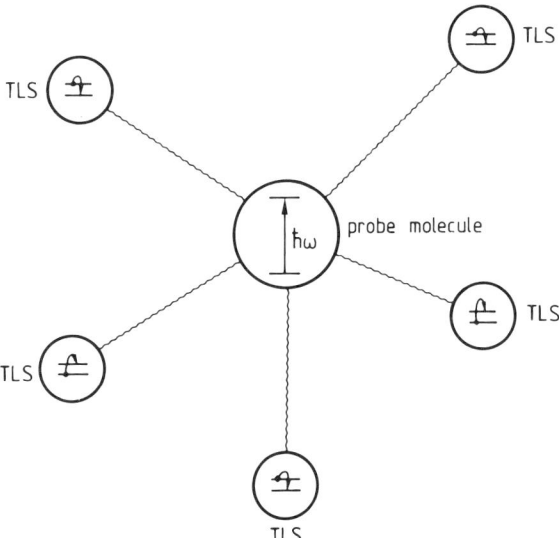

Fig. 11. Scheme of the coupling of an optical transition ($\hbar\omega$) to various two-level systems (TLSs) of the glass.

be solved analytically under the restrictive conditions that there is only an interaction between the probe molecule and the TLS ensemble, but not within the TLSs. Additionally it has to be assumed that this interaction is of the dipole–dipole type. In this case one finds the surprising result that the line shape is Lorentzian with a width γ_D which is proportional to the spatial density n of interacting spins:

$$\gamma_D \equiv [\langle \varepsilon^2 \rangle]^{\frac{1}{2}} = C \frac{\Delta}{E} n$$

Δ and E represent the asymmetry and the energy values of the involved TLS. The corresponding parameters of the probe TLS are contained in the coupling constant C which is of tensor character. The appearance of Δ/E in the coupling parameter follows from the TLS Hamiltonian in the pseudo-spin representation [42]. The coupling can be purely electronic or electronic-elastic, in case it is mediated via strain fields [23]. It is worthwhile noting that the line shape, in spite of being inhomogeneous in nature, is Lorentzian. Now we introduce a time dependence into the above concept. Suppose, we consider a probe molecule at a certain time t_1, at which it may be found somewhere in frequency space at a well-defined frequency ω_1. Then, as time goes on, the pseudo-spins (or TLSs) start to flip and the number of new configurations increases with the number of spins having undergone a transition within time interval $(t - t_1)$. Hence $\gamma_D(t - t_1)$ increases with time in the same way in which $n(t - t_1)$ increases with time

$$\gamma_D(t - t_1) = C \frac{\Delta}{E} n(t - t_1) \tag{30}$$

with $n(t - t_1)$ being the number of spins per volume which have flipped an odd number of times within the interval $(t - t_1)$. If we again express the experimental time in terms of the inverse relaxation rates of the amorphous system considered — that is, $t = 1/R$ — then, for $t_1 = 1/R_{max}$ Equation (30) represents (approximately) the full diffusion width. To get the explicit time dependence of the optical width one has to find the time dependence of $n(t - t_1)$. This can be achieved by solving the corresponding rate equation [42, 43]. For the optical experiments (Section 8) a simpler model seems applicable [30–32]. First we calculate the number of TLSs having flipped in the interval $t - 1/R_{max}$ by integrating the rate distribution function from $R = 1/t$ to R_{max}:

$$\int_{\Delta E} \int_{R=1/t}^{R_{max}} p(R', E) \, dR' \, dE = -\int_{\Delta E} \frac{1}{2} p_\Delta p_\lambda \ln \left\{ \frac{1 + [1 - R/R_{max}(E)]^{\frac{1}{2}}}{1 - [1 - R/R_{max}(E)]^{\frac{1}{2}}} \right\} \Bigg|_{R=1/t}^{R_{max}(E)} dE. \tag{31}$$

ΔE is the whole energy range of the TLS states involved. We assume that the time scale is slow compared to R_{max}. In this case a linear expansion of the square root is a good approximation and one gets

$$\int_{\Delta E} \int_{R=1/t}^{R_{max}} p(R', E) \, dR' \, dE = \int_{\Delta E} \frac{1}{2} p_\Delta p_\lambda \ln 4 R_{max}(E) t \, dE. \quad (32)$$

Thus, the number of TLSs which change configuration at least once in the time interval considered increases logarithmically with time. We point out that $\ln 4 R_{max}(E) t$ depends only weakly on energy and, hence, can be put in front of the integral without making to large an error. We further stress that with hole-burning experiments the fastest relaxation times which can be measured are on the order of seconds or minutes and, hence, it is clear that, the tunneling contribution to the TLS-energy is extremely small. Hence, we may put

$$E = \Delta. \quad (33)$$

To calculate the linewidth according to Equation 30 we have to modify Equation 32 a little bit further in order to take into account the possibility that a certain fraction $(1 - \kappa)$ of the TLS ensemble may have flipped an even number of times within the interval considered and hence contribute nothing to the number of configurations determining the linewidth. Therefore, to calculate the diffusion width according to Equation (30) the distribution $p(R, E)$ in Equation (32) has to be multiplied with a factor $\kappa(E) \leq 1$. We assume that κ is time independent but depends, of course, on temperature and energy.

As to the time independence of κ our arguments are as follows: For long times, the number of relaxed TLSs is almost constant, i.e. depends only logarithmically on time; hence, we can, for the majority of the centers having already relaxed on a time scale $t' < t$, let t' approach infinity. This procedure is equivalent to the assumption that one fraction of the TLS systems, namely those with rates $R > 1/t$, can be considered as being, at time t, in thermal equilibrium. The other fraction of the TLS systems with rates $R < 1/t$ must be considered as being, at time t, still far off thermal equilibrium. Similar arguments have been discussed in the field of photoconduction in amorphous solids by Tiedje and Rose [44]. As far as the energy dependence of κ is concerned we have to take into account that the initial population of the TLS systems contributing to the linewidth will, in general, not be in thermal equilibrium. We characterize the pseudo-spin states of the systems involved by population numbers p_- and p_+. Then $p_-\kappa_-$ and $p_+\kappa_+$ represent the fraction of the TLS systems being in the lower $(-)$ or in the higher $(+)$ state having changed from $(-)$ to $(+)$ or from $(+)$ to $(-)$, respectively. The parameters $\kappa_\pm$ are solely governed by the Boltzmann factor:

$$\kappa_\pm = 1 / \left(1 + e^{\pm \frac{E}{kT}}\right) \quad (34)$$

The above quantities $\kappa_\pm$ vary between 1 and $\frac{1}{2}$. The former case holds for photo-induced TLS systems (the TLS$_p$) where E is generally assumed to be large compared to kT (see below).

An integration of the distribution function $p(E, R)$ from R_{min} to R_{max} and over the whole range ΔE must yield the total number $\bar{N}$ of TLS states per volume. From this normalization condition it follows that

$$\frac{1}{2} p_\Delta p_\lambda = \bar{N}/\Delta E \ln\left(\frac{4R_{max}}{R_{min}}\right) \tag{35}$$

Hence, combining Equations (32)–(35) with Equation (30) we derive for the diffusional part of the hole width the expression

$$\gamma_D(t - R_{max}^{-1}) = \frac{C\bar{N}}{\Delta E \ln(4R_{max}/R_{min})} \ln 4R_{max} \cdot t \int_{\Delta E} (p_-\kappa_- + p_+\kappa_+)\, dE \tag{36}$$

In order to eliminate the unknown quantities C, $\bar{N}$ and ΔE one can normalize $\gamma_D(t)$ to some value $\gamma_D(t_1) \equiv \gamma_1$

$$\frac{\gamma_D(t - R_{max}^{-1})}{\gamma_1} = \ln\left[\frac{4R_{max}}{R_1}\right]^{-1} \ln 4R_{max} t, \tag{37}$$

where we have again used $R_1 = 1/t_1$.

We see that the reduced as well as the absolute spectral diffusion width evolve linearly on a logarithmic time scale. While the slope factor of the absolute width is governed by several system parameters such as C, N, ΔE, R_{max}, R_{min} the slope factor s of the reduced width depends just on one parameter, namely R_{max}, the maximum relaxation rate, which, therefore, can be easily determined if one succeeds in measuring γ_D/γ_1.

At this point it seems worthwhile to point out that we did not specify the TLS ensemble considered, either concerning their energy or their rate parameters. The model holds for times far away from R_{max}^{-1}. However, since the singularity at R_{max} in the distribution function is integrable and contains only a very small fraction of the total TLS population, the error made in determining R_{max} from Equation (37) is of minor importance compared to the very large range of relaxation rates between R_{max} and R_{min}.

We finally point out that similar arguments as have been used above to derive the time dependence of the hole width can be used to derive the time dependence of the sepcific heat. For the specific heat experiments one has to consider the number of TLS states which relax into thermal equilibrium with a time $t - R_{max}^{-1}$ [23, 26]. This number is also given by Equation (32), and, hence, the time dependence for both quantities C_p and γ_D is the same.

If one measures the relative change of $C_p(t)$ with respect to some value $C_p(t_1)$ one can derive exactly the same logarithmic law as for the time evolution of the optical width by integrating the distribution from $R = 1/t$ to R_{max}.

$$C_p(T, t) = \frac{\pi K^2 p_\Delta p_\lambda \cdot T}{12} \ln 4R_{max}t \tag{38}$$

$$\frac{C_p(t)}{C_p(t_1)} = 1 + \left[\ln \frac{4R_{max}}{R_1}\right]^{-1} \ln 4R_{max}t \tag{39}$$

In deriving Equation (38) it was assumed that, as in the optical experiments, the time scale of the experiment is long compared to the fastest process characterized by R_{max}, i.e.

$$t \gg R_{max}^{-1} \tag{40}$$

Since one does not know, *a priori*, if the same TLS ensemble determines the thermodynamic and the optical relaxation processes, it would be very interesting to compare the time dependence of the reduced specific heat (Equation (39)) with the reduced diffusional width (Equation (37)). From heat release experiments it is known that the specific heat may evolve on rather long time scales [45], comparable to the optical experiments. It is also of great interest to compare the time scales of the hole area reduction with the time scale of the hole broadening. In principle, both processes can be governed by totally different TLS systems. While the hole-filling is, by definition, due to the photochemically induced TLS_p, the line-broadening might be due to the normal TLS ensemble of the glass.

8. The Logarithmic Decay Law and its Relation to Other Dispersive Time Dependencies

The derivation of the logarithmic decay law is based on the distribution of rates, as given by Equation (16), which was derived from evaluating specific heat experiments of disordered materials. However, the problem which arises in case of highly non-exponential relaxation phenomena with long time tails, is the fact that a reasonable fit to the experimental data can be achieved with a variety of model decay laws. This is a consequence of the very small time window in which these decay laws can be probed experimentally.

In this section we consider the interrelation between the various forms of decay laws [46] which are widely used to interpret the relaxational phenomena in disordered systems. We list them in the order of increasingly slower decay characteristics.

(a) The Kohlrausch [47] or Williams–Watts [48] stretched exponential

$$\Phi(t) = \exp\left[-\left\{\frac{t}{\tau}\right\}^{\alpha_1}\right] \quad (0 < \alpha_1 < 1, t > \tau). \tag{41}$$

(b) The exponential-logarithmic form [49–52]

$$\Phi(t) = \exp\left[-a_2 \ln^\beta\left(\frac{t}{\tau}\right)\right] \quad (\beta \geq 1, t > \tau). \tag{42}$$

(c) The algebraic decay [50, 51, 53]

$$\Phi(t) \sim \left(\frac{t}{\tau}\right)^{-a_3} \quad (a_3 > 0, t > \tau) \tag{43}$$

a_1, a_2 and a_3 are fitting parameters which determine the dispersion of the system. A point which had been made is that Equations (41)–(43) fail for very short times. This fact follows from their derivation [52] and is evident by an inspection of Equations (42) and (43). Hence, we have to postulate a maximum relaxation rate R_{max}. The logarithmic decay law, on the other hand, also requires a minimum relaxation rate R_{min}.

$$\Phi(t) = 1 - s_a \ln \frac{t}{t_1}$$

$$s_a = \left[\ln \frac{R_1}{R_{min}}\right]^{-1}. \tag{44}$$

One should keep in mind that Equation (44) holds for times $t \geq t_1$; that is, $t_1 \equiv R_1^{-1}$ is determined by the maximum rate which can be probed by the experiment.

It can be shown easily that if the dispersion of rates is large, as is the case for the structural relaxation processes in glasses [30, 33], the above decay laws (41)–(43) may be well approximated by a logarithmic form. For example, for times $\ln t/\tau \ll a_2^{-1}$, Equation (42) is, with $\beta = 1$, well described by a logarithmic form

$$\Phi(t) = 1 - a_2 \ln \frac{t}{\tau}, \quad \left(\ln \frac{t}{\tau} \ll a_2^{-1}\right). \tag{45}$$

The time range in which Equation (45) is valid is, due to its logarithmic form, extremely large if one assumes small a_2-values. If we compare relation (45) with Equation (44) we find [46] that

$$a_2 = \left[\ln \frac{R_1}{R_{min}}\right]^{-1}. \tag{46}$$

Furthermore, Equation (42) with $\beta = 1$ can be rewritten as

$$\Phi(t) = \left(\frac{t}{\tau}\right)^{-\alpha_2} \approx 1 - \alpha_2 \ln \frac{t}{\tau}. \qquad (47)$$

By setting $\alpha_2 = \alpha_3$ we can express α_3 also through the quotient $R_1/R_{\min}$ [46].

These results show that Equations (41)–(44) are mathematically equivalent expressions in the time range $t_{\min} \ll t \ll t_{\max}$. (In Equations (41)–(44), $t_{\max}$ is either given by τ or by t_1.) However, for a large dispersion of rates (i.e. for small α_2 and α_3 values) this range is very large. For example, if $s_a = 0.02$ one has $t_{\max}/t_{\min} = 5 \times 10^{21}$. For such a large dispersion of rates even Equation (41) may be approximated by a logarithmic form, as has been pointed out by Queisser [54]. For small α_1, we get

$$\left(\frac{t}{\tau}\right)^{\alpha_1} = \exp\left[\alpha_1 \ln \frac{t}{\tau}\right] \approx 1 + \alpha_1 \ln \frac{t}{\tau}, \qquad (48)$$

which, inserted in Equation (41), yields $\alpha_1 = \alpha_2 = \alpha_3 = s_a$.

It is an interesting aspect that, by comparing the various decay laws with the logarithmic form (Equation (44)), one can find a physical interpretation of the 'abstract' fit parameters α_1, α_2, α_3. Our interpretation shows that the three parameters are related to the ratios of microscopically interpretable rates. If their dispersion is large, a logarithmic approximation holds over many orders of magnitude in the time domain.

Quite recently, different approaches have been employed to describe relaxation patterns in disordered systems with long time tails [55–58]. In these approaches one starts with a fluctuating quantity of a Gaussian distribution, e.g. the tunneling distance, the activation energy, or the tunneling parameter (λ). The decay law is calculated by assuming first-order kinetics and by integrating over the fluctuating quantity. We note that in all cases, where the rate depends in an exponential fashion on the fluctuating quantity, these models can be very well approximated by a logarithmic law if the dispersion of rates is large enough. To show this, we take as an example the decay law used by Bässler and coworkers to describe the relaxation of photochemical holes [57]:

$$\Phi(t) = \frac{1}{\sqrt{2\pi\sigma^2}} \int_{-\infty}^{+\infty} \exp\left[\frac{-(\lambda - \lambda_0)^2}{2\sigma^2}\right] \exp\left[\frac{-t}{\tau_0} e^{-\lambda}\right] d\lambda. \qquad (49)$$

τ_0^{-1} is a frequency factor on the order of the zero point vibrational frequency of the double minimum potential. We substitute for the Gaussian distribution of the tunnel parameter λ a rectangular function with limits $\lambda_{\max}$ and $\lambda_{\min}$. Then, we transform the integration over λ into an integration over the rates R. We then obtain

$$\Phi(t) \sim -\int_{R_{max}}^{R_{min}} \frac{\exp[-Rt]}{R} \, dR$$

$$= -\left\{ \int_{R_{max}}^{R_1} \frac{\exp[-Rt]}{R} \, dR + \int_{R_1}^{R_{min}} \frac{\exp[-Rt]}{R} \, dR \right\}. \tag{50}$$

We assume 'ideal glass conditions' [59], i.e. all the relaxation processes are either very fast or very slow compared to the experimental time $t = 1/R_1$. (This assumption is equivalent to a large dispersion of rates.) Then the first integral in the above expression yields only a negligible contribution. In the remaining integrals we may, for the majority of centers, approximate the exponential by 1, and thus get a logarithmic decay law.

Quite often the experimental data available span only one or two decades in time, and, hence, the experimentalists can use one or the other of the above mathematical descriptions. From an heuristic point of view, the analysis of the data in terms of a logarithmic law [30, 31, 33, 54] is, however, very informative since it can provide microscopic information on natural bounds of the system (see Section 9).

9. Experimental Investigation of Spontaneous Structural Relaxation Processes

9.1. THE PHOTOCHEMICAL SYSTEMS

So far we have investigated two systems which are very different, namely 1,4-dihydroxyanthraquinone and tetracene both in an EtOH/MeOH glass mixture. In both systems we chose the same concentration of roughly 10^{-4} m. To get an idea about the tunneling mechanism we performed identical experiments in an isotopically substituted (deuterated) glass. The interesting observation is that while 1,4-dihydroxyanthraquinone shows photochemical hole-burning, the hole-burning process in tetracene is of photophysical nature (for a review on photophysical hole-burning see [7]). According to the present understanding, photophysical hole-burning is solely due to a photo-induced structural rearrangement of the amorphous matrix surrounding the probe molecule. In this latter case the 'product' state $|P\rangle$ and the educt state $|E\rangle$ (Figure 10) are considered to be part of the normal TLS ensemble of the alcohol glass, while in the photochemical case the TLS_p states (see Section 7) can only be populated by light induced changes in the structure of the probe molecule. In the photochemical case the TLS_p states can in principle be very different from the ordinary TLS states, for example, with respect to their energies, their symmetries, barriers, etc.

The photoreaction leading to hole-burning in the case of 1,4-dihydroxyanthraquinone in alcohol glass is assumed to be a proton transfer reaction [10, 8]. The absorption of light leads to a breakage of an internal hydrogen bond and to a subsequent formation of an external hydrogen bond between probe molecule and matrix (Figure 10(b)). This reaction model is in agreement with all experiments performed with this system so far: hole-burning is easily achieved in matrices having the capability of forming hydrogen bonds (for instance, alcohol glasses), but occurs with very low yield (if at all) in pure hydrocarbon glasses (for instance, 3-Me-pentane). Furthermore, any hole-burning process is absent in the presence of strong bases such as KOH. Hence, there is a clear picture concerning the hole-burning process of 1,4-dihydroxyanthraquinone in alcohol glasses and consequently the TLS_p states are well characterized, but there is no definite idea about the corresponding processes in the system tetracene in EtOH/MeOH.

9.2. Experimental Results

Figures 12—21 display some of our experimental results concerning persistent spectral holes and on spontaneous structural relaxation processes [30, 31, 33, 60]. A survey spectrum of quinizarin is shown in Figure 12. The hole is the small dip

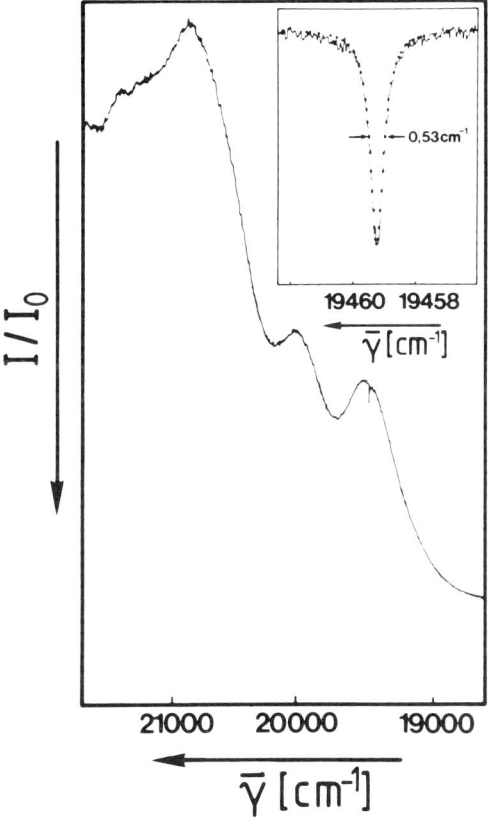

Fig. 12. Survey spectrum of quinizarin in an alcohol glass. The spectral hole is shown, on an enlarged scale, in the insert.

at the wavelength of 5145 Å. We see that the hole area of quinizarin as well as that of tetracene (Figures 13 and 14) in an alcohol glass mixture decays linearly on a logarithmic scale. Hence, these results are in agreement with the simple theoretical model as outlined above. The decay itself is very slow so that, for the protonated alcohol glasses, within one week, only 40% of the total hole area has decayed. The slopes of both systems are rather similar, yielding a ratio of $R_{max}/R_{min} \approx 10^9$ for the quinizarin system and $R_{max}/R_{min} \approx 10^{12}$ for the tetracene system.

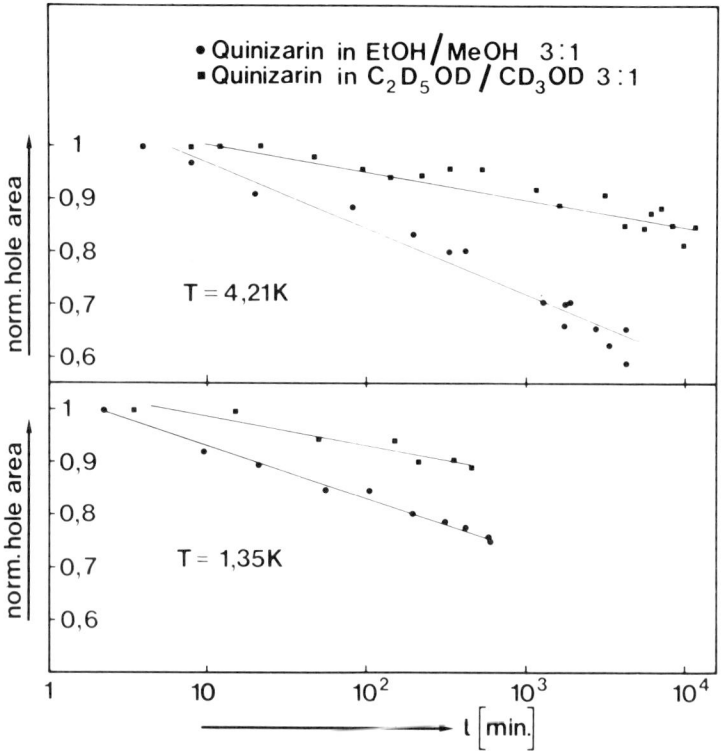

Fig. 13. Decay of the normalized hole area of quinizarin in alcohol glass at 4.2 and 1.3 K respectively. Note the difference between protonated and deuterated quinizarin.

To get more detailed information on the nature of the TLSs involved we, in addition, investigated quinizarin in a perdeuterated glass. As can be seen from Figure 13, deuteration brings about a rather drastic change in the slope factor. The dispersion of rates covers a range of almost 20 orders of magnitude (that is, the hole area decays to 50% in about 2×10^5 years, extrapolated). Hence, in the case of quinizarin, the photochemically induced tunnel states are obviously closely connected to proton tunneling processes. This result is a very strong support for the reaction mechanism proposed above, namely the breakage of an internal H-bond and the formation of an external H-bond to the solvent. Due to this

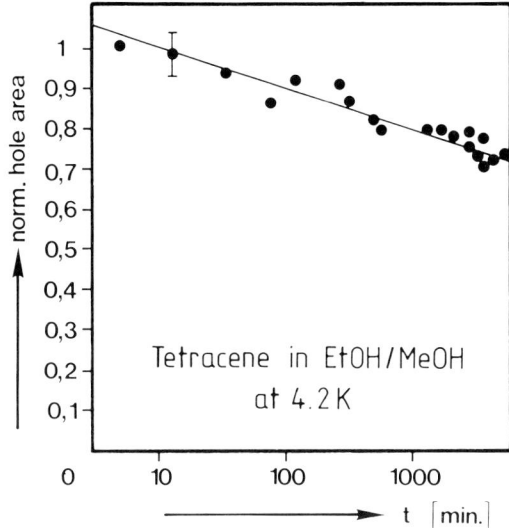

Fig. 14. Decay of the normalized hole of tetracene in an alcohol glass at 4.2 K.

special interaction between probe molecule and solvent we can consider the back relaxation from the product to the educt state as being due to a structural relaxation of the irradiated alcohol glass. From Figure 13 it is evident that the influence of temperature on the slope is rather small.

Figures 15 and 16 show the time evolution of the line shapes of the quinizarin system at 4.2 K. Immediately after burning the line shape is Lorentzian. Even though the width and the area change considerably during the long observation time, the line shape is conserved. It is still Lorentzian, even after a week of observation. The same result holds for the perdeuterated quinizarin system Figures 17 and 18 show the corresponding data series for both systems at 1.35 K. For the protonated system Gaussian shapes are plotted for comparison. The Lorentzian line shape in these experiments is in agreement with the model proposed by Klauder and Anderson for a dipole—dipole-type coupling.

Similar results hold for tetracene in EtOH/MeOH.

A very interesting experiment is the investigation of the linewidth as a function of time [30, 31] (Figure 19). First of all we note that the width, like the area, increases linearly on a logarithmic time scale as predicted by the TLS model. However, what is not predicted *a priori* by the model is the fact that the time scales for the decay at the hole area and the increase of the width are roughly the same. This suggests that for both type of processes the same category of relaxing tunneling systems is responsible. This is a rather surprising result since, in case of quinizarin, the TLS_p states are rather well known and thought to be due to intermolecular hydrogen tunneling processes, while, in principle, for the time evolution of the width all tunneling states could be responsible. In the above model calculation yielding the logarithmic time dependence of the width no specific assumption about the TLSs involved has been made.

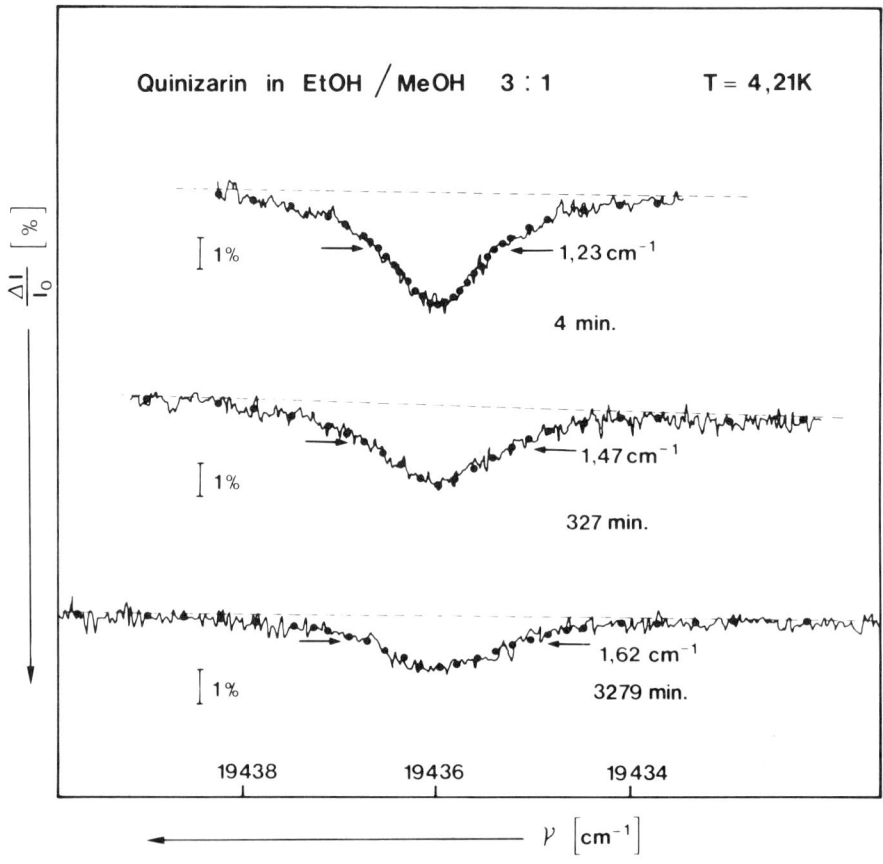

Fig. 15. Time evolution of the hole for protonated quinizarin at 4.2 K.

We summarize the above experimental results as follows. First, if the TLS_p states are responsible for the time evolution of the width then there must be a clearly discernable deuterium isotope effect in the slope factor of the width. As can be seen from Figure 19, this is indeed the case. The absolute magnitude of the change in the slope factor of the width is about the same as the change in the slope factor characterizing the hole area.

Second, as one can infer from the symmetry of the quinizarin molecule (Figure 10(b)) the TLS_p states are asymmetric; hence the singularity in the distribution function (Equation (16)) at R_{max} is absent. The distribution seems to be of the R^{-1} type even for the faster processes. Consequently, there is no deviation from the logarithmic law over the whole range of involved rates. As will be shown below, this can also be verified experimentally. It is also clear that in the case of asymmetric TLSs with slow tunneling relaxation the approximation $\Delta \approx E$, which was made in deriving Equations (36) and (37), holds for all times.

As to the question how the TLS_p may influence the optical width, we assume that the relaxing product states create strain or electric fields which, in turn, influence the energy levels of the probe molecules thereby leading to a diffusion in

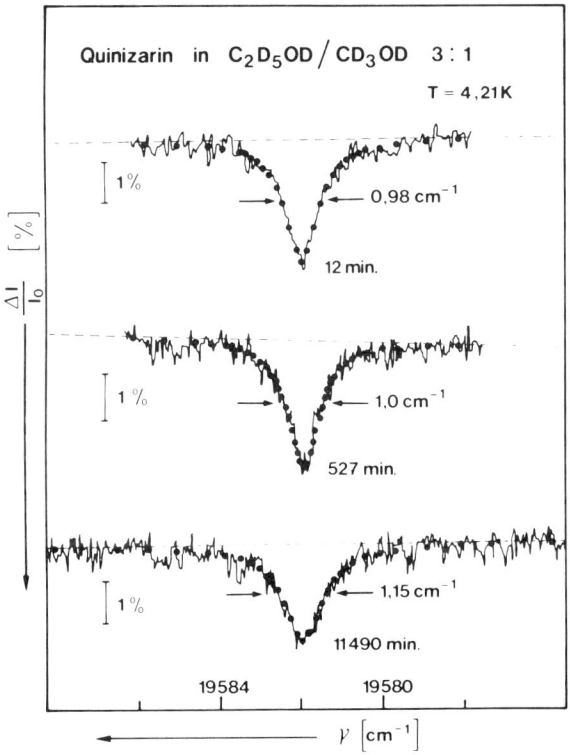

Fig. 16. Time evolution of the hole for deuterated quinizarin at 4.2 K.

ω-space. We finally note that, at least as far as the decay function of the hole is concerned, the difference between the photochemical and the photophysical systems seems to be small.

9.3. Investigation of the Microscopic Rate Parameters of the Logarithmic Law: The Deuteration Effect

As we have pointed out in the foregoing sections, the logarithmic laws should be of rather general nature as far as the magnitude of the slope factors s_a and s_w for the area and the width of the holes is concerned.

$$s_a = \left[\ln \frac{R_1}{R_{\min}}\right]^{-1}$$

$$s_w = \left[\ln \frac{R_{\max}}{R_1}\right]^{-1} \tag{51}$$

(The factor 4 in s_w (Equation (37)) is absent in case the distribution of rates obeys a pure $1/R$ law.) s_a as well as s_w are expected to have roughly the same magnitude

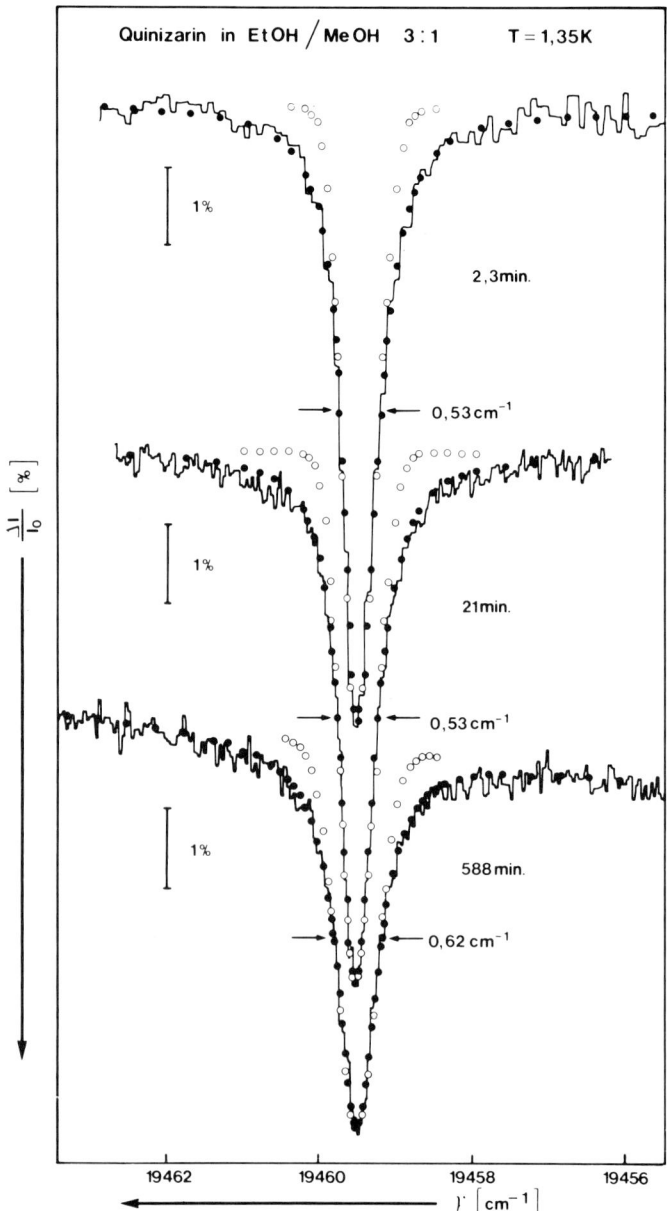

Fig. 17. Time evolution of the hole for protonated quinizarin at 1.3 K.

for a wide class of amorphous materials due to their logarithmic form. Any dependence on parameters like temperature, pressure, external fields, etc., should be suppressed by the logarithmic function except in cases in which these parameters enter into the rates in an exponential fashion (see Equation (51)). This is obviously the case for experiments with isotopic substituents, like, for instance,

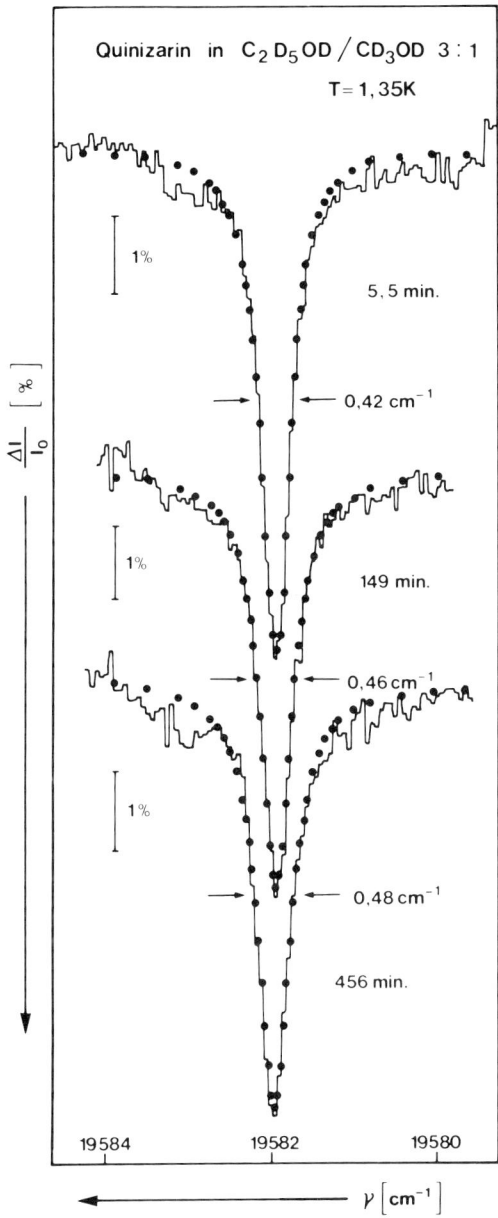

Fig. 18. Time evolution of the hole for deuterated quinizarin at 1.3 K.

deuteration of the amorphous matrix. The tunneling matrix element Δ_0 depends in an exponential fashion on the square root of the mass of the tunneling particle

$$\Delta_0 = \hbar\Omega \exp[-\sqrt{m} \cdot \lambda']. \tag{52}$$

λ' is determined by the barrier height and by the tunneling distance (see

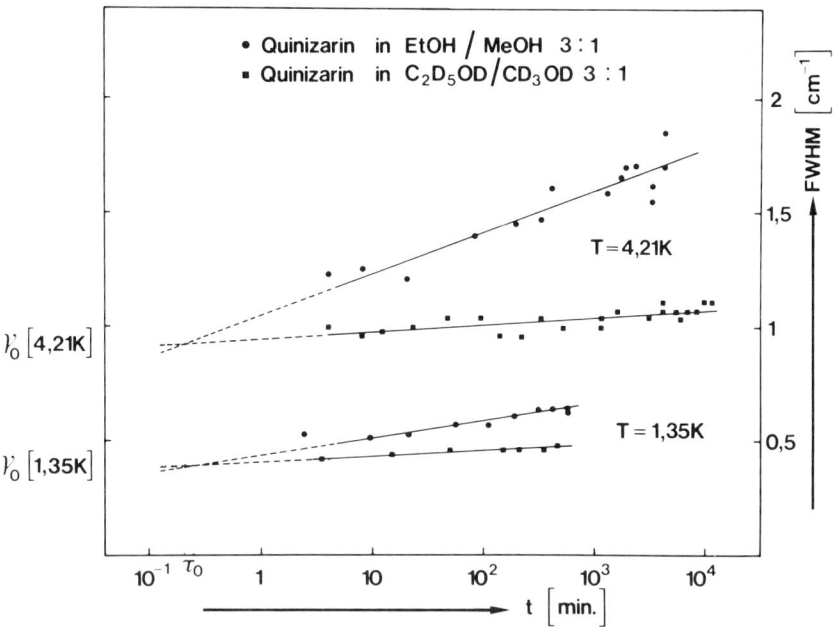

Fig. 19. Change of the linewidth of protonated and deuterated quinizarin at 4.2 and 1.35 K respectively. The dotted lines are extrapolations of the experimental data based on the logarithmic time law.

Equations (11) and (12))

$$\lambda' \simeq \left[\frac{V_0 d^2}{2\hbar^2}\right]^{\frac{1}{2}} \tag{53}$$

It is obvious that if λ' is large enough, Δ_0 may change many orders of magnitude in case m increases by a factor of two by deuteration. If, on the other hand, λ' is sufficiently small the change in Δ_0 upon deuteration will be small. λ' is large for those tunneling centers which have large barriers and (or) large tunneling distances. These centers have very slow relaxation rates. As can be seen from Equation (51), the slope factor s_a of the hole decay is governed by the slowest process in the system considered, i.e. by the rate constant $R_{\min}$. Following the above arguments this slow process shows the largest possible deuteration effect. It amounts to about 10 orders of magnitude and, hence, shows up clearly even in the logarithmic slope factor. Since the TLS_p states in the case of the quinizarin system are rather well characterized we can try to estimate the maximum barrier heights in the TLS_p ensemble from the observed isotope effect. If we assume that isotopic substitution leads mainly to changes in the tunneling matrix element, then

$$\frac{R_{\min}(H)}{R_{\min}(D)} = \frac{\Delta_0^2(H)}{\Delta_0^2(D)} = \exp[2\lambda'(\sqrt{m_D} - \sqrt{m_H})]. \tag{54}$$

From the measured slopes one can determine the isotope effect using Equation (54). This procedure yields a well-defined value λ'. Assuming proton tunneling distances of a few Ångstroms (e.g. 5 Å) one estimates maximum barrier heights of several 1000 cm^{-1}. This is a rather interesting result because it indicates that there are indeed barrier heights which exceed the glass transition temperature ($kT_g \approx 150$ cm^{-1}) considerably. Such a behavior is, however, quite consistent with the Vogel–Fulcher law (see, for example, [18]), which describes the relation between the viscosity η of the glass and its temperature

$$\eta = \eta_0 \exp\left[\frac{A}{k(T-T_0)}\right] \tag{55}$$

with parameters η_0, A and T_0. The above equation can be interpreted within the frame of an Arrhenius law with a temperature dependent barrier

$$V_0 = A(1 - T_0/T) \tag{56}$$

which the system has to overcome for thermally activated structural relaxation processes. At $T = T_0$ the barrier heights diverge formally, indicating that, on average, the thermally activated relaxation processes become infinitely slow. T_0 cannot be defined in a straightfoward fashion. Within the free volume model of amorphous solids it can be interpreted as the temperature at which the free volume vanishes [18]. We argue that if the system approaches the glass transition temperature there may be a large reorganization of free volume which leads to a lowering of the very high barriers of the system. As a consequence, all points in the configuration space are accessible to the glass at temperatures above T_g. Hence, we learn from the experiments on the isotope effect that in glasses, at low temperatures, very high barriers do indeed exist [33, 31]. We conclude that barrier formation and reduction is a collective phenomenon which renders ergodicity to the system at $T > T_g$.

Equation (36) shows that the slope factor of the hole width, like the slope factor of the area, is governed by the logarithm of the ratio R_{max}/R_{min}. When the TLS systems responsible for the evolution of the width are the same as those determining the area decay, we expect a deuteration effect of the same order of magnitude for both phenomena. This is caused by the rate R_{min} in the logarithmic ratio. Figure 19 shows that this is indeed the case. However, in contrast to Equation (27), the slope factor in the optical width (Equation (36)) still contains some experimental parameters. That is, we can explain the qualitative behavior but are unable to extract any system parameters from the measured hole width. However, the situation would be different if one would succeed in measuring separately the diffusional part of the total width; in this case one could calculate the reduced width γ_D/γ_1 (Equation (37)) and would be able to determine R_{max} from the corresponding slope [30, 31].

In order to determine γ_D from the measured width one has to know γ_0, the diffusion independent contribution — that is, one has to determine the onset of the spectral diffusion process. In the case considered, this is possible by evaluating the measured isotope effect. Figure 19 shows that the extrapolated data series of the protonated and the deuterated system crosses over at a certain time τ_0. For $t < \tau_0$ the width of the deuterated system would be larger as compared to the protonated system. Since this cannot be physically meaningful we conclude that the crossover point τ_0 determines the onset of the spectral diffusion process. Consequently

$$\gamma(\tau_0) \approx \gamma_0. \tag{57}$$

Hence, we can calculate from the measured data and the estimated value of γ_0 the reduced width $\gamma_D(t)/\gamma_1$ with γ_1 being an arbitrary reference value:

$$\frac{\gamma_D(t)}{\gamma_1} = 1 + \left[\ln \frac{4R_{max}}{R_1}\right]^{-1} \ln R_1 t. \tag{58}$$

This rescaling procedure has a very interesting consequence. Instead of R_1/R_{min} the ratio R_{max}/R_1 appears in the slope factor. Since, however, R_{max} is characterized by the fastest rate, the corresponding tunneling matrix element will be large and consequently λ' will be rather small. As a result there will be only a minor isotope effect in the ratio of the rates which will be suppressed by the logarithmic behavior. Figure 20 shows the result. Here we have rescaled our data according to Equation (58). All the data points of the four independently taken series (Figure 19) fall on a unified plot. There is indeed no isotope effect according to the theoretical model. It is gratifying to see how these differences are scaled out, and, a rather uniform time law holds. From (58) we find for $t = \tau_0$

$$R_{max} = \frac{1}{4\tau_0}. \tag{59}$$

We get a value of approximately 50 s^{-1}, which is surprisingly small. Hence, hole-burning experiments and investigation of deuteration effects allow, for the first time, a complete determination of the rate dispersion of a special TLS system, which is dominated by photochemical parameters.

9.4. POLARIZATION DIFFUSION

From the time scales of the investigated relaxation processes and from the observed isotope effects we concluded that in the quinizarin system the same kind of TLS states, namely the photoactive TLS$_p$, are responsible for both the broadening and the decay of the hole. Even though the TLS$_p$ in the quinizarin system are rather well characterized, there is no clear picture yet as to the nature of the corresponding photoactive TLS states of the photophysical system tetracene in EtOH/MeOH. To get some information on the nature of these states we carried out long-time investigations of the anisotropy of the corresponding

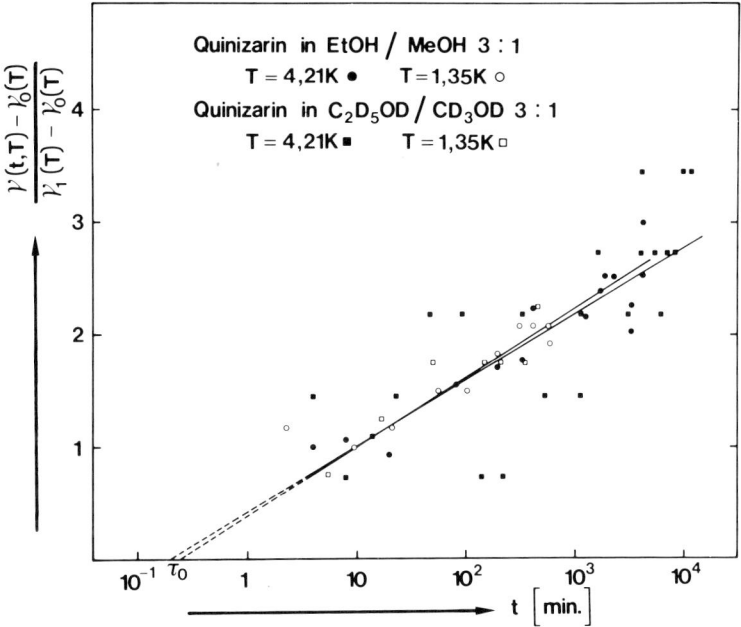

Fig. 20. Reduced spectral width of both protonated and deuterated quinizarin samples (see text).

hole spectra. In case the holes are burnt with polarized light, and in case the photoreactive dye molecules have well-defined non-degenerate dipole transitions, the spherical symmetry of the sample gets lost due to the polarized hole-burning process and the hole can be characterized by a definite degree of polarization which depends, as a rule, on the laser power and the burning time. The quantities which are measured are the degree of polarization ρ, defined by

$$\rho(t) = \frac{L_\parallel(t) - L_\perp(t)}{L_\parallel(t) + L_\perp(t)} \qquad (60)$$

and the hole areas $A_\parallel(t)$ and $A_\perp(t)$. L is the hole depth. The indices $\parallel$ and $\perp$ refer to whether the hole is scanned parallel or perpendicular to the direction of the burning field. ρ depends on the frequency within the hole spectrum. All our experiments were done with respect to the line center.

In the photochemical case one can, in principle, distinguish between two different mechanisms, which might occur on rather different time scales: (i) Relaxation of the photoproduct molecules to the educt state. If this process is connected with reorientation of the transition dipole as compared to the state before burning, the anisotropy of the hole may change. (ii) Polarization diffusion of those dye molecules which were not affected by the laser photochemistry. It is clear that such processes may also lead to a loss in anisotropy.

While process (i) is related to the special TLS$_p$ systems, process (ii) would reflect transitions in the ordinary TLS system of the doped glass. In principle, both tunneling systems may be characterized by rather different TLS parameters. It is obvious, however, that such a distinction breaks down in the case of non-photochemical hole-burning systems. In this case both tunneling systems belong to the same category.

Based on the above considerations we want to answer two questions. First, does a dye molecule change its orientation during the complete photochemical cycle as shown in Figure 10(a) and, second, is there any reorientation of the molecules which are left in the educt state $|E\rangle$ on the time scale of the experiment?

Figures 21 and 22 show the results. For both systems the degree of polarization remains constant throughout the whole experimental observation time. The hole does not lose its anisotropy. An extrapolation of the logarithmic decay of $A_\|$ and $A_\perp$ yields, in both cases, a crossover point which lies exactly on the time axis. This means that, apart from the degree of polarization, the ratio $A_\|/A_\perp$ is also constant in time.

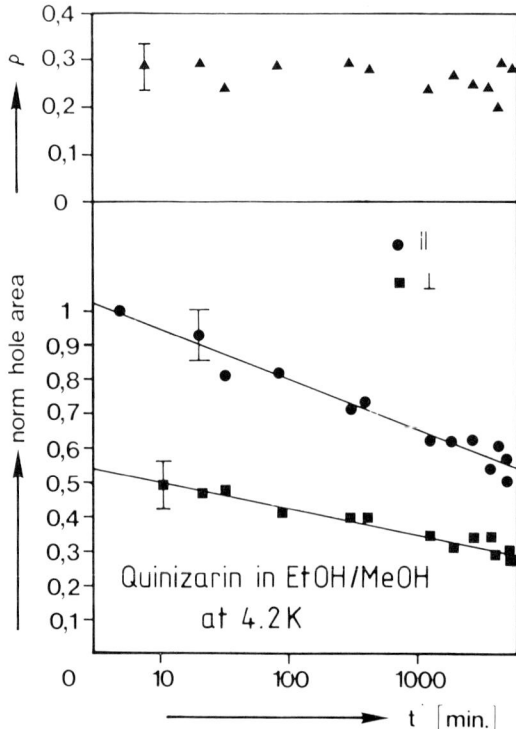

Fig. 21. Time dependence of the polarization of holes in the quinizarin system (upper plot). Time dependence of the corresponding photochemical holes (lower plot). Notice, that the polarization ratio does not change. Temperature 4.2 K.

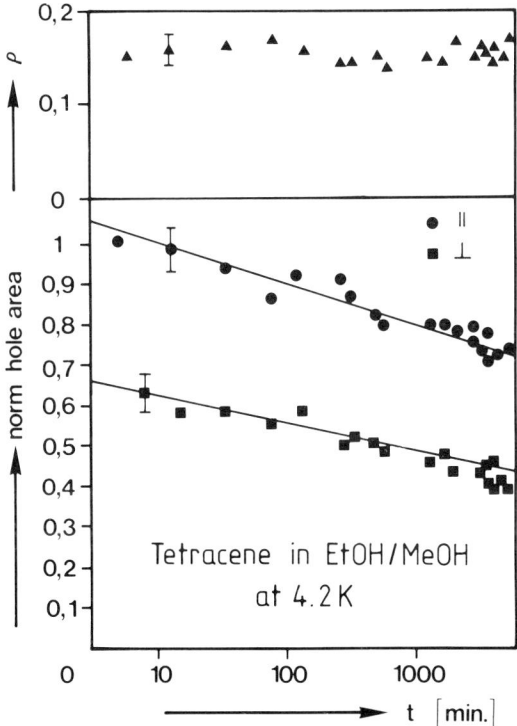

Fig. 22. Time dependence of the polarization of holes in the tetracene system (upper plot). Time dependence of the corresponding photochemical holes (lower plot). Notice, that the polarization ratio does not change. Temperature 4.2 K.

From these results we draw several conclusions [60]. Since $A_\parallel/A_\perp$ and $L_\parallel/L_\perp$ are constant in time, it follows that the number of molecules relaxing into a special spatial configuration with excitation frequency ω is exactly proportional to the number of molecules which were burnt in this configuration. Hence, the molecules relaxing from the product to the educt state return exactly to the positions in space which they had occupied before burning. If we would assume, for example, that the relaxing molecules are randomly distributed, then we would have to conclude that $A_\parallel$ and $A_\perp$ are reduced at the same absolute rate. In this case the degree of polarization would have to increase with time, which is not the case. As to the positions of the relaxing molecules in the frequency domain, we note that they are spread out within a narrow range around the frequency which they had before burning. This spread is due to spectral diffusion processes, as discussed in the foregoing sections.

The above conclusions are remarkable because we can conclude that the solvent cage seems to be a rather well-defined species which does not change in an irreversible fashion during the complete photochemical cycle indicated in

Figure 10(a). We can further conclude that a double minimum potential seems to be a very good description for the phototransformation and the related relaxation processes of both the photochemical and the non-photochemical system. The probability that the system relaxes to a third minimum seems to be of minor importance. Finally, we can infer from the data that a dye molecule is not coupled to several photochemically active TLSs. Otherwise it would be very difficult to understand why a relaxing molecule resumes its place in the frequency and space domain which it had before burnig. We have to assume that it couples mainly to one photoactive TLS (strong coupling); this, however, does not rule out a weak coupling to the other TLSs of the system. We consider this type of coupling as being responsible for the spectral diffusion.

As to the second question on the reorientation dynamics of the molecules left in the educt state, we can definitely say that, as long as the reorientation leads to a population redistribution within the contours of the hole, the anisotropy should fade as the hole is smoothed out. Since ρ remains constant over the whole experimental period, we can safely exclude such reorientation processes of the dye molecules on the time scale of the experiment.

Summarizing the results of the polarization experiments, we draw the following conclusions.
  (i) Molecules relaxing from the product to the educt state return into the position (frequency and space) which they had before burning. The solvent cage seems to be a well-defined species and does not change during a complete photochemical cycle.
  (ii) The relaxation process of a glass is well described in terms of a double minimum potential. Relaxation between several minima seems to be absent.
  (iii) A dye molecule seems to be strongly coupled just to one photoactive TLS.

10. Field Effects and Spectral Diffusion Phenomena

As has been pointed out in the foregoing sections, various external parameters enter into the experimentally determined shape of a photochemical hole. It was shown that there are two main contributions to the linewidth, namely the site memory function $\delta N(\omega')$ and the dynamical linewidth function $z(\omega - \omega')$. The site memory function had a 'quasi-static' character, i.e. is characterized by slowly varying processes, whereas the dynamic linewidth function reflects the non-adiabatic relaxation processes of the host—guest system.

If we introduce into our considerations external field parameters q, then one has to rewrite Equation (7) in a way which allows the q-parameters to enter into both the hole-burning part of the experiment $(\hat{q})$ and into the detection part of the hole via absorption or fluorescence experiments (q). If one attributes frequency shifts $\Delta\omega(q)$ to the various linewidth contributions, then one can express the linewidth by modifying Equation (7).

$$L^{\tau}(\omega, T, T_b, t, q, \hat{q})$$

$$= \sigma \int \delta^{\tau} N[T_b, t, (\omega' - \Delta\omega(\hat{q}))] \, z \, [T, \{\omega - (\omega' - \Delta\omega(q))\}] \, d\omega'. \quad (61)$$

Let us first consider the q-variable of the electric field, which is understood on a more quantitative basis before introducing strain fields, which can also be considered in a formally similar fashion. The first quantitative data on the Stark effect were documented by the Personov group [62]. The authors have introduced a general electric field term of the following form:

$$\Delta\omega(q) \equiv \Delta\omega(E) = \frac{1}{\hbar} (f \cdot \Delta\boldsymbol{\mu} \cdot \mathbf{E} + \tfrac{1}{2} f^2 \mathbf{E} \cdot \Delta\hat{\alpha} \cdot \mathbf{E}), \quad (62)$$

where f is a Lorentz field correction factor $f = (\varepsilon + 2)/3$ and $\Delta\boldsymbol{\mu}$ and $\Delta\hat{\alpha}$ are the changes in the dipole moment and the polarizability upon creating an excited state of the molecule under consideration. Note, that for evaluating the experimental data two field values $\hat{E}$ and E enter the expressions for the hole-burning and the hole detection experiments respectively.

The first experimental data [62] and subsequent experiments [63–65] have shown that, even in the cases in which molecules with inversion symmetry were investigated, the linear Stark effect yielded the dominating term for explaining the experimental data.

10.1. ELECTRIC FIELD EFFECTS FOR MOLECULES WITH INVERSION SYMMETRY

Let us consider the case of a molecule with inversion symmetry. In this case the individual induced dipole can, to a good approximation, be considered as not being correlated to the transition dipole of the hole-burning molecule. The analytical [64] and numeric data [62, 65] show that a symmetric line-broadening is expected. It can be readily shown that if one calculates the number of molecules per Stark-shift interval, one expects, for a given $\Delta\boldsymbol{\mu}$ value, a step function extending from $-(\Delta\boldsymbol{\mu} \cdot \mathbf{E})/\hbar$ to $+(\Delta\boldsymbol{\mu} \cdot \mathbf{E})/\hbar$. This function is shown in Figure 23 for a given, constant value of $\Delta\boldsymbol{\mu}$. If one introduces a distribution of $\Delta\boldsymbol{\mu}$-values obeying a Gaussian statistic, as has been done by Bogner *et al.* [64], the character of the step function is smoothened. Figure 24(a) and (b) show numerical line shape calculations using a constant $\Delta\boldsymbol{\mu}$ and a statistically varying $\Delta\boldsymbol{\mu}$, respectively [65].

Analytical data describing the hole intensity at the hole-burning frequency have been presented by [64]. As has been pointed out above, the intensity dependence is only a function of the difference between the burning field $\hat{E}$ and the measuring

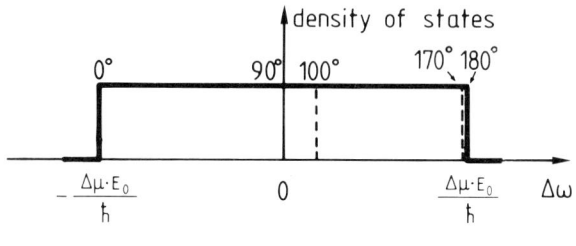

Fig. 23. Density of states due to Stark shifts of the molecular levels (see text).

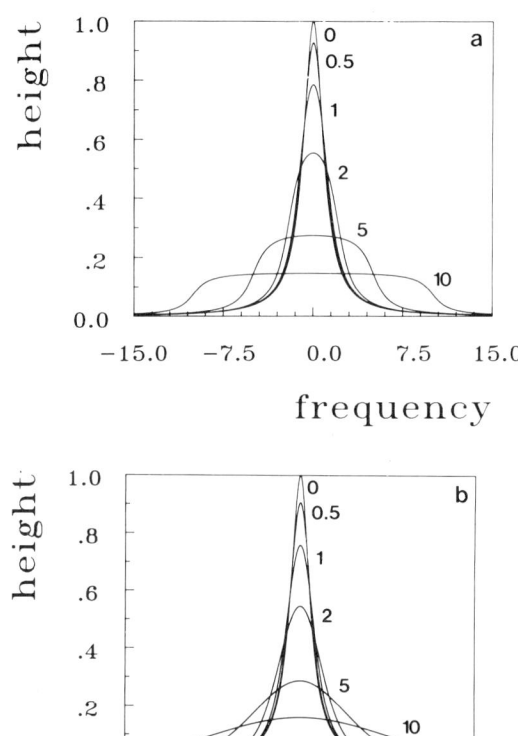

Fig. 24. (a) Line shape of a photochemical hole of an ensemble of molecules with inversion symmetry and a constant $\Delta\mu$-value [65]. (b) Line shape as in (a), but with a statistically varying $\Delta\mu$-value [65]. (a + b) The frequency is normalized to a half width of 1 at zero field. The parameters in the plots represents dimensionless field strengths. A field strength of 1 is reached when the dipolar energy spread (Figure 23) corresponds to the hole width at zero field.

field E. The field dependence, as given by [64], can be written as:

$$L(\hat{E}, E) \approx \left\{ 1 - \frac{r\Delta\omega_N}{\Delta\omega_N + \Delta\omega_z} \pi^{\frac{3}{2}} \hbar \frac{(\Delta\omega_N + \Delta\omega_z)}{f\Delta\bar{\mu}|\hat{E} - E|} \times \right.$$

$$\left. \times \exp\left[\frac{\hbar^2(\Delta\omega_N + \Delta\omega_z)^2}{f^2 \Delta\bar{\mu}^2 |\hat{E} - E|^2}\right]\left[1 - \mathrm{erf}\left(\frac{\hbar(\Delta\omega_N + \Delta\omega_z)}{f\Delta\bar{\mu}|\hat{E} - E|}\right)\right]\right. . \quad (63)$$

In the above equation r is a factor characterizing the photochemical condition of the hole. $\Delta\bar{\mu}$ is the average change in dipole moment; $\Delta\omega_N$ and $\Delta\omega_z$ are the widths of the site memory function $\delta N(\omega)$ and the spectral line shape function $z(\omega)$, respectively, taken in the 'short burning time approximation'. In this approximation both line shape functions can be approximated by Lorentzians. Figure 25 shows a comparison between the model calculations and the experimental data of [64]. Since the hole width $\Delta\omega_{\mathrm{hole}}$ is, under the above restrictions, given by $\Delta\omega_{\mathrm{hole}} = \Delta\omega_N + \Delta\omega_z$, Equation (63) contains $f \cdot \Delta\bar{\mu}$ as sole fitting parameter. $\Delta\mu$-values for various molecules have been reported [62, 64, 65, 71]; they range between $\Delta\mu = 0.1$ and 6 Debye units.

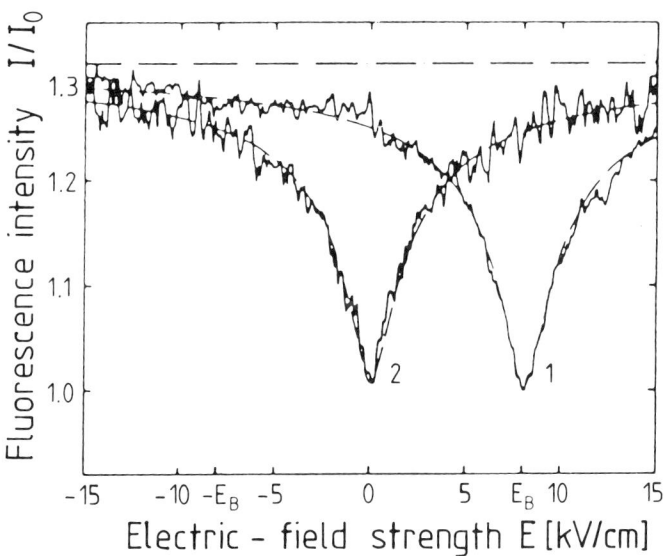

Fig. 25. Hole depths at the laser frequency ω_L (hole center) as a function of the external electric field (taken from reference 64). The broken line is calculated with Equation (63).

It should be pointed out that the above experimental findings are rather surprising. The measured, induced dipole moments are solely due to the molecule—matrix configuration, i.e. due to the lack of inversion symmetry by the guest—host configuration, since the guest molecule itself does have inversion symmetry. This means that an amorphous matrix, which is usually being considered as 'isotropic' matrix, can only be called isotropic in a macroscopic sense. In a microscopic picture, the matrix has a rather low symmetry as to render a dipole moment to a molecule which does have inversion symmetry. As far as organic molecules in glasses are concerned there are little systematic experiments on local matrix effects which are responsible for removing molecular symmetries.

10.2. ELECTRIC FIELD EFFECTS FOR MOLECULES WITHOUT INVERSION SYMMETRY

We have assumed, in the above section, that the molecules involved have inversion symmetry. Therefore, we could, with good reasons, consider the matrix induced $\Delta\mu$ as being uncorrelated to the transition dipole of the molecule. The reason for this approximation is the fact that the induced dipole moment is considered as a matrix effect, whereas the transition dipole is linked to a molecular axis and, for allowed optical transitions, is not coupled to the symmetry of the host matrix.

If a molecule lacks inversion symmetry, then $\Delta\mu$ and the transition dipole moment can be assumed to have a well-defined orientation with respect to each other. For simplicity two cases have been considered in the past. They are sketched in Figure 26a. If we call the transition dipole moment μ_{01}, then $\Delta\mu$ can either be parallel or perpendicular to μ_{01}. A case in which this condition is not fulfilled can be considered as superposition.

In Figure 26 the change in dipole moment is parallel to μ_{01}. It is also assumed that the electric field vector is parallel to the k-vector of the incident light (spectroscopy through two semi-transparent electrodes perpendicular to the light beam). In this case the $\Delta\mu$ ensemble, which couples to the light field is in the $\cos^2\theta$ distribution, as shown in Figure 26, and has its main contributions perpendicular to the external electrical field thus yielding no appreciable Stark shift. Hence, the line will broaden symmetrically, as has been shown by Personov et al. [62] in the first numerical calculations.

The opposite case of $\Delta\mu$ perpendicular to μ_{01} is shown in Figure 26(b). Here the $\sin^2\theta$ distribution has strong contributions in both the 'up field' and the 'down field' direction, leading to a double peaked line shape, as has also been verified by the first numerical calculations [62].

The presently available mathematical models can only be considered as first approximations since, in principle, the induced dichroism of the sample, as a consequence of the hole-burning, has to be included in the model [61]. Most likely, however, those calculations can only be performed numerically.

In summary, electric field effects can be rather well described by extending the pertinent equations, which describe hole-line shapes with the appropriate Stark terms. Up to now, it had been possible to extract average numbers for the change in dipole moment $\Delta\bar{\mu}$; this holds for both the induced moments and the molecular moments [66]. These numbers can be taken as numerical parameters, describing the 'site properties' of molecular centers in amorphous materials. The microscopic

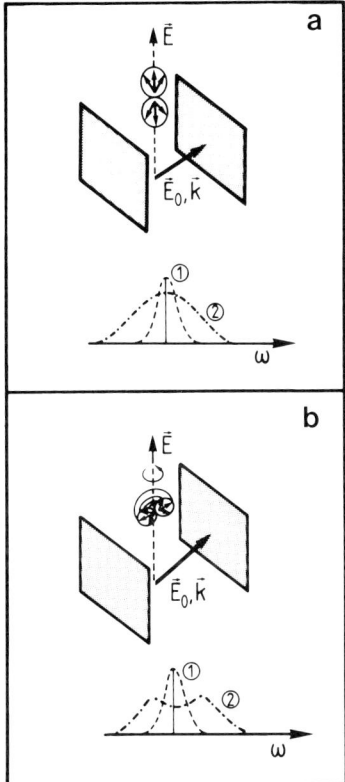

Fig. 26. (a) Stark broadening of a zero-phonon line with the change in dipole moment $\Delta\mu$ being parallel to the transition dipole moment μ_{01}. ① shows the line without the static electric field $\mathbf{E}_0$ and ② shows the line with a finite electric field $\mathbf{E}_0$. $\mathbf{k}$ is the wavevector of the incident light beam, which is also perpendicular to the electrodes (shaded areas). (b) Stark broadening with conditions as in (a) but with $\Delta\mu$ perpendicular to μ_{01}. Note, that the line develops a double-peak structure for large electric fields.

understanding of site induced asymmetries, which remove the inversion symmetry of symmetric molecules in amorphous media, is far from being understood.

10.3. HOLE-BURNING EXPERIMENTS UNDER EXTERNAL PRESSURE

Instead of using electric fields, one can use external uniaxial or hydrostatic pressure to achieve a frequency shift $\Delta\omega(q)$ of the sites which are involved in the hole-burning experiments. These experiments seem to be promising, because very small spectroscopic changes can be observed due to the high resolution features of the experiments. However, few experimental data for uniaxial [67] and hydrostatic [68] pressure are available.

In spite of some similarities in the formal description of field effects, there are some basic differences between external electric fields and external strain fields.

As has been shown, a linear Stark effect tends to broaden the photochemical hole symmetrically to both sides of the spectrum, since an equal amount of dipoles points, in a statistical average, in field direction and against the field direction. In contrast to this general situation, uniaxial or hydrostatic pressure generally shifts the center of the line to higher or lower energies (a shift to lower energies for increasing pressure is the common situation). This energy shift has, in the past, been described by various models. One model uses the approach of a spectral shift due to the pressure-induced changes in the dielectric properties of the medium [69]. Other models are based on parametrized configuration coordinate models [70, 71 and references]. For both descriptions one expects a line shift, which can be described by a linear and a quadratic term in pressure (with the linear term being, for low p values, the dominating term):

$$\Delta\omega(p) = A \cdot p + B \cdot p^2 \tag{64}$$

Superimposed on the line shift there is usually a line-broadening of comparable magnitude. Figure 27 gives a schematic view of both terms contributing to the line

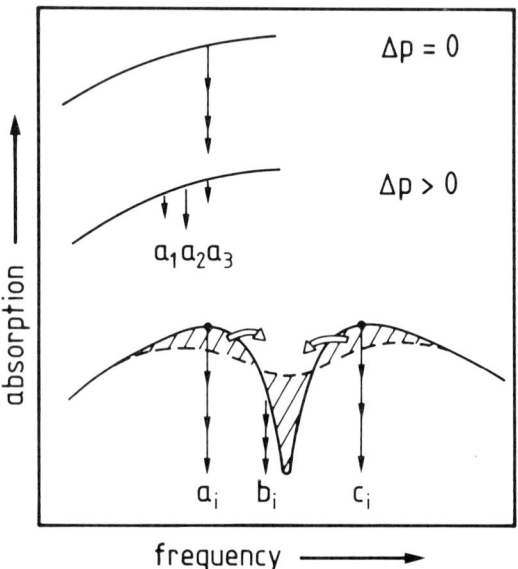

Fig. 27. Scheme for pressure-induced hole broadening. The figure shows the individual site shifts of the molecular absorbers a_i-c_i. The mechanism for the broadening of each individual site group is symbolically depicted in the upper graph [67].

shift and the line broadening. In a perfect single crystal, in which each site corresponds to an identical center, the shift term would be the only contribution to the pressure-induced change. In an amorphous medium, however, there is a concomitant broadening of the spectral line due to the dispersion of 'shift parameters', characterizing the various sites of the sample. One can symbolize this behavior as being due to a variation of the 'local compressibility', which is different from site to site.

Figure 28 shows experimental data of the hole spectrum for free base phthalocyanine in a PMMA matrix in the very low pressure regime [67]. Note, that the observed broadening is of the same magnitude as the observed line shift. The pressure shift values are on the order of $d\bar{\nu}/dp \simeq 10^{-2}$ cm$^{-1}/10^3$ hPa; they are known from high-pressure experiments on dyes and color centers using straightforward optical spectroscopy.

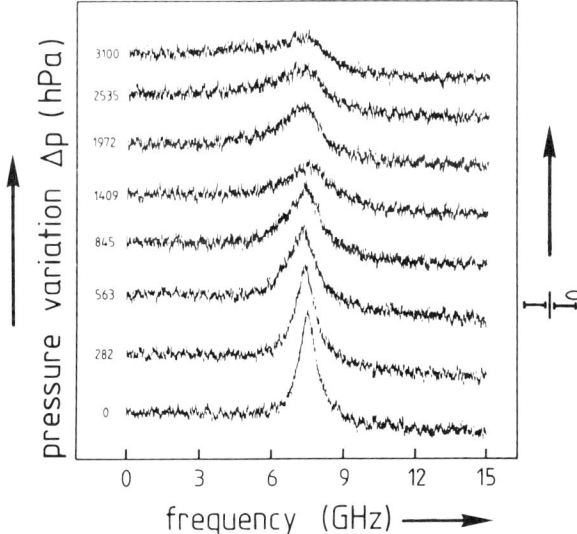

Fig. 28. Experimental line shift and line broadening of free base phthalocyanine in a PMMA matrix [67].

Whereas there are quite detailed theories on the Stark effect of molecular spectra, the understanding of the pressure-induced line-broadening of holes is, for the time being, limited to an empirical understanding. First data indicate that one can describe the pressure-induced line-broadening $\Delta\Gamma$ by the following empirical equation:

$$\frac{\Delta\Gamma}{\Gamma} = \left(\frac{\Delta p}{p}\right)^\alpha. \tag{65}$$

The measured pressure exponents α range between $\alpha = 1.5$ and 2.5.

A phenomenon which is worth mentioning and which, if more data can be produced, may shed light on barrier heights and barrier height distributions is the observed irreversible spectral line-broadening. This broadening has been observed at pressures exceeding 3×10^4 hPa (i.e. at comparatively small pressures). It has been tentatively attributed to pressure-induced spectral diffusion. Here one would have to assume that the external pressure induces a barrier crossing of a TLS system, which couples to the molecular center under investigation. As one can see from Figure 29, this pressure-induced change can be irreversible when the involved TLS systems are asymmetric and, hence, the removal of the external pressure will not bring the system back to its original configuration. It should, however, be pointed out, that the existence of non-equilibrium configurations is necessary to describe the observed irreversible effects.

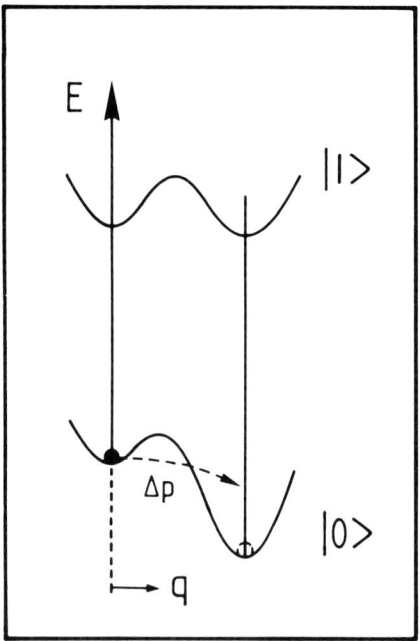

Fig. 29. Pressure-induced, irreversible barrier crossing. If the particle performs the pressure-induced change, which is depicted by the dotted arrow, it will not return to its original position if the pressure is removed due to the asymmetry of the potential. $|0\rangle$ and $|1\rangle$ label the electronic ground and excited states. q is a tunneling coordinate.

In summary, pressure data on hole-burning systems are, at the present time, scarce. They may, however, provide a new approach for the understanding of some details of the TLS configurations which are responsible for reversible and irreversible line changes.

Acknowledgment

The authors acknowledge the support of their own research on hole burning through the Stiftung Volkswagenwerk. Some of the early work on hole burning was performed at the IBM Research Laboratory in San Jose, California and was supported by the ONR.

References

1. A. A. Gorokhovskii, R. K. Kaarli and L. A. Rebane, *JETP Lett.* **20** (1974) 216.
2. B. M. Kharlamov, R. I. Personov, and L. A. Bykovskaya, *Opt. Commun.* **12** (1974) 191.
3. A. Szabo, *Phys. Rev. Lett.* **25** (1970) 924.
4. R. I. Personov, E. I. Al'shits, and L. A. Bykovskaya, *Opt. Commun.* **6** (1972) 169.
5. L. A. Rebane, A. A. Gorokhovskii, and J. Kikas, *Appl. Phys.* **B29** (1982) 235.
6. R. I. Personov in *Spectroscopy and Excitation Dynamics of Condensed Molecular Systems* (eds V. M. Agranovich and R. M. Hochstrasser), North Holland, Amsterdam (1983).
7. G. J. Small in *Spectroscopy and Excitation Dynamics of Condensed Molecular Systems* (eds V. M. Agranovich and R. M. Hochstrasser), North Holland, Amsterdam (1983).
8. J. Friedrich and D. Haarer, *Ang. Chemie* **96** (1984) 96, Int. Ed. Engl. **23** (1984) 113.
9. S. Völker, R. M. Macfarlane, and J. H. van der Waals, *Chem. Phys. Lett.* **53** (1978) 8.
10. F. Graf, H. K. Hong, A. Nazzal, and D. Haarer, *Chem. Phys. Lett.* **59** (1978) 217.
11. I. I. Abram, R. A. Auerbach, R. R. Birge, B. E. Kohler, and J. M. Stevenson, *J. Chem. Phys.* **63** (1975) 2473.
12. W. C. McColgin, A. P. Marchetti, and J. H. Eberly, *J. Am. Chem. Soc.* **100** (1978) 5622.
13. J. Friedrich, J. D. Swalen, and D. Haarer, *J. Chem. Phys.* **73** (1980) 705.
14. J. Friedrich and D. Haarer, *J. Chem. Phys.* **76** (1982) 61.
15. J. Friedrich, H. Wolfrum, and D. Haarer, *J. Chem. Phys.* **77** (1982) 2309.
16. A. R. Gutierrez, G. Castro, G. Schulte, and D. Haarer in *Electronic Excitations and Interaction Processes in Organic Molecular Aggregates* (eds P. Reineker, H. Haken, and H. C. Wolff), Springer, Berlin (1983).
17. J. Friedrich, D. Haarer, and R. Silbey, *Chem. Phys. Lett.* **95** (1983) 119.
18. G. Crest and M. Cohen, *Adv. Chem. Phys.* **48** (1981) 217.
19. J. Jäckle, 'Models of the Glass Transition' (to appear in *Rep. Progr. Physics* 1985).
20. R. Zallen, *The Physics of Amorphous Solids*, Wiley, New York (1983).
21. S. Alexander and R. Orbach, *J. Phys. Paris Lett.* **43** (1982) L-625.
22. S. K. Lyo and R. Orbach, *Phys. Rev.* **B22** (1980) 4223.
23. W. A. Philips, *Amorphous Solids*, Springer, Berlin 1981.
24. S. Hunklinger and A. K. Raychaudhuri in "Progress of Low Temperature Physics", ed. by D. F. Brewer, Elsevier Science Publ. Co. 1986, p. 265.
25. R. O. Pohl in *Amorphous Solids*, Springer, Berlin (1981).
26. P. W. Anderson, B. I. Halperin, and C. M. Varma, *Philos. Mag.* **25** (1971) 1.
27. A. S. Davydov, *Quantenmechanik*, VEB Deutscher Verlag der Wissenschaften, Berlin (1970).
28. J. Jäckle, *Z. Phys.* **257** (1972) 212.
29. V. Hizhnyakov and J. Tehver, *Phys. Stat. Sol.* **B95** (1979) 65.
30. W. Breinl, J. Friedrich, and D. Haarer, *J. Chem. Phys.* **81** (1985) 3915.
31. J. Friedrich and D. Haarer, *J. Physique* C7 (1985) 357.
32. D. Haarer and J. Friedrich, *Physikalische Blätter* **41** (1985) 363.
33. W. Breinl, J. Friedrich, and D. Haarer, *Chem. Phys. Lett.* **106** (1984) 487.
34. W. Breinl, J. Friedrich, and D. Haarer, *J. Chem. Phys.* **80** (1984) 3496.
35. L. W. Molenkamp and D. A. Wiersma, *J. Chem. Phys.* **83** (1985) 11.
36. S. Völker, R. M. Macfarlane, and J. H. van der Waals, *Chem. Phys. Lett.* **53** (1978) 8.
37. H. P. H. Thijssen, S. Völker, M. Schmidt, and H. Port, *Chem. Phys. Lett.* **94** (1983) 537.
38. H. P. H. Thijssen, R. van Berg, and S. Völker, *Chem. Phys. Lett.* **97** (1983) 295.
39. S. Völker, in *Photoreaktive Festkörper* (ed. H. Sixl), Wahl Verlag, Karlsruhe (1984), p. 749.
40. D. A. Fushman, *Phys. Stat. Sol.* B **127** (1985) 679.

41. J. R. Klauder and P. W. Anderson, *Phys. Rev.* **125** (1962) 912.
42. J. L. Black and B. I. Halperin, *Phys. Rev.* **B16** (1977) 2879.
43. S. Hunklinger and M. Schmidt, *Z. Phys. B., Condensed Matter*, **54** (1984) 93.
44. T. Tiedje and A. Rose, *Solid State Commun.* **37** (1980) 49.
45. M. Schwartz, F. Pobell, M. Kubota, R. M. Mueller, *J. Low Temp. Phys.* **58** (1985) 171.
46. J. Friedrich and A. Blumen, *Phys. Rev.* **B32** (1985) 1434.
47. R. Kohlrausch, *Ann. Phys. (Leipzig)* **12** (1847) 393.
48. G. Williams and D. Watts, *Trans Faraday Soc.* **66** (1970) 80.
49. M. Inokuti and F. Hirayama, *J. Chem. Phys.* **43** (1965) 1978.
50. H. Scher and M. Lax, *Phys. Rev.* **B12** (1975) 2455.
51. H. Scher and E. W. Montroll, *Phys. Rev.* **B12** (1975) 2455.
52. A. Blumen, *Nuovo Cimento* **B63** (1981) 50.
53. A. Blumen, J. Klafter, and G. Zumofen, *Phys. Rev.* **B27** (1983) 3429.
54. H. J. Queisser, *Phys. Rev. Lett.* **54** (1985) 234.
55. T. Doba, K. U. Ingold, W. Siebrand, and T. A. Wildman, *Chem. Phys. Lett.* **115** (1985) 51.
56. R. Richert and H. Bässler, *Chem. Phys. Lett.* **116** (1985) 302, **118** (1985) 534.
57. R. Jankowiak, R. Richert, and H. Bässler, *J. Phys. Chem.* **89** (1985) 4561.
58. C. Aubert, J. Fünfschilling, I. Zschokke-Gränacher, T. A. Wildmann, and W. Siebrand, *Chem. Phys. Lett.* **122** (1985) 465.
59. R. J. Roe, *J. Chem. Phys.* **79** (1983) 936.
60. W. Köhler, W. Breinl, and J. Friedrich, *J. Chem. Phys.* **82** (1985) 2935.
61. W. Köhler, W. Breinl, and J. Friedrich, *J. Phys. Chem.* **89** (1985) 2473.
62. V. D. Samoilenko, N. V. Razumova, and R. I. Personov, *Opt. Spectrosc.* **52** (1982) 346.
63. F. A. Burkhalter, G. W. Suter, U. P. Wild, V. D. Samoilenko, N. V. Rasumova, and R. I. Personov, *Chem. Phys. Lett.* **94** (1983) 483.
64. U. Bogner, P. Schätz, and M. Maier, *Chem. Phys. Lett.* **102** (1983) 267.
65. Kador D, Haarer, and R. I. Personov, to be published.
66. A. J. Meixner, A. Renn, S. E. Bucher, and U. P. Wild, *XIth Molecular Crystal Symposium*, Lugano 1985.
67. W. Richter, G. Schulte and D. Haarer, *Opt. Commun.* **51** (1984) 412.
68. T. Sesselmann, Diplomarbeit 1985, Bayreuth; T. Sesselmann, W. Richter and D. Haarer to be published.
69. N. S. Bayliss, *J. Chem. Phys.* **18** (1950) 292.
70. D. Curie, D. E. Berry, and F. E. Williams, *Phys. Rev.* **B22** (1980) 4109.
71. C. P. Slichter and H. G. Drickamer, *Phys. Rev.* **B22** (1980) 4097.
72. A. Gorokhovskii, V. Korrovits, U. Palm, M. Trummal, *Chem. Phys. Lett.* **125** (1986) 355.

MODELS FOR REACTION DYNAMICS IN GLASSES

A. BLUMEN

Max-Planck Institut für Polymerforschung, D-6500 Mainz and Lehrstuhl für Theoretische Chemie, Technische Universität, D-8046 Garching, West Germany.

J. KLAFTER

Corporate Research Science Laboratories, Exxon Research and Engineering Company, Clinton Township, Annandale, N.J. 08801, U.S.A.

and

G. ZUMOFEN

Laboratorium für Physikalische Chemie, ETH-Zentrum, CH-8092 Zürich, Switzerland

1. Introduction

Glasses differ from crystals in that they lack long-range spatial order. Furthermore, due to their high viscosities, glasses are less favorable to the internal rearrangements displayed by liquids. Hence glasses show a multitude of microscopic patterns around each site (microenvironments), which may relax on widely different time scales. Any local probe of the glass structure, be it an elementary particle (neutron), a trapped charge carrier (electron or hole) or an impurity (ion or sensibilizing molecule) senses not only the different geometry of its surroundings, but, because of it, also changes in the local potentials. These may then reflect themselves in the rates of the processes under investigation. Thus geometrical disorder also implies energetic and temporal disorder. On the other hand, the long lifetime of the local microenvironments precludes an efficient internal averaging; in glasses one is forced to deal with the full complexity of a disordered medium.

Let us note from the start that there is little hope in describing such a complexity in terms of perturbations from ideal, non-random patterns. This point has been forcefully stressed by P. W. Anderson in the Les-Houches lectures on ill-condensed matter [1]. He remarked that multiple scattering theories (effective-medium or coherent-potential approximations) are paradigms [2] of an old attitude, which tries to model random systems through regular ones; the new attitude is exemplified by localization and percolation, which typify disordered situations as genuinely distinct problems [1].

In this contribution we adopt this line of thought and present several new classes of models for disorder. These models have come into close scrutiny only during the last decade. Each class may be viewed as arising due to a particular aspect of randomness. Thus fractals [3, 4] exemplify the spatial disorder, continuous-time processes [5—9], the temporal and ultrametric structures [10—12], the energetic disorder. We focus on dynamical processes, illustrated by reaction schemes, such as obtain for electron—hole recombination, electron scavenging, and energy transport, trapping and annihilation. We feel that in disordered systems aspects related to dynamics and relaxation have been treated less extensively than the static aspects, such as percolation, on which a series of reviews exists [13—16]. Also, the related dynamical problems on spin glasses will not be analyzed, the only connections to our subject matter being made in discussing ultrametric (hierarchical) structures [10—12]. For spin glasses we refer to a recent proceedings volume [17]. Furthermore, we will center on newer developments of the continuous-time random-walk approaches, the former works being summarized in references [5] to [9].

Basic to our understanding of disorder is the temporal evolution of the systems under investigation. We will thus start in the next section by reviewing reaction patterns of widespread use, such as pseudo-unimolecular and bimolecular decays. A quick comparison to actually observed relaxation patterns will reveal that the long-time decays which are observed in disordered materials have seldom kinetical chemical counterparts, the reason being the implicit underlying assumption in the kinetic scheme of a 'well-stirred' chemical reactor. The lack of such a homogenization for disordered media leads then to interesting deviations from the kinetic picture.

To display decay laws in possibly simplest form, we focus in Section 3 on the direct transfer to randomly distributed centers. The problem is of much interest in energy transfer from sensitizing molecules to acceptors in glasses, but appears in a mathematically equivalent form also when treating the scavenging of electrons. The theoretical advantage of the direct transfer laws resides in their mathematical simplicity; for many systems exact expressions for the decay are known, and easy-to-use approximate forms may be derived in a straightforward manner. In Section 3 these forms, which belong to the so-called parallel relaxation, will be investigated for several geometries. We concentrate both on regular lattices and on fractals, and we take several interaction types into account. Since fractal geometries may be thought to arise from restrictions on the regular lattices, we will also discuss the influence of restricted geometries on the direct transfer, a point which we believe may have important applications.

After showing in Section 3 the decay forms obtainable for fixed molecular configurations and one-step processes, we proceed in Section 4 to investigate the influence of random walks on the decay laws. After an overview on random walks on regular lattices we discuss the influence of a fractal geometry on the relaxation process. As in Section 3 we focus on pseudo-unimolecular reactions (in which one of the species is in very low concentration, and is hence reaction-determinant). Two extreme models may be distinguished: in the target problem the majority species moves and the minority species is immobile, whereas in the trapping problem the minority species moves among static traps. Also intermediate cases, in which both species move will be considered. In low-dimensional spaces there are large qualitative differences in the relaxation patterns for the target and the trapping problems, fact distinct from the predictions of the standard kinetic descriptions, in which only the relative motion of the two species enters.

In Section 5 we extend our approaches to disorder by including the temporal randomness in form of continuous-time random walks (CTRW). The CTRW picture is of importance, both in connection with electronic conduction properties and also as a model for spin relaxation in glasses. We will show that depending on the waiting-time distributions used, the trapping and the target decays are different. To display the flexibility of the CTRW approach, we will also apply it to fractals, thus including in a generalized model both the temporal and the spatial disorder aspects.

Section 6 will be devoted to the influence of energetic disorder on the relaxation behavior. Two schemes have proven to be fruitful here: the one is the multiple trapping (MT) model, whereas a more recent approach is based on the idea of a hierarchical distribution of barrier heights such as for instance postulated for spin glasses. This distribution of different separation scales leads to the concept of ultrametric spaces (UMS). In Section 6 we will present results for both the target and the trapping model, obtained by letting the motion of the particles involved take place on ultrametric spaces.

In Section 7 we make connection to another class of chemical reactions, these of $A + A \to 0$ or $A + B \to 0$ type (the latter under the assumption of an equal number of A and B particles) such that in the chemical kinetic schemes the relaxation at long times follows a $1/t$ behavior. In the presence of disorder, the relaxations obey more general power law patterns. In this section we analyze such chemical reactions under several aspects: we monitor motions and fractals, we employ the CTRW model and we also display the influence of UMS.

2. Relaxation Viewed as Chemical Reaction: the Kinetic Approach

In this section, we remind the reader of chemical reaction schemes of widespread use and show their relaxation patterns. As will become obvious in the second part of this section, the decay laws which follow from the kinetic formalism are not suitable for describing many of the intriguing relaxation forms found in glasses. The main reason for this dichotomy is to be found in the important role played by the disorder.

We start by noting that many of the methods used to locally probe glasses may be viewed in terms of an underlying chemical reaction. Thus the capture of an electron by a scavenger may be considered as being a reaction in which at the microscopic level one free electron and one active scavenger disappear. The same holds true for the electron–hole recombination in glasses: in each recombination step one electron and one hole are annihilated. Also sensitized energy transfer in glasses, by which the energy is exchanged between different molecular species may be envisaged as a reaction which changes the numbers of excited molecules of different kinds. Another example may be found in the triplet–triplet annihilation in mixed molecular solids, in which two triplets are annihilated and one excitation in the singlet state appears; the whole process may be monitored through the delayed fluorescence of the singlet. Furthermore, chemical analogs are to be found in several models for dielectric relaxation, which assume that local strains may relax only when free defects (or volumes) which meander through the glass happen to be in their vicinity. Evidently, the reactions invoked here are of $A + B \rightarrow P$ or $A + A \rightarrow P$ type.

General irreversible reactions have the form:

$$A_1 + A_2 + \cdots + A_n = \sum_{i=1}^{n} A_i \xrightarrow{k} P. \tag{2.1}$$

In the classical kinetic scheme these reactions are modeled through a system of (in general) nonlinear differential equations [18–22]:

$$\frac{dA_i(t)}{dt} = -k \prod_{i=1}^{n} A_i(t). \tag{2.2}$$

Here the $A_i(t)$ denote the concentrations (or the number of particles) for the ith molecular species. One looks now for the general solution of Equation (2.2), the integration constants being the initial values of the A_i, i.e. $A_i(0) = A_{i0}$.

The simplest case obtains in (2.1) for $n = 1$, and leads to a unimolecular reaction:

$$A \xrightarrow{k} P \tag{2.3}$$

whose solution is exponential in time:

$$A(t) = A_0 e^{-kt}. \tag{2.4}$$

Bimolecular reactions ($n = 2$) are only slightly more complex:

$$A + B \xrightarrow{k} P. \tag{2.5}$$

To this reaction corresponds in the kinetic scheme, Equation (2.2):

$$\frac{dA(t)}{dt} = -kA(t)B(t) = \frac{dB(t)}{dt}. \tag{2.6}$$

From the left- and right-hand sides of Equation (2.6) one observes that the difference $A(t) - B(t)$ stays constant during the reaction. We set $A(t) - B(t) = -C = A_0 - B_0$ and separate the variables in Equation (2.6), obtaining after integration:

$$\frac{1 + C/A(t)}{1 + C/A_0} = e^{Ckt}. \tag{2.7}$$

From Equation (2.7) we infer for $B_0 \gg A_0$, that $C \simeq B_0$ and thus $C/A(t) \gg 1$:

$$A(t) \simeq A_0 e^{-B_0 kt}. \tag{2.8}$$

Thus the decay of the minority species is quasi-exponential. It may also be viewed as being pseudo-unimolecular (one may compare it to Equation (2.4)), the decay rate being now $B_0 k$.

On the other hand, if $A_0 = B_0$ then $C = 0$ in Equation (2.7). An expansion in small C leads to the decay form valid for this case:

$$A(t) = \frac{A_0}{1 + A_0 kt} \tag{2.9}$$

from which at longer times, $t \gg (A_0 k)^{-1}$, an algebraic time-dependence emerges:

$$A(t) \sim \frac{1}{kt}. \tag{2.10}$$

We pause to note that a very similar behavior also obtains for the $A + A \to P$ reaction, with the kinetic equation:

$$\frac{1}{2}\frac{dA(t)}{dt} = -k[A(t)]^2. \tag{2.11}$$

Separation of variables and integration lead to:

$$A(t) = \frac{A_0}{1 + 2A_0 kt} \tag{2.12}$$

a form very akin to Equation (2.9). The long-time behavior obeys here

$$A(t) \sim \frac{1}{2kt}. \tag{2.13}$$

Up to now we thus find as long-time behaviors either an exponential decay or an algebraic time-dependence. These two functional forms do not, by any means, exhaust the rich panoply of relaxation patterns found in glasses, which we will consider in the second part of this section. Let us also point out that increasing the number of reactions partners in Equation (2.1) does not yield readily other temporal dependences. One may envisage a general equation, with several reactants and different initial conditions. We will call minority species that reaction component or components, whose supplies are exhausted at the end of the reaction. Clearly, just before completion, the concentrations of the non-minority species practically do not change, and may be viewed as constant. For only one minority species present, its long-time decay is exponential, if its order of reaction is one, and is algebraic, if its order of reaction is n, $n \neq 1$, going as $t^{1/(1-n)}$ at longer times. If two minority species are present, say A and B, with orders n and m, the decay at longer times is always algebraic, going as $t^{1/(1-n-m)}$.

What is the reason behind the simple findings of exponential decays for pseudo-unimolecular and for $1/t$ decays for the strictly bimolecular kinetics? The basic assumption underlying the general kinetic scheme of Equation (2.2) is the 'well-stirred reactor' model in which all spatial dependences in the number of particles are neglected. Thus the use of Equation (2.2) implies a *homogeneous* spatial distribution of particles during the *whole* course of the reaction. That such an assumption is untenable in general form was discussed by us in previous works [23, 24]. There we pointed out that non-homogeneous conditions are widespread. First, the initial conditions may be already non-homogeneous (surface catalysis [25], pair-creation in irradiation tracks [26], energy transfer in doped systems [27, 28]). Second, even under homogeneous initial conditions, the *microscopic* reactions create by themselves non-homogeneities and enhance already existing density fluctuations. Diffusion can only partly wipe out such effects [24, 29, 30] but only when the diffusion length is large compared to the mean interparticle distance. At low particle densities, diffusion (or stirring) cannot create a homogeneous background.

These nonhomogeneity effects become even more accute, when random materials such as glasses are considered. The relaxation behavior in glasses may be arranged in the following way [31] the slower decays appearing lower in the list:

(I) $$\Phi(t) = \exp\left[-\left(\frac{t}{\tau}\right)^\alpha\right]. \tag{2.14}$$

This is for α between zero and one the so-called stretched exponential. This slower-than-exponential decay is known in the theory of relaxation processes in disordered media under the names of Kohlrausch [32] and of Williams and Watts [33] and shows up repeatedly at the glass transition (see Jäckle [34]). Several relaxation patterns in glasses display this ubiquitous behavior [35, 36]. The form has been used with much success in fitting dielectric [33], NMR [37] and dynamical light-scattering data [38–40]. Measurements by optical bleaching of the reversible transformation of spiropyran into merocyanine in a polymer display the stretched-exponential form, which is explained as arising from the frozen-in disorder of the system [41]. The same form is also present when monitoring abstraction reactions of hydrogen atoms from matrix molecules [42, 43]. Reactions of trapped hydrogen atoms in γ-irradiated sulfuric acid glasses have been shown [44] to be describable via time-dependent rate coefficients of the form $k(t) \sim t^{\alpha-1}$, which correspond to the decay law (2.14).

In the field of intermolecular energy transfer stretched-exponential behavior is widespread (Förster [45], Inokuti and Hirayama [46]). Thus using picosecond light pulses singlet–singlet energy transfer between rhodamine 6G and malachite green in a glycerol environment was analyzed by Rehm and Eisenthal. Their findings confirm that the Förster expression (which gives a stretched exponential) describes the transfer well [47]. The same decays are also found in the transfer of energy among rare earth ions in glasses and in mixed crystals [48, 49], on surfaces [50, 51], and even in porous glassy media, such as Vycors [52]. In the following sections, we will present several models which are related to Equation (2.14), and there we will again refer to the experimental situation.

(II) $$\Phi(t) = \exp\left[-C \ln^\beta\left(\frac{t}{\tau}\right)\right]. \tag{2.15}$$

This exponential—logarithmic relaxation pattern is slower than the stretched exponential but, for $\beta > 1$, more rapid than the algebraic one. Equation (2.15) is of considerable use in describing electron scavenging and electron—hole recombination, and it also appears in the analysis of relaxation phenomena related to hole-burning in glasses. In the next section we will show how Equation (2.15) arises naturally, when tunneling over energy barriers or transfer due to exchange interactions are involved. Experimentally, a large body of results concerning electron-transfer rates in glasses could be fitted to Equation (2.15). The techniques employed were pulsed radiolysis and optical detection, and the glasses used ranged from aqueous NaOH [53] to 2-methyltetrahydrofuran [54] for which dozens of different electron acceptors were investigated. Another example for Equation (2.15) is the recombination of free electrons from the conduction band to valence states [55]; in Ge_xSi_{1-x}:H the recombination is due to direct tunneling and follows an exponential—logarithmic behavior.

Furthermore, the direct energy transport also shows an exponential—logarithmic behavior [46, 56]. Recently Morgan and El-Sayed have found this behavior in 1,4-dibromonaphthalene [57]. Interestingly, the experiments show a crossover at longer times from an exponential-logarithmic form (due to *intra*chain interactions) to a stretched exponential (due to *inter*chain interactions).

For small values of C in Equation (2.15) one may expand the exponential in a series:

$$\Phi(t) \simeq 1 - C \ln^\beta \left(\frac{t}{\tau} \right). \tag{2.16}$$

The temporal range of validity of Equation (2.16) can be very wide if the parameter C is small; indeed, in such cases the differences between decay laws are hard to detect [31]. We observe that logarithmic dependences such as Equation (2.16) have been often reported. Thus the recovery after hole-burning of quinizarin embedded in ethanol—methanol glass follows a logarithmic pattern, as determined by Friedrich and Haarer using optical methods [58]. Equation (2.16) is related to differences in the local geometries of solvent cages in glasses, which lead to hyperbolic-type distributions of relaxation rates [58, 59].

(III) $\quad \Phi(t) \sim \left(\dfrac{t}{\tau} \right)^{-\gamma}. \tag{2.17}$

This is the prototype form of algebraic decays. Equation (2.17) is meaningful only at longer times, since it diverges for $t \to 0^+$. Thus one has to specify a certain cut-off (say τ), above which the expression is adequate. In general, the form arises from more complex patterns, such as $[1 + (t/\tau)]^{-\gamma}$, which are well behaved at $t = 0$. Equation (2.17) decays more slowly than the stretched exponential and than the exponential—logarithmic forms with $\beta > 1$. In fact (III) is a special case of (II) for $\beta = 1$ and $C = \gamma$. To obtain a form suitable for relaxation only $\gamma > 0$ is required in Equation (2.17), but the most interesting patterns occur for $0 < \gamma < 1$, where the transport displays long-time tails [5—9].

Forms related to Equation (2.17) have been reported for the relaxation processes of photogenerated carriers, which occur after electron—hole pair creation in amorphous Si : H [60]. In the field of energy transfer algebraic decays have been observed by Evesque and by Klymko and Kopelman in substitutionally mixed (protonated and deuterated) naphthalene crystals [28, 61, 62]. The transfer is monitored here via delayed fluorescence from a singlet, created by the annihilation of two triplet excitations. Another example for algebraic behavior is provided by the carrier diffusion in xerographic films, which shows algebraic long-time tails [7, 8]. A different application, which is mathematically similar, concerns the effect of chromatographic tailing [63], in which dissimilar chemical species display different trapping-release patterns during their hindered diffusion through chromatographic columns.

Thus, Equation (2.14) to (2.17) display forms which are not readily reconciled with the exponential or the $1/t$ decays which are found in chemical kinetics for pseudo-unimolecular and for bimolecular reactions. In the following sections we will systematize our knowledge of models which lead to Equations (2.14), (2.15) and (2.17). For this we will first focus on pseudo-unimolecular schemes in Sections 3 to 6, and will monitor deviations from exponentiality. Then, in Section 7 we will discuss bimolecular reactions and deviations from the $1/t$ behavior. As stressed, the decay forms of interest to us arise because of the presence of disorder. Thus the models are tailored corresponding to different aspects: one has fractals for geometric randomness, processes in continuous time for temporal and ultrametric spaces for energetic disorder.

3. A Parallel Relaxation Scheme: the Direct Transfer

In this section we begin our analysis of mechanisms that display non-exponential decay forms with a model in which the reaction involves a single dynamical step, as for instance found in the transfer of a trapped electron to a nearby scavenger. In the energy-transfer field this model is known as direct transfer: an excited donor molecule may transfer its excitation to several acceptor molecules surrounding it, which differ chemically or energetically from the donor. We stay in the framework of pseudo-unimolecular reactions, and we assume the number of donors to be considerably less than the number of acceptors. The disorder is introduced through the different positions which may be occupied by the acceptors; one averages the decay patterns over the different acceptor configurations, and follows thus a parallel [64] relaxation scheme.

3.1. REGULAR LATTICES WITH IMPURITIES

To fix the ideas we consider the energy transfer, and take the donor and acceptors to occupy substitutionally sites on a regular lattice. In this classical problem

[45, 46, 56, 65, 66] one considers an excited donor molecule, at position $\mathbf{r}_0$, surrounded by acceptors which occupy some of the sites $\mathbf{r}_i$ of a given structure. The transfer rates $w(\mathbf{r})$ depend on the mutual distance $\mathbf{r}$ between each donor–acceptor pair. Neglecting back-transfer, the probability of the decay of the donor due to the presence of an acceptor at $\mathbf{r}_i$ is thus

$$f(t, \mathbf{r}_i, \mathbf{r}_0) = \exp[-tw(\mathbf{r}_i - \mathbf{r}_0)]. \tag{3.1}$$

One assumes the acceptors to act independently, which means that they contribute multiplicatively to the decay. Let $g(j)$ be the probability of having j acceptors at one site. The decay of the donor, averaged over all possible distributions of acceptors around it, is given by

$$\Phi(t, \mathbf{r}_0) = \prod_{\mathbf{r}_i}{}' \left\{ \sum_j g(j) [f(t, \mathbf{r}_i, \mathbf{r}_0)]^j \right\}, \tag{3.2}$$

where the product extends over all structure sites with the exception of $\mathbf{r}_0$. Thus, for a binominal distribution, $g(j) = (1-p)\delta_{0,j} + p\delta_{1,j}$ ($\delta_{i,j}$ being the Kronecker delta) one has

$$\Phi(t, \mathbf{r}_0) = \prod_{\mathbf{r}_i}{}' \{1 - p + p \exp[-tw(\mathbf{r}_i - \mathbf{r}_0)]\}. \tag{3.3}$$

Equation (3.3) is exact, and thus theoretically very valuable. In the past, Equation (3.3) has been derived in several ways [67–70], but from the above procedure it is obvious that only combinatorial aspects matter. Basic ingredients for it are: (a) the exponential form of the decay of a single donor surrounded by acceptors, (b) the necessary average over all acceptor configurations, which mirrors the *parallel* relaxation of isolated donors embedded in these configurations, and (c) the fact that the acceptors act independently. Because of this generality, Equation (3.3) may be readily extended to several acceptor species [70] and allows us to treat anisotropic situations exactly [71, 72].

For a small acceptor concentration, $p \ll 1$, one may replace Equation (3.3) by

$$\Phi(t, \mathbf{r}_0) = \exp\left(-p \sum_{\mathbf{r}_i}{}' \{1 - \exp[-tw(\mathbf{r}_i - \mathbf{r}_0)]\}\right) \tag{3.4}$$

which is, in fact, the exact decay of Equation (3.2) under the Poisson law, $g(j) = e^{-p} p^j/j!$.

Usually, in treating energy transfer on infinite, regular lattices the dependence of $\Phi(t, \mathbf{r}_0)$ on $\mathbf{r}_0$ is irrelevant, since there all sites are equivalent. The situation is different in confined geometries and for irregular objects such as fractals, for which translational symmetry does not hold. In these cases the dependencce on $\mathbf{r}_0$ may be important, as will be demonstrated in the next subsections. Staying in the general case and introducing a site-density function $\rho(\mathbf{r})$ we can transform the sum in Equation (3.4) to the usual integral form. We set

$$\rho_0(\mathbf{r}) \equiv {\sum_{\mathbf{r}_i}}' \delta(\mathbf{r} - \mathbf{r}_i), \tag{3.5}$$

where the index 0 in $\rho_0(\mathbf{r})$ acts as a reminder that $\mathbf{r}_0$ is excluded from the sum of the right-hand side. With $\rho_0(\mathbf{r})$ one obtains

$$\check\Phi(t, \mathbf{r}_0) = \exp\left(-p \int d\mathbf{r}\, \rho_0(\mathbf{r})\{1 - \exp[-tw(\mathbf{r} - \mathbf{r}_0)]\}\right). \tag{3.6}$$

From Equation (3.6) one now arrives at the Förster-type decays by taking $\rho_0(\mathbf{r}) = \rho = \text{const}$ and extending the integration over the whole space:

$$\check\Phi(t) = \exp\left(-p\rho \int d\mathbf{r}\{1 - \exp[-tw(\mathbf{r})]\}\right). \tag{3.7}$$

To obtain decay patterns it is now only necessary to specify the interaction $w(\mathbf{r})$ and to perform the integration in Equation (3.7). Often encountered forms for $w(\mathbf{r})$ are:

$$w(\mathbf{r}) = ar^{-s} \tag{3.8}$$

for isotropic multipolar interactions [45], and

$$w(\mathbf{r}) = a\, e^{-\kappa r} \tag{3.9}$$

for isotropic exchange interactions [65]. The parameter s in Equation (3.8) equals 6 for dipole, 8 for dipole–quadrupole and 10 for quadrupole–quadrupole interactions, and the parameter κ in (3.9) is a measure of the range of the exchange interaction. Evidently, Equations (3.8) and (3.9) are only approximations to realistic transfer rates, which may have a considerably more complex structure [73]. Inserting now Equation (3.8) into Equation (3.7) we obtain

$$\check\Phi(t) = \exp(-At^{d/s}). \tag{3.10}$$

Here d is the spatial dimension of the underlying lattice and A is a time-independent constant

$$A = V_d p \rho \Gamma(1 - d/s) a^{d/s}, \tag{3.11}$$

where V_d is the volume of the unit sphere in d dimensions and $\Gamma(z)$ is the Euler-gamma function. One should note that Equation (3.10) is a stretched-exponential form, Equation (2.14) with $\alpha = d/s$. Here the stretched exponential is obtained for a parallel relaxation to acceptors, when the different configurations determine a hierarchy of distances [74].

Considering now the exchange interactions, insertion of Equation (3.9) into Equation (3.7) gives

$$\check\Phi(t) = \exp[-Bg_d(at)], \tag{3.12}$$

where B is again a time-independent constant

$$B = V_d \rho \kappa^{-d} \tag{3.13}$$

and $g_d(z)$ is an analytical function of z (see [75] for details). For longer times, $at \gg 1$, one has [75]

$$g_d(at) \simeq \ln^d(at) \tag{3.14}$$

and one obtains from Equation (3.12)

$$\tilde{\Phi}(t) \simeq \exp[-B \ln^d(at)]. \tag{3.15}$$

Expression (3.15) corresponds exactly to the exponential–logarithmic decay, Equation (2.15), with $C = B$ and $\beta = d$. This example shows how this reaction pattern obtains readily from a parallel reaction scheme.

It is perhaps worthwhile to stress that Equations (3.10) and (3.15) are non-analytical at $t = 0$. Evidently, this is due to the approximations involved in Equation (3.17). The exact expression for the decay, Equation (3.3), does not show any problems at $t = 0$, as amply discussed by us in [70] and [75]. Even in the continuum model of Equation (3.7) this situation (non-analiticity at $t = 0$) may be amended, by introducing a lower cut-off to the integral. This approach is best handled in the framework of geometrical restrictions, to which we now turn.

3.2. RESTRICTED GEOMETRIES

In this subsection we display the role of excluded volume effects and the role of confined geometries on the direct transfer. The main result here is the appearance of crossover effects in the decay patterns [76, 77].

It is convenient to rewrite Equation (3.6) as

$$\tilde{\Phi}(t, \mathbf{r}_0) = \exp[-pI(t, \mathbf{r}_0)] \tag{3.16}$$

by setting

$$I(t, \mathbf{r}_0) = \int d\mathbf{r} \, \rho_0(\mathbf{r}) \{1 - \exp[-tw(\mathbf{r} - \mathbf{r}_0)]\}. \tag{3.17}$$

We start with an infinite volume and consider the influence of a minimal distance b of approach between donor and acceptors [70]. Thus:

$$\rho_0(\mathbf{r}) = \begin{cases} 0 & \text{for } |\mathbf{r} - \mathbf{r}_0| < b \\ \rho & \text{otherwise.} \end{cases} \tag{3.18}$$

For simplicity we exemplify the situation in three dimensions where from (3.17) and (3.18) and for multipolar interactions, Equation (3.8), one has

$$I(t, \mathbf{r}_0) = 4\pi\rho \int_b^\infty dr \, r^2 [1 - \exp(-tar^{-s})] \equiv I(t), \tag{3.19}$$

which, because of the infinite volume, no longer depends on $\mathbf{r}_0$.

It is now a simple matter [70] to consider the behavior of $I(t)$ at short $t \ll b^s/a$ and at long $t \gg b^s/a$ times. At short times one may expand the exponential inside the integral, obtaining to first order in t:

$$\check{\Phi}(t) = \exp[-4\pi(s-3)^{-1}\rho_0 b^{3-s} at], \tag{3.20}$$

i.e. an exponential decay at short times. At long times the lower cut-off in (3.19) does not play any role, and one recovers the Förster decay in three dimensions:

$$\check{\Phi}(t) = \exp\left[-\left(\frac{4\pi}{3}\right)\rho_0 \Gamma\left(1 - \frac{3}{s}\right)(ta)^{3/s}\right]. \tag{3.21}$$

Both results are well borne out by the exact decay [70].

For exchange interactions the situation is similar; one finds an exponential decay at short times, followed by an exponential–logarithmic form at long times [71].

As next example, we consider the acceptors to be distributed in a sphere of radius R, and calculate for multipolar interactions the decay for several donor positions. Taking first the donor at the center of the sphere $\mathbf{r}_0 = \mathbf{0}$, one has:

$$\rho_0(\mathbf{r}) = \begin{cases} \rho & \text{for } r < R \\ 0 & \text{otherwise.} \end{cases} \tag{3.22}$$

Inserting Equations (3.22) and (3.8) into (3.17) one obtains

$$I(t, \mathbf{0}) = 4\pi\rho \int_0^R dr\, r^2 [1 - \exp(-tar^{-s})]. \tag{3.23}$$

Setting $x_0 = taR^{-s}$ it follows:

$$I(t, \mathbf{0}) = \left(\frac{4\pi\rho}{3}\right)(ta)^{3/s}\left[\Gamma(1-3s) + x_0^{-3/s}(1 - e^{-x_0}) - \int_0^{x_0} dx\, x^{-3/s}\, e^{-x}\right]. \tag{3.24}$$

For fixed R and at small times, $t \ll R^s/a$, and thus $x_0 \ll 1$, for which we again obtain the Förster-type behavior, the second and third terms in the brackets of Equation (3.24) being small. On the other hand, for large t, $t \gg R^s/a$, one finds directly from Equation (3.23)

$$I(t, \mathbf{0}) = \left(\frac{4\pi\rho}{3}\right)\rho R^3. \tag{3.25}$$

Accordingly, at long times,

$$\check{\Phi}(t, \mathbf{0}) = \exp\left[-p\rho\left(\frac{4\pi R^3}{3'}\right)\right]. \tag{3.26}$$

Consequently, the decay law shows a crossover behavior at intermediate times $t \sim R^s/a$ from the Förster-type form to a constant non-zero value.

To display this crossover, we present in Figure 1 the decay laws for several values of a, such that $w(\mathbf{R})$ varies between 0.1 and 10. The decays are plotted as $-\ln[-\ln \Phi]$ vs. $\ln t$. In these scales the short- and long-time forms appear as two different straight lines, with a crossover region between them, as is evident from Figure 1. All decays scale with ρ, and we have thus set $\rho = 1$.

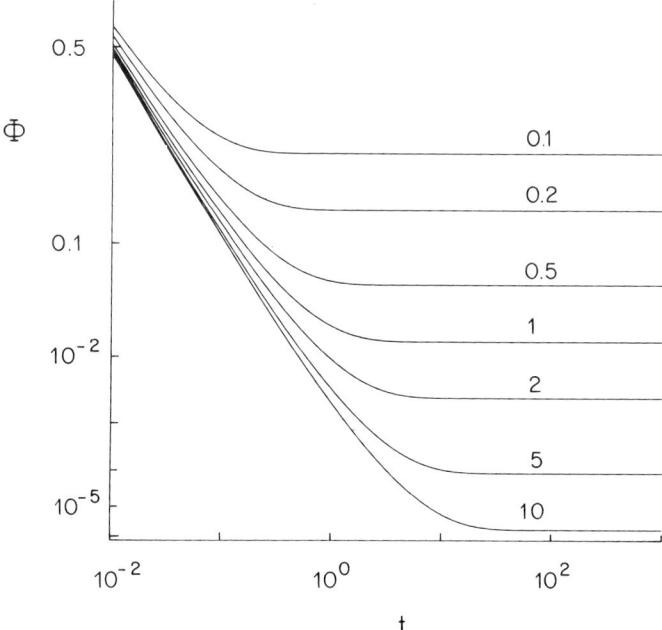

Fig. 1. Decay law Φ of an excited donor located in the center of a sphere of radius R. The sphere contains randomly distributed acceptors. The interactions are dipolar $w(\mathbf{r}) \sim r^{-6}$, and the decays are parametrized according to the value of $w(R)$ on the sphere, going from 0.1 to 10. Note the $-\ln[-\ln \Phi]$ vs $\ln t$ scales.

Assuming the donor to be placed on the surface of the sphere, say at $\mathbf{R}_0 = (0, 0, R)$, changes the situation so that at short times $I(t, \mathbf{R}_0)$ now differs from $I(t, 0)$ by a factor of 2 [76]:

$$I(t, \mathbf{R}_0) = \frac{I(t, \mathbf{0})}{2} = \left(\frac{2\pi\rho}{3}\right)(ta)^{3/s}\Gamma\left(1 - \frac{3}{s}\right). \quad (3.27)$$

This is due to the fact that at short times the donor sees only one-half of the space occupied by acceptors. The long-time behavior again follows Equation (3.26). Thus also here there is a crossover between a Förster-form at short times and a constant at longer times.

As a third type of decay which can be obtained by considerations on $\rho_0(\mathbf{r})$ we now treat the direct transfer to acceptors distributed in a cylinder of radius R. Let the donor be located on the *surface* of the cylinder at the origin $\mathbf{0} = (0, 0, 0)$. The axis of the cylinder lies at a distance R from the origin and points along the z-direction, and thus:

$$\rho_0(r)\,dr\,dz = 2r\rho \arccos\left(\frac{r}{2R}\right) dr\,dz. \qquad (3.28)$$

Inserting Equations (3.28) and (3.8) into Equation (3.17) we obtain

$$I(t, \mathbf{0}) = 4\int_0^{2R} dr \int_0^{\infty} dz\, r\rho \arccos\left(\frac{r}{2R}\right)\{1 - \exp[-ta(r^2 + z^2)^{-s/2}]\} \qquad (3.29)$$

In the cylindrical case, as in the spherical volume example, for short times $t \ll R^s/a$ one recovers the Förster-type decay, where the donor sees only half the space occupied by acceptors, Equation (3.27). For the long-time decay, $t \gg R^s/a$, Equation (3.29) yields the Förster decay corresponding to one dimension

$$\Phi(t, \mathbf{0}) = \exp\left[-\rho\rho 2\pi R^2 \Gamma\left(1 - \frac{1}{s}\right)(ta)^{1/s}\right]. \qquad (3.30)$$

Thus, the cylindrical geometry leads to a crossover between a three-dimensional and a one-dimensional Förster-type behavior. Here again the geometrical restriction induces a slower relaxation of the donor. The corresponding decays are given in Figure 2.

If the donor is placed on the *axis* of the cylinder then the crossover [76] is from the three-dimensional Förster-decay given by Equation (3.21) (full space) to Equation (3.30). Thus, as shown here and in the previous example, $\Phi(t, \mathbf{r}_0)$ may indeed depend, for restricted volumes, on the donor position $\mathbf{r}_0$. Hence, if the position of the donor is not fixed, an additional averaging procedure is required in order to relate to experimental data:

$$\Phi(t) = \langle \Phi(t, \mathbf{r}_0)\rangle_{\mathbf{r}_0} = \int d\mathbf{r}_0\, \tilde{\rho}_0(\mathbf{r}_0)\Phi(t, \mathbf{r}_0). \qquad (3.31)$$

In many simple cases, however, the qualitative pattern of the decay is not strongly changed by this average.

Summarizing this subsection we have shown deviations from Förster-type behavior for the direct energy transfer on restricted geometries. The decay laws show crossovers and a richer pattern than that which obtains for regular infinite lattices. In all the cases studied the smooth site-density function $\rho_0(\mathbf{r})$ accounts well for the geometrical restrictions which enter the direct energy transfer decay laws. While for the finite spherical volume the long-time decay Equations (3.21) and (3.27) crosses over to constant value (zero dimension), in the case of

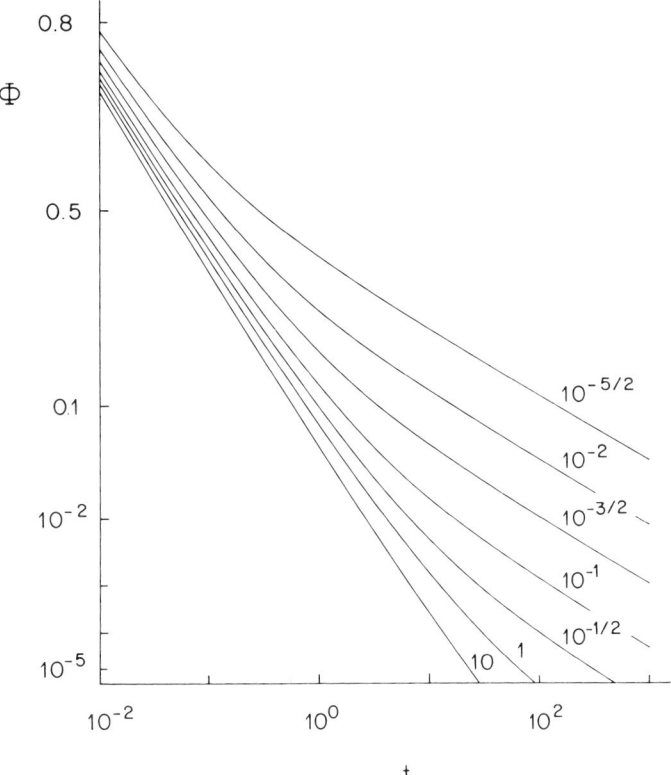

Fig. 2. Decay law Φ of an excited donor placed on the surface of a cylinder of radius R. The symbols are as in Figure 1.

cylinders the crossover is from three to one dimension, Equation (3.30). Extensions to other, more complex, situations are readily envisaged, by remarking that each decay stage mirrors faithfully the dimension of the space probed by the acceptors, which in that time regime contribute most to the decay. As we proceed to show, the decay patterns recover again a simple form, when the geometry scales with distance, a basic idea which leads us to the concept of fractals [3, 4].

3.3. FRACTALS

The term fractal was coined by Mandelbrot [78] to describe a wide class of objects which display scale-invariance: The patterns are self-similar, i.e. basically unchanged under dila(ta)tion operations. As topological objects, some fractals have come into mathematical scrutiny already early in this century [79, 80]. It is, undoubtedly, the merit of Mandelbrot, to have pointed out the ubiquitous presence of fractal patterns in Nature [3]. On physical grounds, the *dynamical* aspects of fractals have turned out to be very important, as stressed by Alexander and Orbach [81]. These aspects are connected to scaling and renormalization-group ideas [81—85]. Examples for fractals are linear and branched polymers [86—88], amorphous and porous materials [52, 89—91], epoxy resins [92], diffusion-limited aggregates (DLA) [93—95] and percolation clusters at criticality [15, 16, 95]. Reference [95] presents the 'state of the art' in the display of fractals in aggregation and gelation processes.

It is now of much importance to assess in how far glasses may be described by fractal concepts. In this subsection we discuss the decay laws which follow for the direct transfer on fractals [52, 96] after presenting a few selected fractal structures.

Perhaps the simplest fractal is the Sierpinski gasket [3, 79, 80] embedded in the two-dimensional space, as displayed in Figure 3. The structure can be generated

Fig. 3. Sierpinski gasket at the 6th iteration stage.

by a prescription which clearly renders the underlying symmetry. One starts from a triangle of sidelength 2, which includes three smaller, upwards-pointing triangles of sidelength 1. This basic pattern is called generator in the nomenclature introduced by Mandelbrot [3].

A dilatation by a factor of 2 from the upper corner transforms the upper small triangle into the larger one, and creates two additional, larger triangles. The procedure is then iterated M times, and leads to the Mth stage structure. Thus, the portion of the Sierpinski gasket depicted in Figure 3 is at the 6th stage.

Sierpinski gaskets can be generated in embedding spaces of arbitrary dimension d, by starting with the corresponding hypertetrahedrons [3, 97]. Also, related fractals of more general shape may be constructed along similar lines of thought [98].

Distinct from regular lattices, whose random-walk properties are mainly determined by their embedding spaces, properties of fractal lattices are describable through several, mostly non-integer parameters, which play roles similar to the dimension. For us here two major parameters are of concern: the fractal [3] (Hausdorff) dimension $\bar{d}$, which is related to the density of sites, and the spectral [85] (fracton [81]) dimension $\tilde{d}$, which appears in connection with dynamical properties (such as heat conduction, wave propagation and also reaction and diffusion) on fractals. Thus denoting by N the number of lattice points inside a sphere of radius R, one defines $\bar{d}$ as being $\lim_{R \to \infty} (\ln N/\ln R)$, i.e. one requires

$$N \sim R^{\bar{d}}. \tag{3.32}$$

Hence for Sierpinski gaskets embedded in d-dimensional spaces one has [3, 81]

$$\bar{d} = \frac{\ln(d+1)}{\ln 2}, \tag{3.33}$$

and furthermore, based on a scaling argument [81],

$$\tilde{d} = \frac{2\ln(d+1)}{\ln(d+3)}. \tag{3.34}$$

Therefore, for the Sierpinski gasket of Figure 3, $\bar{d} = 1.584$ and $\tilde{d} = 1.365$. In general, the relation $\tilde{d} \leq \bar{d} \leq d$ holds; furthermore, for Sierpinski gaskets $1 < \tilde{d} < 2$, as may be seen from Equation (3.34).

Not all fractal systems display the strong symmetry of the Sierpinski gaskets, which are 'deterministic' fractals [97], designated as such since for each point in space it is unambiguously clear whether it belongs to the structure or not. The distribution of points in a fractal pattern may be random, as exemplified by percolation clusters or DLA (*vide infra*). Nevertheless even for such structures one has scaling with distance according to Equation (3.32), i.e. $N \sim R^{\bar{d}}$. One calls such random, scaling patterns 'stochastic' or 'statistical' fractals [97, 99].

Two examples of stochastic fractals are given in Figures 4 and 5. Figure 4 shows the infinite cluster at the critical concentration for site percolation $p_c = 0.593$. The concentration p_c is that value of p at which for the first time a connected path through the whole (in theory, infinite) sample is achieved. All

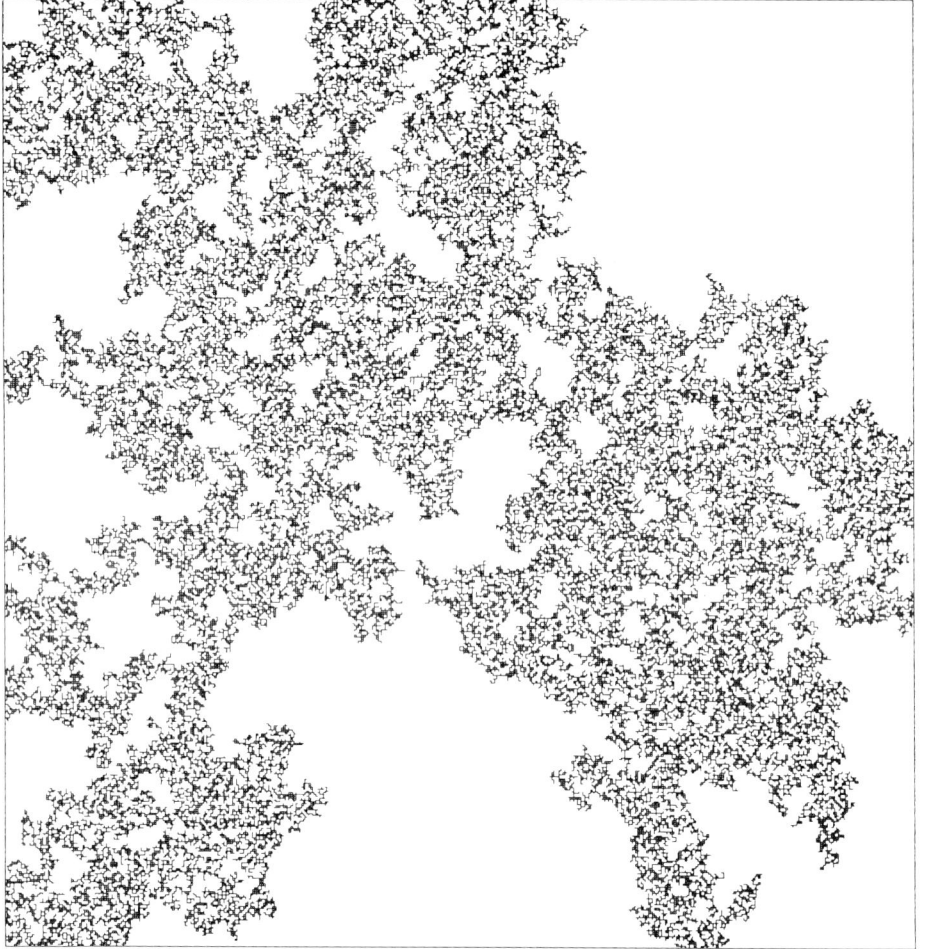

Fig. 4. Percolation cluster at the critical concentration for site percolation $p_c = 0.593$, on a 500 × 500 square lattice.

points connected via nearest-neighbor relations to this path form the displayed critical cluster. As a result of criticality, the cluster scales with distance and its fractal dimension $\bar{d}$ is around 1.9 [16].

Figure 5 shows a diffusion-limited aggregate (DLA), constructed by allowing particles to diffuse one-by-one to a centrally placed seed and letting them stick to the already formed object [93, 94]. Typical is the appearance of a tree-like shape which is due to the fact that latter arriving particles are less prone to reach the central region and thus attach themselves rather at the periphery of the cluster. This fact is borne out by the figure, in which different arrival times are denoted by different shading. Again, the structure is fractal, with $\bar{d} \simeq 1.67$.

For modeling purposes it is not always adequate to work with stochastic

Fig. 5. Diffusion-limited aggregate (DLA) on a 600 × 600 square lattice. The different shading shows stages in the evolution of the DLA.

fractals. Since, as mentioned, many properties of fractal objects depend only on $\bar{d}$ and $\tilde{d}$, it is advisable to find deterministic fractals whose dimensions come close to preset values. In fact, by changing both the dimension of the embedding space and the generator used, one may attain predesigned values with high accuracy [96, 97, 100].

In Figure 6 we display a fractal structure obtained by combining two species of triangular generators called B and C in reference [97]. When infinitely repeated, the structure leads to a fractal of dimensions $\bar{d} = 1.49$ and $\tilde{d} = 1.26$.

Fig. 6. Fractal structure obtained by combing two simple generators.

Figure 7 shows a diamond-shaped sponge in two-dimensions [3], again a fractal object with a dimension of $\bar{d} = 1.893$.

Figure 8(a) is a photograph of a model for the three-dimensional Sierpinski gasket, at the 4th iteration stage. Its dimensions are, from Equations (3.33) and (3.34), $\bar{d} = 2$ and $\tilde{d} = 1.547$.

Fig. 7. Diamond-shaped Sierpinski carpet.

Another possibility of constructing fractals of predetermined dimensions is the direct (set-) multiplication [3, 83, 84]. Multiplying a two-dimensional Sierpinski gasket with a linear chain leads to an object of dimensions $\bar{d} = 2.584$ and $\tilde{d} = 2.365$. For obvious reasons, the structure was named Toblerone [98, 101] and is shown in Figure 8(b).

Let us now focus on optical properties related to fractals. In subsection 3.1 we have seen that the Förster-type decays are determined by the spatial dimension d of the region accessible to acceptors, see Equations (3.10) and (3.15). In fact, dimensionality features are so essential that they impose crossover behaviours, as exemplified in subsection 3.2. To which dimension are the Förster-decays related

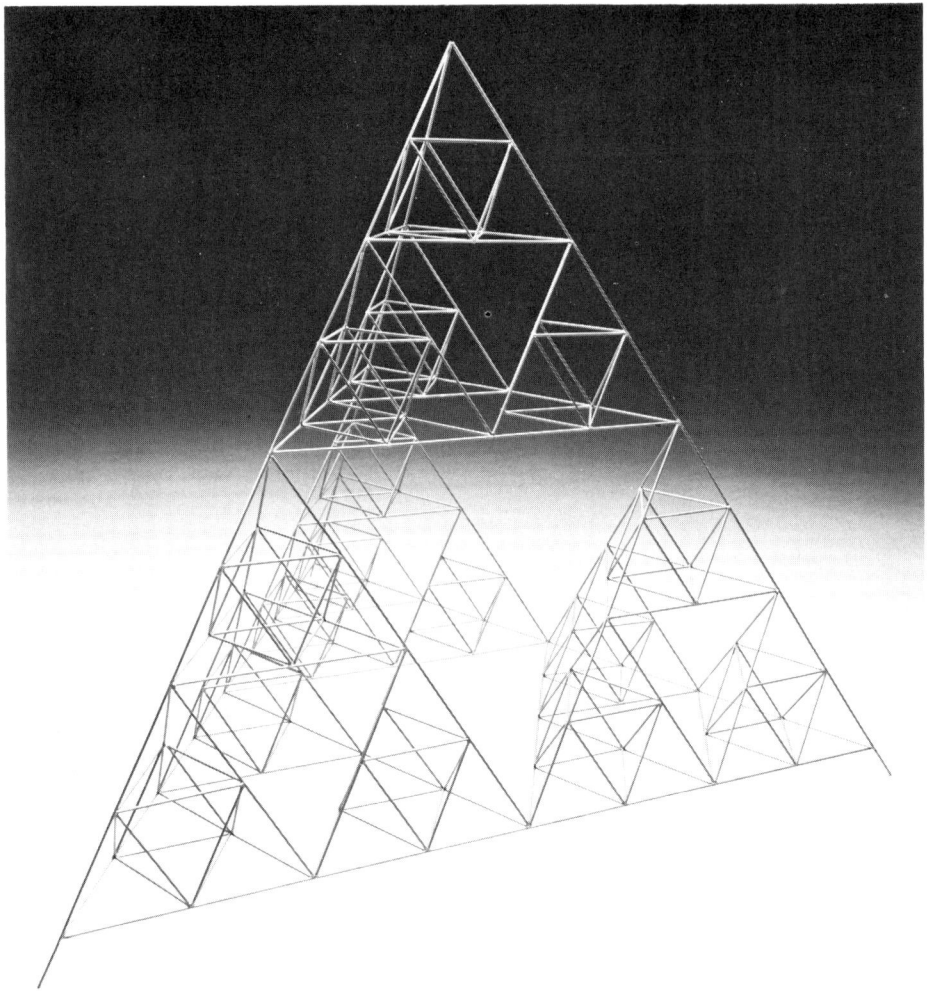

Fig. 8(a). Photograph of a three-dimensional model of a Sierpinski gasket.

when the accessible space happens to be fractal? This question was addressed in reference [96], whose line of thought we now follow.

The natural starting point is Equation (3.3), since it is exact and depends only on combinatorial arguments, see the discussion in subsection 3.1. For small acceptor concentrations the continuum approximation again leads to Equation (3.6) for which the density $\rho_0(\mathbf{r})$ has to be specified. Now, in determining the density a few remarks are in order. First, a fractal structure is not translationally symmetric, so that the decay may depend on the location of the donor. For relatively homogeneous fractals, such as the Sierpinski gaskets, this is not a very serious matter, since on such objects the local densities around the sites are quite

Fig. 8(b). The Toblerone fractal, the direct (set-) product of a two-dimensional Sierpinski gasket with a linear chain.

similar. Here we remark that in general a fractal does not fill its embedding space, so that the density $\rho_0(\mathbf{r})$ is a decreasing function of the distance. Second, $\rho_0(\mathbf{r})$ being a density, it must fulfill

$$N = \int_V \rho(\mathbf{r})\,d\mathbf{r}, \tag{3.35}$$

where N is the number of lattice sites in the volume V. Combining Equation (3.35) with Equation (3.32) one sees that for homogeneous fractals one has to a good approximation

$$\rho_0(\mathbf{r}) \simeq \hat{\rho} r^{\bar{d}-d}, \tag{3.36}$$

where $\bar{d}$ is the fractal dimension and $\hat{\rho}$ is a proportionality constant. Now Equation (3.36) no longer depends on the origin. Inserting it into Equation (3.6) one obtains

$$\tilde{\Phi}(t) = \exp\left(-p\hat{\rho} \int d\mathbf{r} \, r^{\bar{d}-d}\{1 - \exp[-tw(\mathbf{r})]\}\right). \tag{3.37}$$

For isotropic multipolar, (Equation (3.8)) and exchange (Equation (3.9)) interactions, the integration is immediate [96]. One obtains as long-time behaviors for multipolar-type interactions:

$$\tilde{\Phi}(t) = \exp(-\hat{A} t^{\bar{d}/s}) \tag{3.38}$$

as exemplified in Figure 9, and for exchange

$$\tilde{\Phi}(t) = \exp[-\hat{B} \ln^{\bar{d}}(at)], \tag{3.39}$$

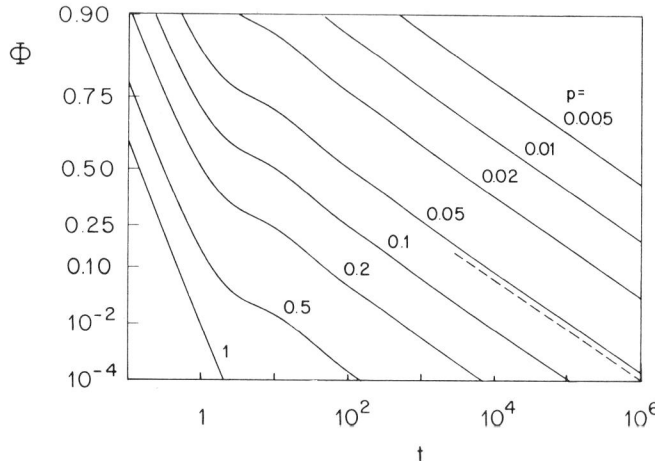

Fig. 9. Förster-type decay on a two-dimensional Sierpinski gasket for dipolar interactions $w(\mathbf{r}) \sim r^{-6}$ and various acceptor densities (indicated by p). The dashed line gives the slope of $\bar{d}/s = 0.264$.

where $\hat{A}$ and $\hat{B}$ are time-independent. Comparison of Equations (3.38) and (3.39) with Equations (3.10) resp. (3.15) shows that the only change in the Förster-type decays brought about by the fractal space consists in the replacement of d by the fractal dimension $\bar{d}$. Remarkably, the fractal spaces leave invariant the stretched-exponential and the logarithmic—exponential form of the Förster-decay laws. Due to their simple functional form Equations (3.38) and (3.39) could be very useful in the interpretation of experimental data, and they allow, for fractals, an efficient determination of $\bar{d}$. Following our approach recent measurements on energy transfer from rhodamine 6G to malachite green in a Vycor glass were interpreted as arising from a fractal-like arrangement of pores [52], whose dimension lies around $\bar{d} = 1.74$.

4. Parallel-Sequential Schemes: Random Walks

After having studied in the last section the relaxation behavior determined by direct transfer processes, we now turn our attention to such relaxation mechanisms, for which a series of steps (mostly randomly taken) is necessary for the completion of a reaction. Typical for such behavior are generalized diffusion models, under which random walks over discrete geometrical structures are very prominent [102—105]. We stay here in the framework of the pseudo-unimolecular reactions, in which we have a minority and a majority species, and we again monitor deviations of the relaxation pattern from exponentiality. In the simplest models one has one A and several B particles, and the A particle is annihilated at the encounter of a B particle. Depending on which of the species performs the motion one distinguishes between the trapping model [103—110] (only the A moves), the target (scavenging) model [111—114] (only the B moves) and the moving targets [115] (both species move). We monitor the survival probability of the A particle averaged over all possible realizations of particle distributions and motions. Because of this, the models have both sequential (random-walk) and parallel (ensemble average) characteristics [64]. We consider first random walks on regular lattices and extend then our treatment to fractals.

4.1. RANDOM WALKS ON REGULAR LATTICES

We start our discussion of reaction patterns with the trapping problem, and focus on the fate of a single A molecule which performs a random walk over the lattice. The B molecules are assumed to be distributed randomly, with probability p over the structure. For a finite, relatively large system of N_T sites, $p \simeq B_0/N_T$. For simplicity, we let at each step the walker move, with equal probability, to one of its neighboring sites, the extension to longer-ranged steps being discussed by us in [116] and [117]. The reaction is assumed to occur instantaneously when the A molecule lands on a site occupied by a B molecule. Thus the B molecules act as traps.

In order to obtain the survival law of the A molecules, an average over all random-walk realizations and all distributions of B molecules is needed. We proceed as in our previous works [23, 106]. For a particular realization of the random walk on the structure we let R_n denote the number of distinct sites visited in n steps, and set $R_0 = 1$. Here the starting point is irrelevant, but it may matter if fractals are considered. For the same realization of the walk we let $\tilde{F}_n$ denote the probability that trapping has not occurred up to the nth step:

$$\tilde{F}_n = (1-p)^{R_n-1} \tag{4.1}$$

assuming the origin of the walk not to be a trap. The measurable survival probability after n steps is Φ_n, the average of $\tilde{F}_n$ over all realizations of random walks [106, 108, 109] and over all starting points

$$\Phi_n = \langle \tilde{F}_n \rangle = \langle (1-p)^{R_n-1} \rangle \tag{4.2}$$

Introducing $\lambda = -\ln(1-p)$, Equation (4.2) turns into

$$\Phi_n = e^\lambda \langle e^{-\lambda R_n} \rangle \tag{4.3}$$

which allows the following expansion in the cumulants $K_{j,n}$ of the distribution of R_n:

$$\Phi_n = e^\lambda \exp\left[\sum_{j=1}^{\infty} K_{j,n} \frac{(-\lambda)^j}{j!}\right] \tag{4.4}$$

The first two cumulants are, for instance,

$$K_{1,n} = \langle R_n \rangle \equiv S_n \tag{4.5}$$

and

$$K_{2,n} = \langle R_n^2 \rangle - \langle R_n \rangle^2 \equiv \sigma_n^2, \tag{4.6}$$

where S_n and σ_n^2 denote the mean and the variance of R_n.

Apart from the case $d = 1$, for which both Φ_n and the distribution R_n are known, in general not much information is available on R_n. Truncating the sum in Equation (4.6) one obtains

$$\Phi_{N,n} \simeq e^\lambda \exp\left[\sum_{j=1}^{N} K_{j,n} \frac{(-\lambda)^j}{j!}\right] \tag{4.7}$$

The expression for $N = 1$

$$\Phi_{1,n} = \exp[-\lambda(S_n - 1)] \tag{4.8}$$

has been advanced in many areas [118–120]: In the random-walk field it corresponds to the first-passage-time approximation; in the fractal field it was recently used by de Gennes [120]. For $N = 2$ one obtains from Equation (4.7) the form

$$\Phi_{2,n} = \exp\left[-\lambda(S_n - 1) + \frac{\lambda^2 \sigma_n^2}{2}\right] \tag{4.9}$$

Now, for regular lattices one has for not too small n [103]:

$$d = 1: \quad S_n = a_1\sqrt{n} + a_2/\sqrt{n} + \cdots ; \sigma_n^2 \sim 4\left(\ln 2 - \frac{2}{\pi}\right)n \quad (4.10)$$

$$d = 2: \quad S_n = \frac{a_1 n}{\ln(a_2 n)} + \cdots ; \sigma_n^2 \sim \frac{n^2}{\ln^4 n} \quad (4.11)$$

$$d = 3: \quad S_n = a_1 n + a_2\sqrt{n} + \cdots ; \sigma_n^2 \sim n \ln n, \quad (4.12)$$

where the a_i are constants which depend on the lattice structure.

With these expressions for S_n and σ_n^2 we found for three-dimensional regular systems, that Equation (4.8) was a good approximation over the main portion of the decay. Furthermore, Equation (4.9), which includes the variance, approximates the survival probability of the A molecule very well over several orders of magnitude in its decay, both for three- and two-dimensional systems [106]. For the linear chain, Equation (7) converges very slowly, so that in order to describe two orders of magnitude in the decay, at least four terms ($N = 4$) in Equation (7) are needed [106].

We exemplify these findings in the figures. In Figure 10 we show the survival probability of a walker on a simple cubic lattice, the steps being taken to nearest-neighbor sites only [106]. The concentration of traps is as indicated in the figure and varies from $p = 0.01$ to $p = 0.5$. The dots are the result of the simulation, the dashed lines correspond to the Rosenstock form, $\Phi_{1,n}$, Equation (4.8), whereas the full lines are $\Phi_{2,n}$, Equation (4.9). The survival probabilities are practically exponential over four orders of magnitude in the decay. As stated, $\Phi_{1,n}$ is a good description of the simulated trapping form, whereas $\Phi_{2,n}$ is, in the range considered, indistinguishable from the simulation results.

In Figure 11 we present the decay behavior for trapping in $d = 2$. Here the walk takes place over a square lattice [106]. The survival probabilities are clearly non-exponential, which demonstrates the difference between random walks in $d = 2$ from walks in $d = 3$. Thus the expression $\Phi_{1,n}$ approximates Φ_n only qualitatively, while $\Phi_{2,n}$ is again, in the decay range of the figure, hardly distinguishable from the exact form.

The situation changes drastically in one dimension [106], as exemplified in Figure 12 for a trap density of $p = 0.01$. Here the dots again denote the decay law obtained by simulation, and several approximating forms, $\Phi_{N,n}$ ($N = 1, 2, 3$ and 4) are considered. Now $\Phi_{1,n}$ is qualitatively incorrect after less than one order in the decay, and one has to take into account the first four cumulants in order to obtain a description valid over two orders of magnitude.

Remarkably, in $d = 1$, one is in the fortunate position that an exact analytic solution for trapping is known [103, 121]. This expression allows to determine an asymptotic expansion for Φ_n, valid for large n, or, equivalently, for large t in continuous time [121–123]. The leading term of this expansion is [123]

$$\Phi_n \sim \lambda n^{1/2} \exp[-(3/2)\pi^{2/3}\lambda^{2/3}n^{1/3}]. \quad (4.13)$$

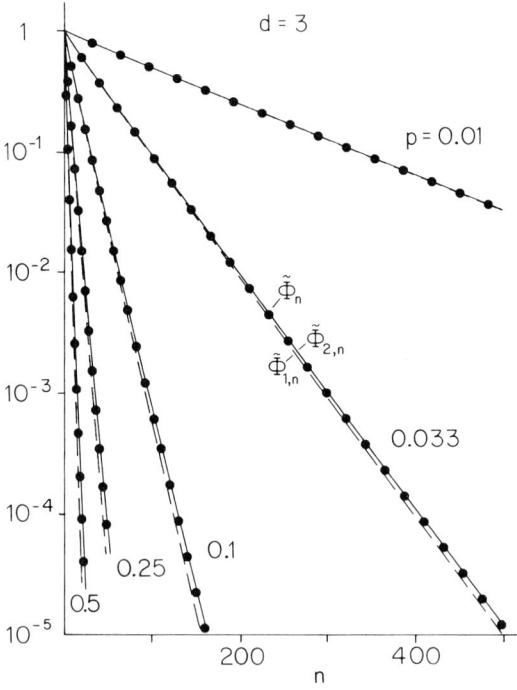

Fig. 10. The decay law due to trapping for nearest-neighbor walks on a simple cubic lattice for different trap concentrations. The number of steps is n. The dots denote the simulated decay $\tilde{\Phi}_n$ (Equation (4.3)); the dashed and the full lines are the $\tilde{\Phi}_{1,n}$ and $\tilde{\Phi}_{2,n}$ approximations (Equation (4.8) and (4.9), respectively). (Reproduced from Ref. 106)

This is a special case of a more general law, proven by Donsker and Varadhan [124] for arbitrary d:

$$\ln \tilde{\Phi}_n \sim -C\lambda^{2/(d+2)} n^{d/(d+2)}. \tag{4.14}$$

Equation (4.14) was discovered by several groups [29, 125, 126], and extended by us to the fractal domain [127]. The corresponding $n^{1/3}$-law [121] is indicated in Figure 12 as a full line, and it obviously fits the long-time data much better than $\tilde{\Phi}_{1,n}$. On the other hand, the convergence of the decay laws to the asymptotic Donsker–Varadhan form, Equation (4.14), is in general quite slow [123, 127–129], so that this form is more of theoretical than of practical interest [127].

Given the difficulties which arise when considering the trapping problem analytically, it is interesting to note that the target annihilation admits an exact solution [114]. As a reminder, in the latter problem, the majority species moves, and the molecules belonging to the minority species act as static targets. We assume the B molecules not to interact with each other, and we focus on the fate of a single A particle, taken to be at the origin of the coordinate system. Several possibilities may now be envisaged in creating the initial distribution of B particles on the remaining lattice sites. In [114] and [130] we have presented the survival

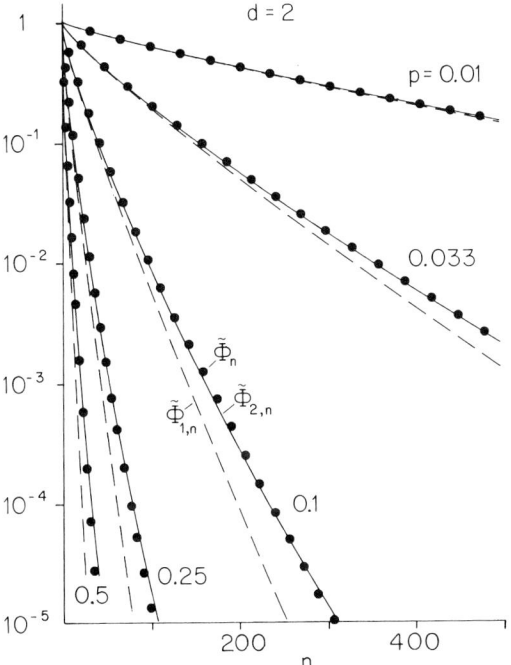

Fig. 11. The decay law Φ for nearest-neighbor walks on a square lattice for various trap concentrations p. The symbols are as in Figure 10. (Reproduced from Ref. 106)

probabilities for B particles which follow binomial, multinomial and Poisson distributions. Here we will exemplify the procedure using the last one, for which the occupancy of a site is taken to be distributed as

$$g(j) = e^{-p} \frac{p^j}{j!}, \qquad (4.15)$$

where $g(j)$ is the normalized probability of having j B particles at one site and p is their average number density.

We now denote by $F_m(\mathbf{r})$ the probability that a random walker starting from $\mathbf{r}$ reaches the origin $\mathbf{O}$ for the first time in the mth step. For regular lattices, because of the symmetry of the walk $F_m(\mathbf{r})$ is also the first-passage time from $\mathbf{O}$ to $\mathbf{r}$, as defined by Montroll and Weiss [131]. In general, we call $H_n(\mathbf{r})$ the probability that a first passage from $\mathbf{r}$ to $\mathbf{O}$ occurred in the first n steps and have:

$$H_n(\mathbf{r}) \equiv \sum_{m=1}^{n} F_m(\mathbf{r}). \qquad (4.16)$$

The probability therefore that a walker from $\mathbf{r}$ did *not* reach $\mathbf{O}$ in the first n steps

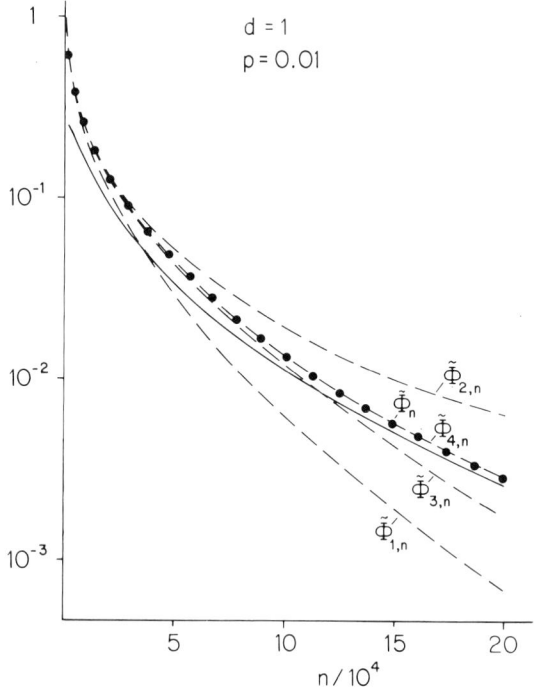

Fig. 12. The decay law Φ for nearest-neighbor random walks on a linear chain for a trap density $p = 0.01$. The dots denote the simulated decay and the dashed lines are the $\tilde{\Phi}_{N,n}$ approximations ($N = 1, 2, 3$ and 4). The full line is the asymptotic form. (Reproduced from Ref. 106)

is thus

$$\Phi_n(\mathbf{r}) = 1 - H_n(\mathbf{r}) \tag{4.17}$$

Using Equation (4.17) we obtain the survival probability of the A molecule by appropriately weighting products of the $\Phi_n(\mathbf{r})$ functions:

$$\Phi_n = \prod_{\mathbf{r}}{}' \left\{ \sum_j g(j) \left[\Phi_n(\mathbf{r})\right]^j \right\}. \tag{4.18}$$

Here the product extends over all structure sites, with the exception of the origin. Inserting now Equations (4.15) and (4.17) into Equation (4.18) leads to

$$\Phi_n = \prod_{\mathbf{r}}{}' \left\{ \sum_j e^{-p} \frac{[p\Phi_n(\mathbf{r})]^j}{j!} \right\}$$

$$= \prod_{\mathbf{r}}{}' \exp[-p + p\Phi_n(\mathbf{r})] = \exp\left[-p \sum_{\mathbf{r}}{}' H_n(\mathbf{r})\right] \tag{4.19}$$

For regular lattices Equation (4.19) may be further simplified since, according to Equations (III.2) and (III.3) of Reference [131], one has

$$\sum_{\mathbf{r}}' F_m(\mathbf{r}) = S_m - S_{m-1} \quad (m > 1). \tag{4.20}$$

Here again S_m is the mean number of distinct sites visited by a random walker in m steps, and $S_0 = 1$. Equation (4.20) may also be derived directly, by noting that the increase of S_m is given by the *total* number of new sites visited in the mth step. Introducing now Equations (4.16) and (4.20) into (4.19), one has *exactly*

$$\Phi_n = \exp[-p(S_n - 1)]. \tag{4.21}$$

We note that Equation (4.21) corresponds to the Rosenstock approximation of the trapping problem, $\check{\Phi}_{1,n}$ in Equation (4.8) with λ replaced by p. Again S_n is given by Equations (4.10) to (4.12) for regular lattices. The long-time behavior of Equation (4.21) is now exponential in three dimensions, but remains non-exponential for $d = 1$ and $d = 2$. It is interesting to see that for $d = 1$ a Williams–Watts relaxation pattern, Equation (2.14), emerges, with $\alpha = \frac{1}{2}$. This result is identical to former findings for the target problem, which used the Glarum model of dipolar relaxation in glasses as experimental application [112, 113, 132–135]. As long as only random walks on regular lattices are envisaged, it is not obvious how to extend the target model to obtain non-trivial Williams–Watts forms with other α values ($\alpha \neq \frac{1}{2}$, $\alpha \neq 1$). In this and the next section we will show that general α values are the rule when fractals, CTRW or ultrametric structures are considered.

To conclude this presentation of target annihilation on regular lattices we stress that the long-time behaviors of the trapping and of the target problems are different. Whereas the target annihilation follows S_n, see Equation (4.21), the trapping decay contains all the higher moments of the R_n distribution, see Equation (4.7), and tends at very long times towards the asymptotic form (4.14). Thus the relaxation patterns differ, depending on which of the two species (A or B) is the mobile one. One has thus a counter-example to the view that in reaction kinetics only the relative motion of the species, but not the individual movements are important. Truly, the solution of a diffusion equation with a sink term leads to time-dependences such as given by Equation (4.21), but these may be only the leading terms of a complex random-walk expression. Here one encounters the fact that a random-walk problem is not necessarily reducible to simple diffusion.

As a proof we display in Figure 13 the decay of A excitations distributed on a surface (square grid). To simulate a realistic situation, we also allow both the A and B species to move, and distribute the mobilities $m_A : m_B$ according to $0 : 1$, $0.5 : 0.5$, $0.75 : 0.25$, $0.9 : 0.1$ and $1 : 0$. The special cases $0 : 1$ and $1 : 0$ correspond to the target and to the trapping problems, respectively, between which the intermediate $m_A : m_B$ interpolate. Hence, in general, one has mobile traps [115]. We obtain the decay Φ_n^A of the A species as a function of the number of steps n, which is distributed according to $m_A : m_B$ over the A and B moieties. Thus, for

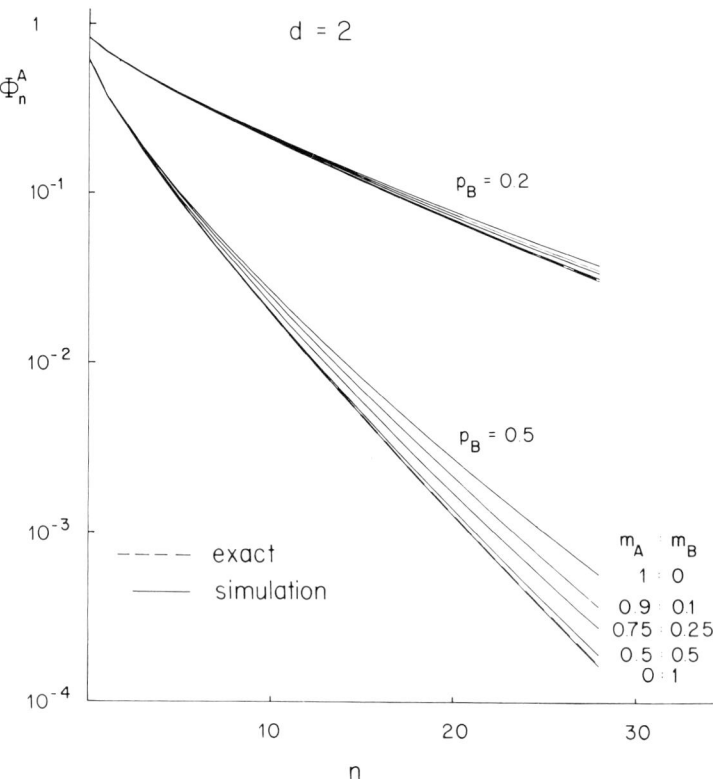

Fig. 13. The decay law Φ_n^A for the A species. Here the A and the B species move on a square lattice. The mobilities are distributed according to $m_A : m_B = 0:1$, $0.5:0.5$, $0.75:0.25$, $0.9:0.1$ and $1:0$. The particular cases $0:1$ and $1:0$ correspond to the target and the trapping problems, respectively; the dashed line is the exact result for the target problem. The concentrations of the B species are $p_B = 0.2$ and 0.5.

each initial concentration p_B of the B species (we chose $p_B = 0.2$ and $p_B = 0.5$) the diffusion picture based on $D_A + D_B$ predicts the same decay, in contrast to the findings of Figure 13. To push the point still further, we show in Figure 14 the corresponding decays for A and B species distributed on a chain ($d = 1$), with $m_A : m_B$ being $0:1$, $0.5:0.5$, $0.75:0.25$, $0.9:0.1$ and $1:0$. Here not only the target but also the trapping solutions are known exactly, as indicated in the figure. Again, the decays are distinct.

After this analysis of simple random walks on regular lattices, we turn in the following subsection our attention to random walks on fractals.

4.2. Random Walks on Fractals

As discussed in the previuos subsection, dynamical models such as the target annihilation show restrictions in the possible range of decay patterns due to

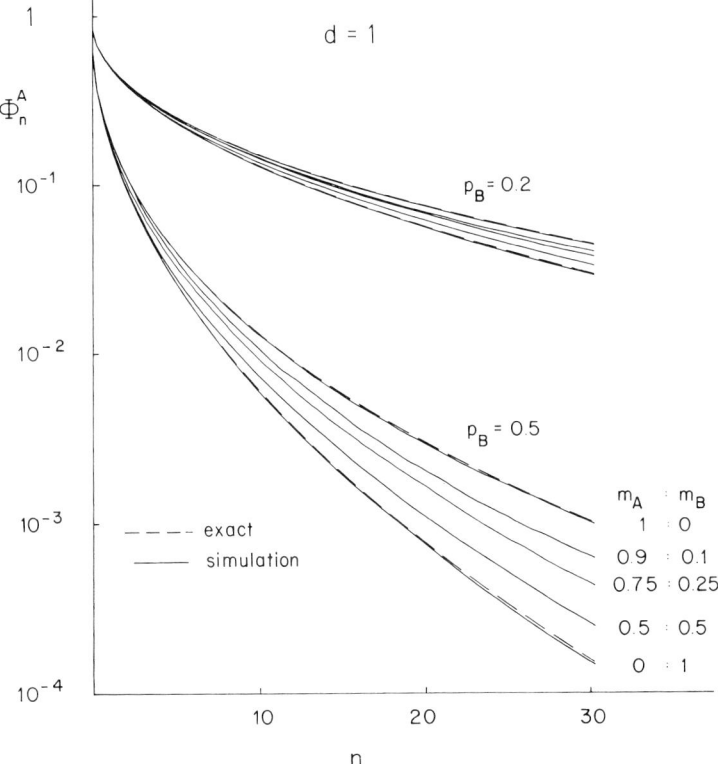

Fig. 14. Same as Figure 13 for the linear chain. The exact results for the target and the trapping problem are given by dashed lines.

limitations of the dynamics to regular lattices. Fractal lattices which are considerably more varied, allow now straightforward extensions of the models considered.

The use of fractals as an underlying space for dynamical processes has been stressed by Alexander and Orbach in a by-now classic paper [81]. A fundamental point which emerged from their analysis is that the density of states of a fractal is connected to an additional parameter, called 'fracton' [81] or spectral [85] dimension $\tilde{d}$, which is in general distinct from d. Interesting, the same quantity has been remarked also in works based on renormalization-group ideas, where it was called 'effective dimensionality' [83, 84]. The dichotomy between d and $\tilde{d}$ is a new aspect for fractals [81], an aspect not found in regular lattices.

The analysis of Alexander and Orbach showed that $\tilde{d}$ is determined by the connectivity of the lattice and that this dimension determines the relevant Hilbert space for the Laplacian equation on the fractal [81]. Thus a whole series of physical problems involve the spectral dimension $\tilde{d}$: diffusion equations for mass and heat transport on fractals, random walks, normal modes of RC circuits and of bead and spring models, and solutions to the Schrödinger equation [81, 85, 137, 138]. We have to remark, however, that fractals are not completely characterized by $\bar{d}$ and $\tilde{d}$, so that for special purposes other parameters have to be introduced: Stanley has listed nine 'extrinsic' fractal dimensions, from which additional 'intrinsic' ones may be computed [99]. Nevertheless, many of these dimensions are rather special, and some do not appear in the main classes of deterministic fractals in current use. In the following we are concerned only with $\bar{d}$ and $\tilde{d}$, which we view as fundamental in applications.

A large body of investigations in disordered media centers on probing the density of normal modes of the materials. Thus Alexander, Entin-Wohlman and Orbach have calculated the interaction between localized electron states and localized vibrational states on fractal structures [138]. In their analysis, in which the energy conservation requirements are observed, they find that the relaxation obeys an exponential—logarithmic decay, Equation (2.15), similar to the Inokuti—Hirayama form [23, 46]. In a following paper [139] they have investigated the spin-relaxation rate in a fractal network, for which two localized modes (fractons) are dominant. The decay patterns show crossovers from stretched exponentials, Equation (2.14) to exponential—logarithmic forms, Equation (2.15).

Stretched-exponential forms appear naturally when the relaxation is due to random-walk motions on fractals. We demonstrate this point by extending the trapping and the target problems of the last subsection to underlying fractal spaces.

As in 4.1 we start with the trapping case, in which the minority species moves. It is now a simple matter to verify that Equations (4.1) to (4.9), which are very general, hold also for fractals, provided that the steps occur with a fixed rate. This requirement is clearly fulfilled for Sierpinski gaskets, for which the coordination number is the same over the whole structure. For other structures, such as symmetric fractals, in which this number varies, one has then to weight the intersite transition rates [97, 98], in order to fulfill this requirement. Furthermore, because of the lack of translational symmetry, the distribution of sites visited, R_n, depends on the origin of the walk. Hence, in Equation (4.3) the average must be taken over both initial sites *and* realizations of the walk, so that an additional average is required in Equations (4.4) and (4.7) to (4.9). Numerical simulations have convinced us that for Sierpinski gaskets, which are fairly homogeneous, this additional averaging — which parallels Equation (3.3) — is not fundamental, so that we include it at the level of the cumulants, Equations (4.5) and (4.6).

In [140] and [141] we have simulated a series of random walks on Sierpinski gaskets embedded in Euclidean spaces of dimensions $d = 2, 3, 4$ and 6, the spectral dimensions being thus from Equation (3.34), $\tilde{d} = 1.36, 1.55, 1.65$ and 1.77, respectively. For the first two moments, S_n and σ_n^2 we found that the relations

$$S_n \simeq an^{\tilde{d}/2} \quad (\tilde{d} < 2) \tag{4.22}$$

and

$$\sigma_n^2 \simeq bn^{\tilde{d}} \quad (\tilde{d} < 2) \tag{4.23}$$

were well obeyed in the range investigated (see Figures 1 and 2 of reference

[141]). The relations:

$$S_n \sim \begin{cases} n^{\tilde{d}/2} & (\tilde{d} < 2) \\ n & (\tilde{d} > 2) \end{cases} \quad \begin{array}{l}(4.24a)\\(4.24b)\end{array}$$

were previously inferred [85] from an argument using the probability of returning to the origin and the compact or non-compact nature [120] of the underlying lattice. On the other hand, information on the higher cumulants is difficult to obtain analytically; from our results on scaling of relaxation patterns on Sierpinski gaskets [128], we expect that the higher cumulants in Equation (4.7) may obey

$$K_{j,n} \sim n^{j\tilde{d}/2} \quad (\tilde{d} < 2), \tag{4.25}$$

a relation which holds [103] in $d = 1$, and of which (4.23) is a special case.

The decay laws $\check{\Phi}_n$ for trapping [140] are given in Figures 15 and 16 for walkers on the Sierpinski gaskets of spectral dimensions $\tilde{d} = 1.36$ and 1.55, i.e. embedded in the two- and three-dimensional Euclidean spaces, respectively. Several trap concentrations are investigated. The full curves correspond to the exact decays, $\check{\Phi}_n$, obtained by performing the average according to Equation

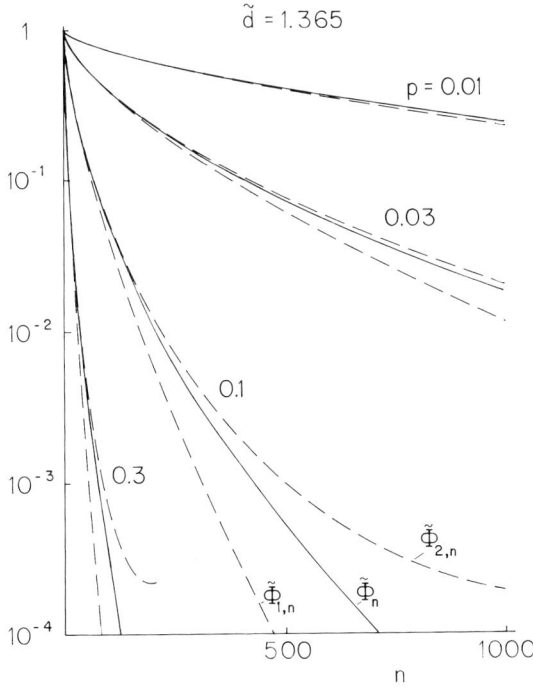

Fig. 15. Decay law due to trapping for random walks on the two-dimensional Sierpinski gasket ($\tilde{d} = 1.365$) where p is the trap concentration. The full lines denote the simulated decay $\check{\Phi}_n$, whereas the dashed lines are the $\check{\Phi}_{1,n}$ and $\check{\Phi}_{2,n}$ approximations. (Reproduced from Ref. 140)

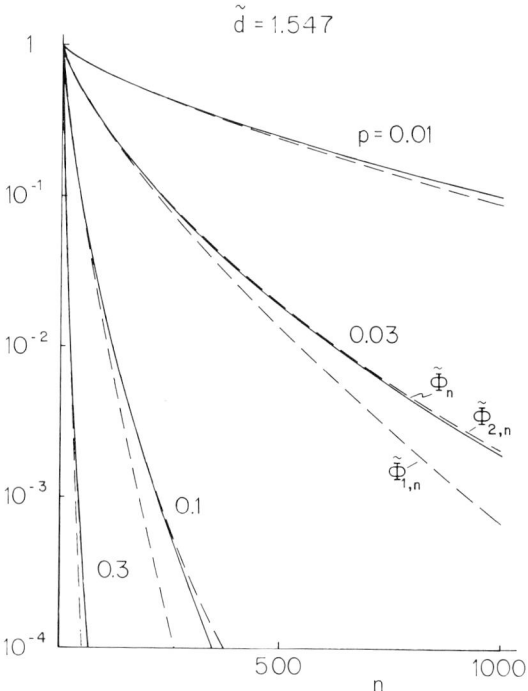

Fig. 16. Decay law due to trapping for random walks on a three-dimensional Sierpinski gasket ($\tilde{d} = 1.547$). The symbols are as in Figure 15. (Reproduced from Ref. 140)

(4.2). The dashed lines are the approximating forms $\check{\Phi}_{1,n}$ and $\check{\Phi}_{2,n}$, Equations (4.8) and (4.9). For all trap concentrations the decays are clearly non-exponential, following for not too-long times

$$\check{\Phi}_{1,n} = \exp(-\lambda n^{\tilde{d}/2}), \qquad (4.26)$$

i.e. a Williams–Watts form. For both gaskets we find a behavior much reminiscent of the square lattice, where $\check{\Phi}_{1,n}$ is only qualitative after an order of magnitude of the decay, whereas the inclusion of the variance in $\check{\Phi}_{2,n}$ considerably improves the agreement with Φ_n. Understandably, in view of the interpolating nature of the spectral dimension, the agreement between $\check{\Phi}_{N,n}$ and Φ_n improves when $\tilde{d}$ is larger.

The long-time behavior of the decay due to trapping on fractals follows a modified Donsker–Varadhan form [124] Equation (4.14):

$$\Phi_n \sim \exp[-C\lambda^{2/(\tilde{d}+2)} n^{\tilde{d}/(\tilde{d}+2)}], \qquad (4.27)$$

where again $\tilde{d}$ (but not $\bar{d}$) enters. We have derived [127] Equation (4.27) from an argument due to Lifschitz [29, 142] and using the concept of compact visitation [120] in $\tilde{d} < 2$. Note the stretched-exponential form of Equation (4.27), with $\alpha = \tilde{d}/(\tilde{d} + 2)$. From our simulations [127] we had to conclude that this slope is probably well outside the experimental range, so that the value of Equation (4.27) is purely theoretical. On the other hand, the slope of $\alpha = \tilde{d}/2$ found from Equation (4.26) should be amenable to experimental observation.

The same Williams—Watts slope, $\alpha = \tilde{d}/2$ follows also for the relaxation behavior in the target problem. The extension of Equations (4.16) to (4.21) for target annihilation due to random walkers on fractals is straightforward, and the same *provisos* as in the trapping problem apply. For Sierpinski gaskets the average over the target positions is a minor problem, and one has, to a very good approximation,

$$\Phi_n \simeq \exp[-pn^{\tilde{d}/2}] \quad (\tilde{d} < 2) \tag{4.28}$$

and

$$\Phi_n \simeq \exp[-pn] \quad (\tilde{d} > 2). \tag{4.29}$$

One may note the Williams—Watts form of Equation (4.28) and the fact that now α may be varied from $\frac{1}{2}$ to unity by a judicious choice of the Sierpinski (or Sierpinski-type) gasket [97, 98] underlying a walk. Hence fractals admit a straightforward extension of the Glarum model [132], discussed in the previous subsection. For $\tilde{d} < 2$, Equation (4.28) also provides a way of experimentally determining the spectral dimension of fractal objects, by monitoring the relaxation due to multiple-step mechanisms in such materials [96, 114].

Summarizing this section, we have shown that random walks on regular lattices can lead to stretched-exponential decays. These findings are even more pronounced for random walks on fractals which extend in a natural way former results to continuously varying dimensions. Different from the one-step parallel relaxation of Section 3, here the basic quantity is the spectral (fracton) dimension $\tilde{d}$ and not the fractal dimension $\bar{d}$.

5. Continuous-Time Random Walks (CTRW)

In continuous-time processes one relaxes the condition that changes, such as the steps of a walker, may occur only at preassigned times. In the CTRW formalism one has thus to introduce a waiting-time distribution $\psi(t)$, which gives the probability density that the time between steps equals t. This method has been implemented in a classic work by Montroll and Weiss [131], who used it to include $\psi(t)$ in the generating-function formalisms for random walks. The approach was elaborated later by Scher and Lax [5, 6] who applied the CTRW to transport in disordered systems, by generalizing the functions $\psi(t)$ to forms $\psi(\mathbf{r}, t)$ which couple the spatial and temporal disorder aspects. The use of a decoupled scheme, i.e. of $\psi(\mathbf{r}, t) = f(\mathbf{r})\psi(t)$ was also proposed by Scher and Montroll [7]. The connection of the CTRW to the master equation, which is the basic equation to describe the transport of a particle in a random medium, was discussed by Bedeaux *et al.* [143] and the connection to the generalized master equation was pointed out by Kenkre *et al.* [144] and by Kenkre and Knox [145]. This later connection was studied only in a decoupled scheme, which corresponds to the above-mentioned Scher—Montroll approximation [7]. Then Klafter and Silbey [146, 147] proved that the original random walk problem on a random lattice can be *rigorously* reduced to a CTRW formulation of the Scher—Lax type. The derivation makes use of projection operator techniques, and $\psi(\mathbf{r}, t)$ is related to a self-energy (or a memory) function. However, the calculation of the self-energy is difficult and one has to look for the best possible approximations.

In more recent works [27, 148] we have discussed ways to approximately obtain $\psi(t)$ for given disordered systems, and we related these functions to temporal decays obtained from parallel processes, such as analyzed here in Section 3. In [27] and [148] we were interested in modeling a given disordered situation as closely as possible. Here we take a somewhat different point of view, in that we start by considering possible $\psi(t)$ forms and discuss which influence these $\psi(t)$ distributions have on random walks on regular lattices and on fractals.

Let us begin by listing a few $\psi(t)$ forms of wide use. The simplest situation obtains for a memoryless process in which the probability of remaining at a given site during the time interval from 0 to t is exponential:

$$\Psi(t) = e^{-bt}. \tag{5.1}$$

If now $\psi(t)$ is the probability density that an event occurs at time t after the previous event has taken place, then obviously

$$\psi(t) = -\frac{d}{dt}\Psi(t) \tag{5.2}$$

and Equation (5.1) leads to the Poisson process

$$\psi^p(t) = b\exp(-bt). \tag{5.3}$$

Slightly more complex forms obtain by using for $\Psi(t)$ Equations (3.10) or (3.15), in order to mimic a transfer due to multipolar or exchange interactions in the presence of substitutional disorder [5, 6, 27, 148]. The corresponding $\psi(t)$ are:

$$\psi^m(t) = \left(\frac{Ad}{s}\right) t^{d/s-1} \exp(-At^{d/s}) \tag{5.4}$$

and

$$\psi^x(t) = \left(\frac{Bd}{t}\right) \ln^{d-1}(at) \exp[-B\ln^d(at)], \quad at > 1 \tag{5.5}$$

We note that Equation (5.4) was also advanced by Ngai and Liu [149] in their CTRW treatment which incorporated low-lying fluctuations for amorphous media.

The waiting-time distributions $\psi^p(t)$, $\psi^m(t)$, and also $\psi^x(t)$ for $d > 1$, are well-behaved in that, for them, the first moments τ_1:

$$\tau_1 = \int_0^\infty t\psi(t)\,dt \tag{5.6}$$

are finite. The distributions $\psi^p(t)$, $\psi^m(t)$ and $\psi^x(t)$ show thus an intrinsic time scale, namely τ_1. As will become evident in the following, many properties of random walks with fixed waiting times will translate to continuous time via $t = n\tau_1$, where n is the number of steps.

Much more interesting are distributions for which the integral in Equation (5.6) is divergent. Thus Scher and Montroll [7] have modeled transport in amorphous media though a $\psi(t)$ which displays a long-time tail:

$$\psi(t) \sim \frac{1}{t^{1+\gamma}}, \quad 0 < \gamma < 1. \tag{5.7}$$

This form was obtained by fitting Equation (5.5) over a large domain to an algebraic form. Interestingly, such an expression is intimately related to a fractal set of event times [9, 150, 151]. From the Poisson process, Equation (5.3), one constructs readily a dilatationally symmetric distribution, by taking into account events occurring on all time scales, in the following way:

$$\psi(t) = \frac{1-N}{N} \sum_{j=1}^\infty N^j b^j \exp(-tb^j), \tag{5.8}$$

where $N < 1$. As is evident, the distribution (5.8) is a normalized sum of Poisson terms and

$$\psi(bt) = \frac{\psi(t)}{(Nb)} - \frac{(1-N)}{N} \exp(-tb) \tag{5.9}$$

For later applications we need $b < N$, so that $b < 1$ and thus at longer times $\psi(bt) = \psi(t)/Nb$. The last expression is equivalent to Equation (5.7) when $\gamma = \ln N/\ln b$ is set. Equation (5.7) shows directly the temporal *scaling* of $\psi(t)$, i.e. its fractal nature in time.

We note that Equation (5.7) obtains readily from the distribution of carrier release times from low-lying traps to the conduction band [8]. This distribution is fundamental in the multiple trapping (MT) formalism. For activated processes the rates depend exponentially on the energy, so that an equidistant level spacing $E_j = j\Delta$ leads to rates proportional to $\exp(-E_j/kT) = b^j$, with $b = \exp(-\Delta/kT)$. Furthermore, the density of states in the energy tail is often itself exponential in energy, $\exp(-E_j/kT_0)$ (where one introduces an effective temperature T_0) so that the density of states follows N^j with $N = \exp(-\Delta/kT_0)$. Thus, in this example $\gamma = \ln N/\ln b = T/T_0$ for $T < T_0$, and one has dispersive transport below T_0.

The form of Equation (5.7) is ill-defined at short times, which is inconvenient, when integrals have to be evaluated. A well-behaved function which is algebraic at large t is the function $\psi_2(t)$, which belongs to the family of functions defined through [7, 152]:

$$\psi_n(t) = c_n a^2 [\exp(a^2 t)] i^n \operatorname{erfc}(at^{1/2}) \tag{5.10}$$

where the $i^n \operatorname{erfc}(z)$ are repeated integrals of the error function and c_n are normalization constants. The function $\psi_2(t)$ has no first moment, $\psi_2(u) = (1 + u^{1/2}/a)^{-2}$, i.e. $\gamma = 0.5$ in Equation (5.7).

Let us now remark that for algebraic $\psi(t)$ forms, which scale with time, this behavior carries over to other quantities related to $\psi(t)$. Let $\chi_n(t)$ denote the probability that exactly n events occurred in time t. This basic quantity of the CTRW formalism is connected to $\psi(t)$ via its Laplace transform:

$$\mathscr{L}[\chi_n(t)] \equiv \chi_n(u) = [\psi(u)]^n \frac{[1 - \psi(u)]}{u}, \tag{5.11}$$

where $\psi(u) = \mathscr{L}[\psi(t)]$.

In Figure 17 we present the $\chi_n(t)$ for the $\psi_2(t)$ distribution, Equation (5.10). As is evident by inspection, the curves scale very well at long times, and their slope is also given by $\gamma = 0.5$. Indeed, for qualitative arguments one may well approximate the $\chi_n(t)$ through $\chi_0(t)$ in the long-time regime [151]:

$$\chi_n(t) = \begin{cases} \chi_0(t) & \text{for } n < n_{\max}(t) \\ 0 & \text{otherwise.} \end{cases} \tag{5.12}$$

On the other hand, for distributions whose first moments exist, scaling does not hold. In Figure 18 we exemplify the situation for $\psi_4(t)$, whose long-time behavior is algebraic with $\gamma = 1.5$, and for which no scaling tendency at longer times is apparent. Finally, in Figure 19 we present the set of $\chi_n(t)$ for the Poisson process, Equation (5.3). The curves show pronounced maxima which, with increasing n shift to longer times. From the log–log plot no scaling is evident.

As an exercise in scaling we may now evaluate $S(t)$, the mean number of sites visited by a walker in continuous time. Qualitatively one expects in situations in which $\tau_1 < \infty$ the pattern to follow Equations (4.10) to (4.12), with n replaced by t/τ. This result is indeed well-fulfilled, as backed by extensive studies on (small) deviations due to higher order terms [144, 148]. On the other hand, when τ_1 is infinite and $\psi(t)$ scales, t/τ_1 is meaningless. Then another argument may be used. From the additional temporal averaging required, $S(t)$ is nothing but

$$S(t) = \sum_{n=0}^{\infty} S_n \chi_n(t). \tag{5.13}$$

We know now that in general in the long-time domain S_n has a power-law dependence on n, both for regular lattices ($d = 2$ excluded), and for fractals.

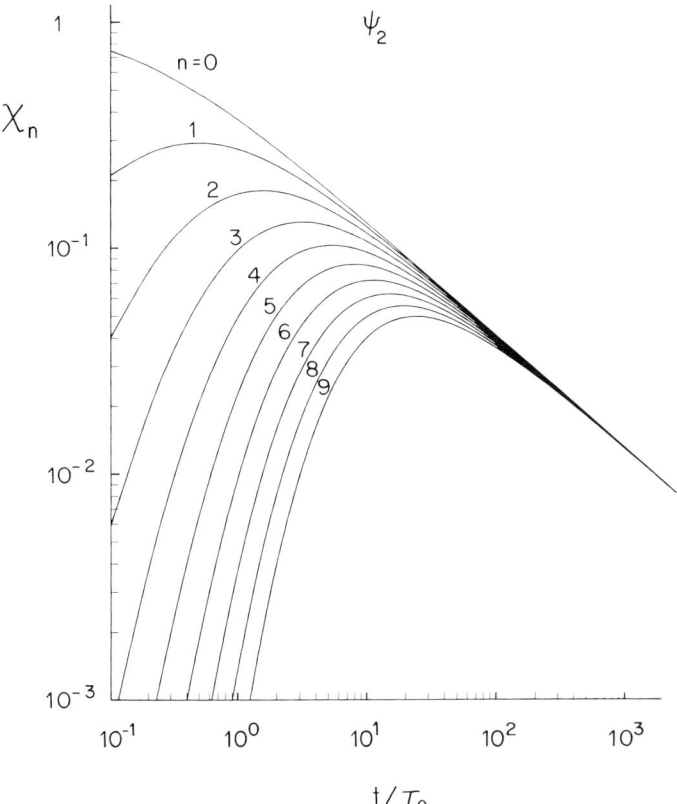

Fig. 17. Probabilities $\chi_n(t)$ that exactly n steps occurred during time t for the waiting-time distribution $\psi_2(t)$ (Equation (5.10)). The time is normalized in terms of τ_e, $\Psi_2(\tau_e) \equiv 1/e$. The long-time decays follow a $t^{-1/2}$ law. (Reproduced from Ref. 151)

Setting $S_n \sim n^\varepsilon$ and using Equation (5.12) it follows that

$$S(t) \sim \sum_{n=0}^{n_{\max}} n^\varepsilon \chi_0(t) \sim \chi_0(t) n_{\max}^{1+\varepsilon}. \tag{5.14}$$

Now the $\chi_n(t)$ are normalized, a fact which determines [151] the time dependence of $n_{\max}(t)$,

$$1 = \sum_{n=0}^{\infty} \chi_n(t) \simeq \sum_{n=0}^{n_{\max}} \chi_0(t) = \chi_0(t) n_{\max}. \tag{5.15}$$

For $\chi_0(t) \sim t^{-\gamma}$ it follows that $n_{\max} \sim t^\gamma$ and hence, from Equation (5.14),

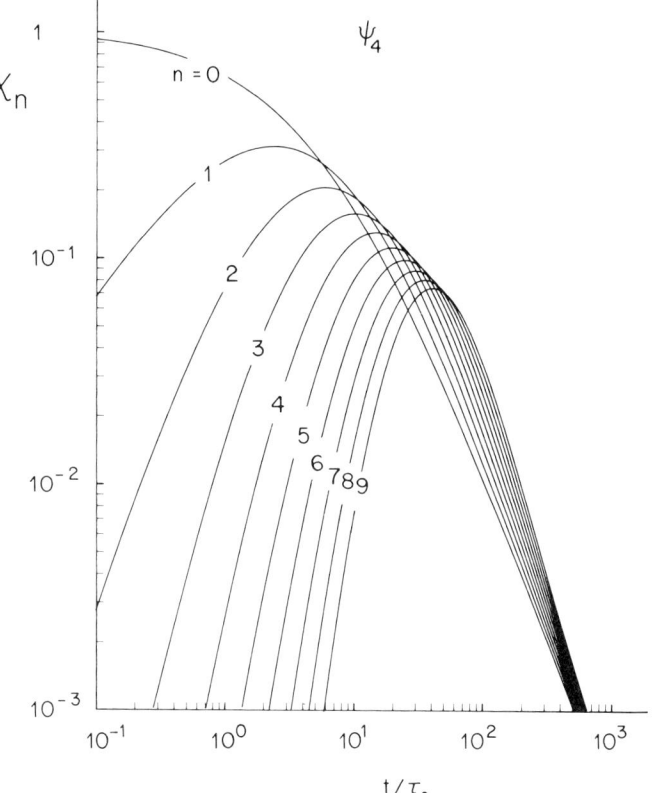

Fig. 18. Same as Figure 17, but for the waiting-time distribution $\psi_4(t)$ (Equation (5.10)). Here $\Psi_4(\tau_e) = 1/e$.

$S(t) \sim t^{\gamma \varepsilon}$, or in fractal notation

$$S(t) \sim \begin{cases} t^{\gamma \tilde{d}/2} & \text{for } \tilde{d} < 2 \\ t^{\gamma} & \text{for } \tilde{d} > 2. \end{cases} \tag{5.16}$$

In Equation (5.16) the two fractal exponents (γ for the temporal and $\tilde{d}/2$ for the spatial aspect) combine *multiplicatively*, i.e. the two processes subordinate [3, 141, 153].

We now consider the relaxation pattern of pseudo-unimolecular $A + B \to 0$, $A_0 \ll B_0$ reactions under continuous-time conditions. Starting with the target problem, which again turns out to be simpler than trapping, we have to extend the formalism of subsection 4.1 to the CTRW situation. From Equation (4.16) we now obtain the probability $H(t; \mathbf{r})$ of a visit from $\mathbf{r}$ to $\mathbf{O}$ in time t:

$$H(t; \mathbf{r}) = \sum_{n=0}^{\infty} \chi_n(t) H_n(\mathbf{r}), \tag{5.17}$$

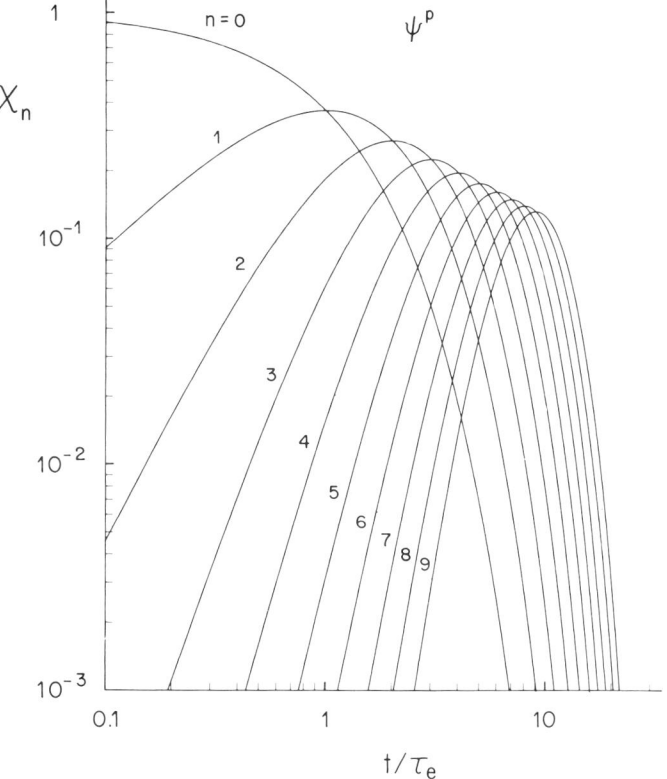

Fig. 19. Same as Figure 17, but for the Poisson waiting-time distribution $\psi^p(t)$ (Equation (5.3)). Here $\Psi^p(\tau_e) \equiv 1/e$. (Reproduced from Ref. 151)

where the $\chi_n(t)$ are defined as in Equation (5.11). The average over times in Equation (5.17) parallels, of course, that of Equation (5.13). Since from Equations (4.16) and (4.20) it follows that

$$\sum_{\mathbf{r}}' H_n(\mathbf{r}) = S_n - 1, \tag{5.18}$$

one has, in fact, using Equation (5.13),

$$\sum_{\mathbf{r}}' H(t;\mathbf{r}) = S(t) - 1. \tag{5.19}$$

Under the assumed independence of motion of the walkers in the target problem, the CTRW transformation of Equations (4.18) and (4.19) carries through, and one obtains for the relaxation of the target:

$$\Phi(t) \sim \exp[-pS(t)], \tag{5.20}$$

where $S(t)$ is given by Equation (5.16).

We stop to note the appearance of a Williams—Watts stretched-exponential form in Equation (5.20). Now the parameter α in Equation (2.14) may take any value between 0 and 1, if one chooses γ accordingly. This is true even for regular lattices in arbitrary dimensions, so that the CTRW provides an adequate extension for the Glarum model of spin relaxation in glasses [132], as pointed out by Shlesinger and Montroll [112, 150].

Notice that due to the multiplicative connection of γ and $\tilde{d}/2$ in Equation (5.16), which stems from subordination, a measured Williams—Watts parameter α may be explained by an infinity of $(\gamma, \tilde{d}/2)$ pairs. Monitoring the target annihilation one cannot distinguish the separate roles played by the spatial disorder exemplified by $\tilde{d}/2$, and by the temporal disorder given by γ. This distinction may be drawn in the trapping problem, since here the CTRW forms lead to quite different decay patterns, as we proceed to show.

Starting point are again the results of subsection 4.1 and, in particular, Equations (4.7) to (4.9), which give the decay due to trapping, $\check{\Phi}_n$, as a function of the number of steps. Now only one particle, the A molecule, moves, so that the transformation to continuous times is [148]

$$\check{\Phi}(t) = \sum_{n=0}^{\infty} \check{\Phi}_n \chi_n(t). \tag{5.21}$$

By viewing Equation (5.21) as an additional average of Equation (4.7), one retrieves in the short-time domain as first term of the cumulant expansion, Equation (4.8):

$$\check{\Phi}(t) \sim \exp[-pS(t)]. \tag{5.22}$$

On the other hand, the long-time behavior of the trapping forms follows from the properties of the algebraic waiting-time distributions $\chi_n(t)$:

$$\check{\Phi}(t) \sim \chi_0(t) \sum_{n=0}^{n_{\max}} \check{\Phi}_n \sim \chi_0(t) \langle n \rangle. \tag{5.23}$$

In Equation (5.23) we made use of the scaling property exemplified in Equation (5.12). Furthermore, here $\langle n \rangle$ is the mean number of steps until trapping for a random walker with fixed waiting times:

$$\langle n \rangle \equiv \sum_{n=1}^{\infty} n(\check{\Phi}_{n-1} - \check{\Phi}_n) = \sum_{n=0}^{\infty} \check{\Phi}_n. \tag{5.24}$$

The last relation of Equation (5.23) holds because $\check{\Phi}_n$ is a decreasing, summable expression, and since for large t, $n_{\max}$ is also large, see Equation (5.15). The remarkable reuslt of Equation (5.23) is that the temporal behavior of $\check{\Phi}(t)$ parallels at long times that of $\chi_0(t) = 1 - \int_0^t \psi(t') \, dt' \equiv \Psi(t)$ (from Equation 5.11), which gives the probability that no step has occurred until t. This effect appeared in our previous numerical analysis of decay laws and leads to concentration-dependent forms which follow avoided-crossing patterns [154].

For algebraic waiting-time distributions, at long times the dependence on the concentration of B particles and on the lattice enters thus only through $\langle n \rangle$, Equation (5.24). For a simple evaluation for small concentrations we again invoke scaling and have

$$\langle n \rangle = \sum_{n=0}^{\infty} \check{\Phi}_n \sim \sum_{n=0}^{\infty} e^{-p(S_n - 1)} \sim \int_0^{\infty} dx \exp(-px^{\varepsilon})$$

$$= p^{-1/\varepsilon} \Gamma\left(1 + \frac{1}{\varepsilon}\right). \tag{5.25}$$

In Equation (5.25) we again made use of $S_n \sim n^{\varepsilon}$, as given by Equations (4.10), (4.12) and (4.24). As before $\Gamma(x)$ is the Euler-gamma function. We furthermore note that in $d = 1$ the relation $\langle n \rangle \sim p^{-2}$ is exact [155] and that it is quite accurately fulfilled in higher dimensions. Putting Equations (5.23), (5.25), (5.15) and (4.10)ff together, one obtains finally:

$$\check{\Phi}(t) \sim \begin{cases} t^{-\gamma}/(p^{2/\tilde{d}}) & \text{for } \tilde{d} < 2 \tag{5.26a} \\ t^{-\gamma}/p & \text{for } \tilde{d} > 2. \tag{5.26b} \end{cases}$$

Thus, at long times the dependence of the trapping decay under algebraic waiting-time forms is algebraic, Equation (2.17), and not of Williams—Watts type, Equation (2.14). The decays of the trapping and of the target problems are thus very different in the long—time regime.

To display graphically these points we present in Figure 20 the relaxation due to trapping for walkers performing a random walk on a Sierpinski gasket [153] embedded in a two-dimensional space, ($\bar{d} = 1.58$, $\tilde{d} = 1.36$), when the steps are taken according to a dipolar waiting-time distribution, $\psi^m(x)$, Equation (5.4) with $s = 6$, so that $\bar{d}/s = 0.26$. This distribution admits a finite first moment τ_1, so that the corresponding $\check{\Phi}_n$ for fixed stepping frequency are also given. Indicated further is the function $\chi_0(t)$, which from Equations (5.2) and (5.11), is identical to $\Psi(t)$. The $\check{\Phi}(t)$ are — except for a very short initial period — bounded from below by $\check{\Phi}_n$, and follow these closely over the first two orders of magnitude in the decay. Hence, for $\psi(t)$ forms, such as Equation (5.4) with $\bar{d}/s > 0.1$, the experimentally relevant portion of the decay $\check{\Phi}(t)$ follows the fractal form $\check{\Phi}_n$.

The situation changes drastically when other $\psi(t)$ are used. In Figure 21 we plot $\check{\Phi}(t)$ for $\psi_2(t)$ and $\psi_4(t)$, Equation (5.10), for the same range of sink concentrations as in Figure 20. For $\psi_2(t)$ the effect of an 'avoided crossing' is clearly evident [153]; after a very short initial decay, the curves are almost parallel to $\Psi_2(t)$, so that $\check{\Phi}(t) \simeq \Psi_2(t)/\lambda$ with $\lambda = -\ln(1 - p)$. Hence, the major part of the decay is determined by $\psi_2(t)$ rather than by the underlying geometrical structure. For $\psi_4(t)$, whose first moment exists, but which also exhibits an algebraic tail, the situation is intermediary between Figure 20 and $\psi_2(t)$. Depend-

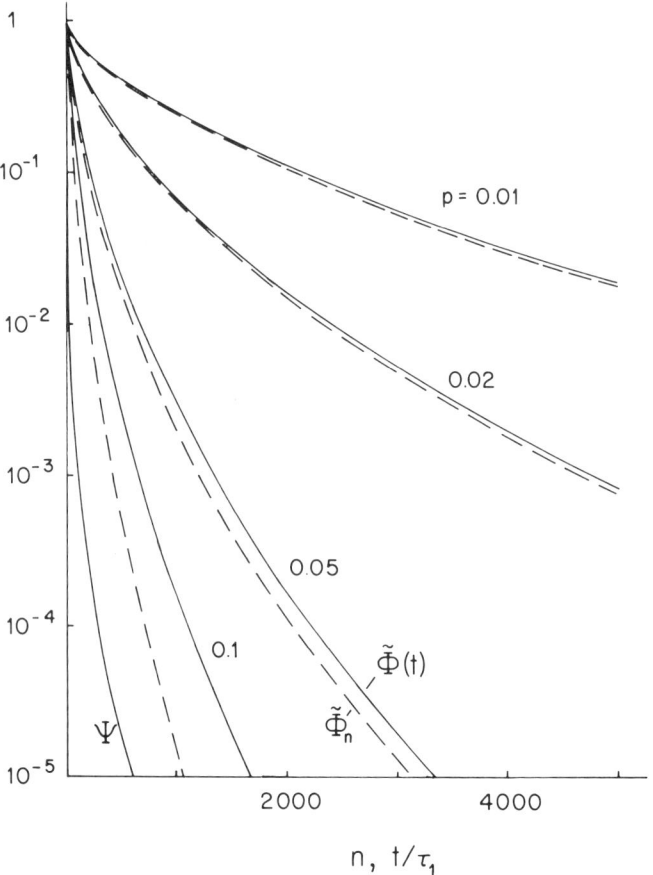

Fig. 20. The decay laws $\Phi(t)$ due to trapping of a random walker on a two-dimensional Sierpinski gasket. The steps are taken according to a dipolar waiting-time distribution, $\psi^m(t)$ (Equation (5.4)) with $\bar{d}/s = 0.26$. The trap densities are as indicated. Included as dashed lines are the corresponding decay laws $\tilde{\Phi}_n$ for fixed waiting times and also $\Psi(t)$ (Equation (5.2)).

ing on p, the short-time decay follows $\tilde{\Phi}_n$ more ($p = 0.01$) or less ($p > 0.1$) closely; the long-time behavior is dictated by $\Psi_4(t)$ and Equation (5.22) is irrelevant. Hence, for waiting-time distributions which decay more slowly than Equation (5.4), it is the CTRW aspect which determines the long-time behavior. This fact can now be used to distinguish between the temporal disorder exemplified by an algebraic $\psi(t)$ form and the spatial disorder typical for fractals: The long-time evolution of $\tilde{\Phi}(t)$ is determined only by $\Psi(t) \equiv \chi_0(t) \sim t^{-\gamma}$, Equations (5.26), whereas the short-time decay depends on $S(t)$, Equation (5.22), and thus subordinates.

As an interesting expression for CTRW on fractals, we mention finally that one may use Equations (5.16) and (5.26) to connect formally $\tilde{\Phi}(t)$ and $S(t)$. One finds

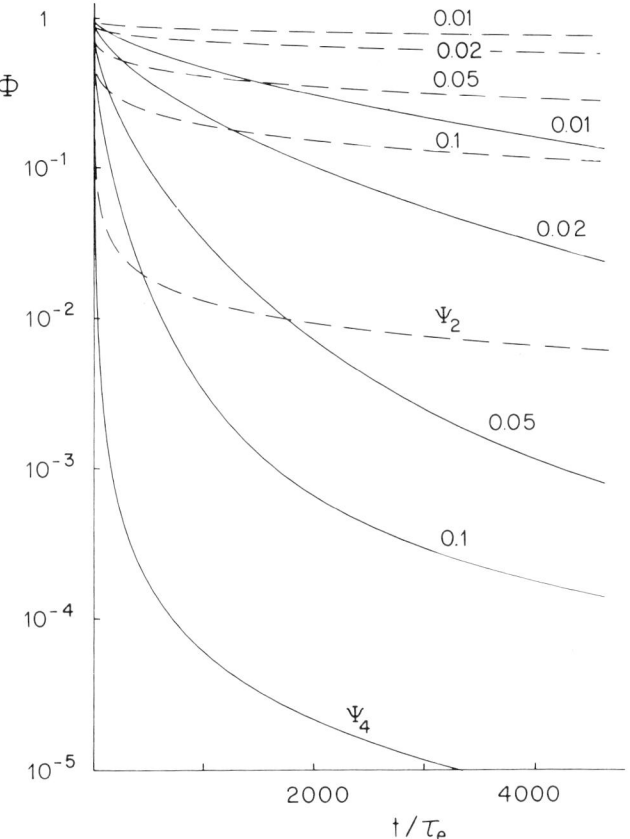

Fig. 21. The decay laws $\Phi(t)$ due to trapping for continuous-time random walks with the waiting-time distributions $\psi_2(t)$, dashed lines, and $\psi_4(t)$, full lines (Equation (5.21)). The parameters and notation are as in Figure 20.

[141, 153]

$$\tilde{\Phi}(t) \sim \begin{cases} [S(t)]^{-2/\tilde{d}} & \text{for } \tilde{d} < 2 \quad (5.27\text{a}) \\ [S(t)]^{-1} & \text{for } \tilde{d} > 2. \quad (5.27\text{b}) \end{cases}$$

Equations (5.27) generalize to fractals a relation established by Scher [156] for $d = 3$.

After having analyzed in this section the inclusion of temporal disorder in dynamical processes through CTRW formalisms, we turn our attention to the energetic randomness. As shown here, waiting time distributions are based on a hierarchy of release times and lead to fractal temporal behavior. The same idea applied to hierarchically arranged energy barriers leads to ultrametric spaces (UMS).

6. Ultrametric Spaces (UMS)

In this section we model energetic disorder through ultrametric spaces. We monitor the relaxation patterns for the trapping and target annihilation problems and demonstrate that the decay forms depend qualitatively on the temperature.

Ultrametric spaces (UMS) occur naturally when classification of objects into distinct clusters is involved [11, 12]. A simple example of a UMS is the *baseline* of Figure 22 which is the space $\mathbf{Z}_2$. Figure 22 may remind one of a Bethe lattice. Here, however, only the baseline points belong to the UMS, the structure above the baseline being subsidiary, in that it serves to indicate connections. If one cuts the structure above the baseline and parallelity to it, one obtains a system of disjoint clusters. The height of the cut determines a generalized distance in the UMS. The distance between two points is the maximal height of a cut which disconnects them.

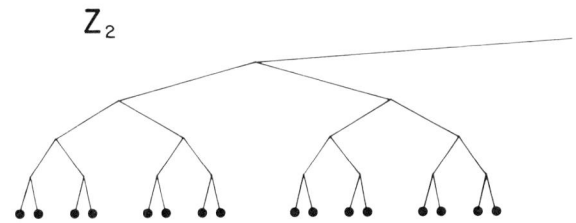

Fig. 22. The ultrametric structure (UMS) $\mathbf{Z}_2$.

More formally, for each hierarchical arrangement as in Figure 22 (dendogram) [11] corresponds a UMS. In this space $\mathbf{X}$ the distance $d(x, y)$ between sites satisfies the strong triangle inequality

$$d(x, y) \leq \max(d(x, z), d(y, z)) \tag{6.1}$$

for all $x, y, z \in \mathbf{X}$ [10–12]. One may readily verify, using Figure 22, that Equation (6.1) is indeed obeyed on $\mathbf{Z}_2$. The name 'ultrametric' stems from the additional constraint on the distance imposed by Equation (6.1), since in a metric space one only requires

$$d(x, y) \leq d(x, z) + d(y, z). \tag{6.2}$$

As a physical picture we take the UMS distance to correspond to the activation energy. Hence the points on the baseline of Figure 1 are separated from each other by energy barriers, which have to be overcome. For simplicity we take all consecutive energy level to differ by Δ. The barrier heights are then of the form $j\Delta$, where Δ is fixed and j is an integer. Then, after an activation by an energy E, a walker may reach only sites which are separated from the original site by barriers not higher than E. To be specific, in Figure 22 a particle activated by E with $\Delta \leq E \leq 2\Delta$ may reach two sites (including the original one) and generally for $j\Delta \leq E < (j+1)\Delta$ may reach 2^j sites. Here the number 2 appears because of the specific geometry of the UMS $\mathbf{Z}_2$. To a general space, say in which the clusters are nested hierarchically in groups of z objects, corresponds the UMS $\mathbf{Z}_z$, and an activation energy of $j\Delta \leq E < (j+1)\Delta$ allows z^j sites to be reached.

We now consider a simple qualitative argument for the temporal dependence of $S(t)$, the mean number of sites visited during t, when the particles are thermally activated, so that the intersite transition rates are proportional to $\exp(-E/kT)$:

$$k_{ji} = w \exp\left(\frac{-j\Delta}{kT}\right). \tag{6.3}$$

Let us focus on the time interval

$$e^{m\Delta/kT} \leq wt_m \leq e^{(m+1)\Delta/kT}. \tag{6.4}$$

During this time interval z^m points of the UMS $\mathbf{Z}_z$ are accessible to the walker, and one has

$$z^m \sim z^{(kT/\Delta)\ln(wt_m)} = e^{\gamma \ln(wt_m)} = (wt_m)^\gamma, \tag{6.5}$$

where we set

$$\gamma = \left(\frac{kT}{\Delta}\right)\ln z \tag{6.6}$$

For $\gamma < 1$, z^m increases more slowly than t_m, and the walker explores practically all accessible points. We are in the case of compact exploration [120]. Therefore,

$$S(t) \simeq (wt)^\gamma \sim t^\gamma \quad (\gamma < 1). \tag{6.7}$$

On the other hand, for $\gamma > 1$, z^m increases more rapidly than t_m, and the mean number of distinct sites visited stays proportional to t_m; hence

$$S(t) \sim t \quad (\gamma > 1). \tag{6.8}$$

From our qualitative argument we have obtained a result much similar to the CTRW expression (5.16), for $\tilde{d}$ above the marginal dimension of 2. This also motivated our use of 'γ' in Equation (6.6). The central role played by γ may also be inferred from an analogy to MT, where b^j is the activation barrier $b^j = e^{-j\Delta/kT}$ and the level density in the hierarchy decreases with increasing level as z^{-j}, i.e. $N \sim 1/z$. Then, from Equations (5.7) and (5.8),

$$\gamma = \frac{\ln(z^{-1})}{\ln(e^{-\Delta/kT})} = \left(\frac{kT}{\Delta}\right)\ln z. \tag{6.9}$$

From this picture we obtain $\psi(t) \sim t^{-1-\gamma}$ and the relaxation pattern is algebraic.

These results can be rendered quantitative by performing simulations on UMS and determining $S(t)$ [157]. Another approach is to evaluate $P_0(t)$, the probability of being at the origin at time t [158—160]; for compact exploration $S(t)$ and $P_0(t)$ are related via $S(t) \sim [P_0(t)]^{-1}$. In [157] we have simulated random walks on the UMS $\mathbf{Z}_2$ and $\mathbf{Z}_3$, where some 1000 hierarchical levels were accounted for. From our calculations we find that for random walks involving some 10^4 steps and 10^4 distinct realizations, the trial function

$$S_n = Cn^\varepsilon \tag{6.10}$$

is a good representation of the behavior over the step range considered. For $\gamma \lesssim 0.5$, ε and γ practically coincide, whereas for $\gamma > 1.5$, ε is practically unity. These results suggest also an analogy to random walks on fractals [158], (Equations (4.24)), by relating γ to $\tilde{d}/2$ and viewing 2γ as a spectral dimension. This analogy also explains our finding that for γ around unity, ε in Equation (6.10) turns out to be systematically lower than γ; since for random walks in $d = 2$ there appear logarithmic corrections in the denominator of S_n, Equation (4.11).

Another quantity which is readily evaluated from simulations is the variance σ_n^2 defined by Equation (4.6). In [157] the results were fitted to

$$\sigma_n^2 = Dn^\delta. \tag{6.11}$$

For small γ values ($\gamma \lesssim 0.5$) δ is practically 2γ, whereas again for γ larger than unity, δ tends to one. Again this finding parallels the behavior for regular lattices, for which the dependence of the variance on d is largest for $d = 2$, see Equations (4.10) to (4.12). Interestingly, for $\gamma < 1$, $2\alpha \simeq \delta$, which seems to imply $\sigma_n^2 \sim S_n^2$, an expression which as indicated after Equation (4.22)ff is exact in $d = 1$, and which holds well [128] for random walks over compact spaces, $\tilde{d} < 2$. It is tempting to conjecture that also for UMS for the higher cumulants $K_{j,n}$ of the distribution on R_n a relation analogous to Equation (4.25) holds, at least for γ around 0.5.

We are now in a position to investigate relaxation phenomena on the UMS. As in the previous sections we study the trapping of walkers and the target annihilation.

Let us start with the target problem, and note that all sites of our standard UMS (fixed branching ratio z and equidistant barriers) are equivalent. We now denote the site on which the target sits as the origin, $r = 0$, and assign integer number $r \in \mathbf{N}$, by counting, to the other sites. Equations (4.16) to (4.19) of subsection 4.2 hold also for UMS, when the position of the site ($\mathbf{r}$ in Section 4) is reinterpreted to be the ordinal integer r. Furthermore, even Equation (4.20) so reinterpreted holds on UMS since $\Sigma'_r F_m(r)$ gives the increase in the total number of visited sites in the mth step, and thus equals $S_m - S_{m-1}$. Consequently, we rederive for UMS, as in Equation (4.21), the exact decay law of A particles annihilated by moving B species:

$$\Phi_n = \exp[-p(S_n - 1)] \tag{6.12}$$

where the B molecules were initially Poisson distributed, Equation (4.15).

From Equations (6.10) and (6.12) one now has:

$$\Phi_n \simeq \exp(-Cpn^\varepsilon) \qquad (6.13)$$

which for $\varepsilon < 1$ shows a stretched-exponential form, Equation (2.14).

In Figure 23 we present the decay due to trapping on the UMS Z_2. The full lines give the decay obtained from S_n in conjunction with Equation (6.12), whereas the dots indicate the direct simulation of the target annihilation. We have taken the density of the walkers to be $p = 0.2$, and the dynamics take place over the UMS Z_2. As expected, in every case the agreement between the two forms is excellent. Note that depending on the temperature one has a crossover from exponential decays for $\varepsilon \simeq 1$, i.e. at higher temperatures ($kT \ln z > \Delta$), to stretched-exponential decays at lower temperatures ($kT \ln z < \Delta$). These aspects stress the point that, in the range of parameters of the figure, random walks on

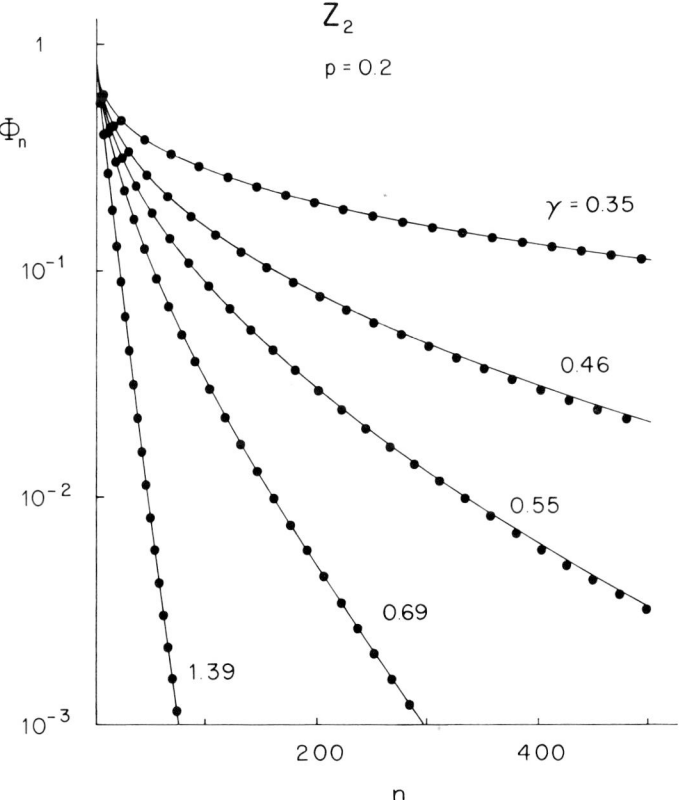

Fig. 23. Decay law due to the annihilation of targets by random walkers on the UMS Z_2. The concentration of walkers is $p = 0.2$ and the decays are monitored as a function of the temperature $\gamma = (\ln 2)/(\Delta\beta)$, for the values $\Delta\beta = 0.5, 1, 1.25, 1.5$ and 2. The full lines are the exact solution (Equation (6.12)), whereas the dots are the results of simulation calculations.

UMS parallel findings for lattices with a spectral dimension $\tilde{d} = 2\gamma$ [158]. Interestingly, one may therefore switch through the marginal behavior at $\gamma = 1$ ($\tilde{d} = 2$) through a simple temperature change.

To pursue this analysis further, we now consider the trapping problem, in which mobile A species get irreversibly trapped by randomly distributed static B molecules. Again we may follow the lines of argument of Section 4: the trapping decay is given by

$$\check{\Phi}_{N,n} \simeq e^{\lambda} \exp\left[\sum_{j=1}^{N} K_{j,n} \frac{(-\lambda)^j}{j!}\right], \tag{6.14}$$

Equation (4.7), where the $K_{j,n}$ are the cumulants (semi-invariants) of the distribution R_n of distinct sites of the UMS visited in n steps.

In Figure 24 we present the decay due to trapping on $\mathbf{Z}_2$ for $\Delta/kT = \frac{1}{2}$, i.e. $\gamma = 1.386$, for several values of p, and compare it to the approximate forms $\check{\Phi}_{1,n}$ and $\check{\Phi}_{2,n}$. We note that over four orders of magnitude in the decay Φ_n is practically exponential, and that $\check{\Phi}_{1,n}$ approximates very well, whereas the agreement with $\check{\Phi}_{2,n}$ is excellent. This situation is very reminiscent of the decay for $d = 3$, see Figure 10.

A decrease in temperature has a very drastic effect on Φ_n. In Figure 25 we present the decay at $\Delta/kT = 1$, i.e. $\gamma = 0.693$. The decay is clearly non-exponential. Furthermore, here neither $\check{\Phi}_{1,n}$ nor $\check{\Phi}_{2,n}$ describes the decay well, and one has to go to higher order in N to obtain a somehow reasonable approximation. The situation is similar to the one encountered for trapping on the linear chain, $d = 1$, Figure 12, and for trapping on the planar Sierpinski gasket, $\tilde{d} = 1.36$, see Figure 15. It is remarkable here that in the range considered a change in temperature lets the relaxation pattern change from exponential to stretched exponential.

To summarize this section, we have analyzed dynamical relaxation behaviors on UMS, and have pointed out the analogies to previous results for random walks on regular lattices and on fractals. All aspects suggest that random walks on UMS with a parameter $\gamma = (kT/\Delta)\ln z$ parallel findings for lattices with a spectral dimension $\tilde{d} = 2\gamma$. Interestingly, one may therefore switch through the marginal behavior at $\gamma = 1$ ($\tilde{d} = 2$) through a simple temperature change. Such a phase transition should be experimentally observable through the qualitative pattern (exponential vs. stretched exponential) of the corresponding relaxation behavior.

This section concludes our exposition on pseudo-unimolecular decays which obtain in disordered systems, when fractals, CTRW or ultrametric spaces are used as models. In the next section we present an overview of bimolecular relaxation patterns.

7. The Bimolecular Reactions $A + A \to 0$ and $A + B \to 0$ ($A_0 = B_0$)

As stressed in Section 2, in which we investigated the chemical kinetic schemes

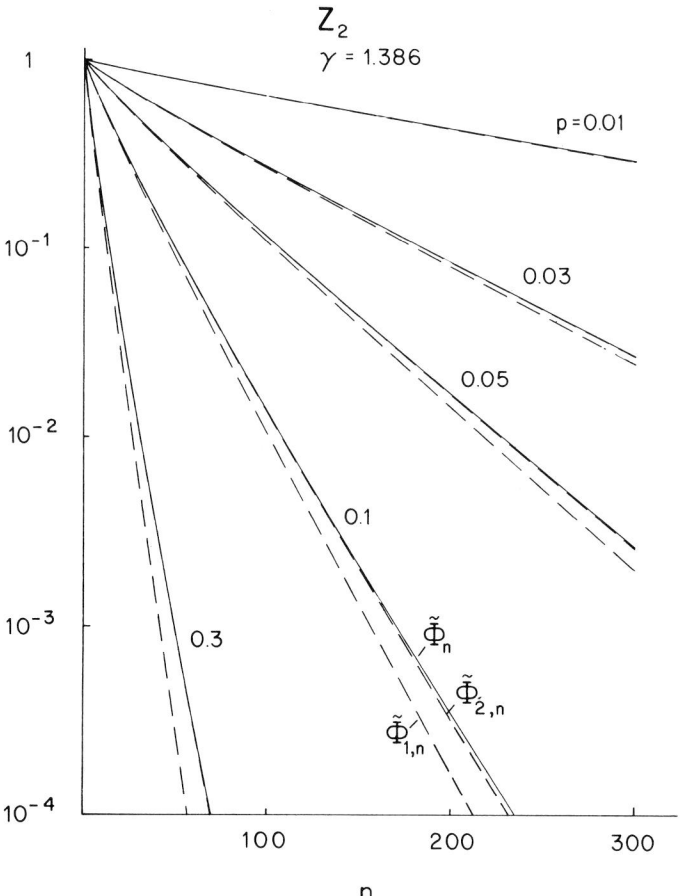

Fig. 24. Decay law Φ_n due to trapping on $\mathbf{Z}_2$, where $\gamma = 1.386$ and the trap density p is as indicated. Also given as dashed lines are $\Phi_{1,n}$ and $\Phi_{2,n}$ (Equation (6.14)).

for the $A + A \to 0$ and for the strictly bimolecular $A + B \to 0$ ($A_0 = B_0$) reactions, in both cases the decay follows the $1/t$ dependence under 'well-stirred' conditions. Our findings, Equations (2.10) and (2.13), emerged from a spatially homogeneous situation. In this section we determine the decay laws which apply in the presence of disorder, which will again be modeled through fractals, CTRW and UMS. The main type of relaxation pattern which will emerge from our studies is algebraic, $\Phi(t) \sim t^{-\gamma}$, Equation (2.17), where γ may take any value between zero and one. In Section 2 we have also encountered more general algebraic forms than $1/t$, when reactions of order higher than 2 were considered, see the discussion after Equation (2.13). On physical and chemical grounds, however, the distinction whether two or more distinct entities are involved in an elementary reaction step is generally easy to draw. The order of the reaction is fixed, and for

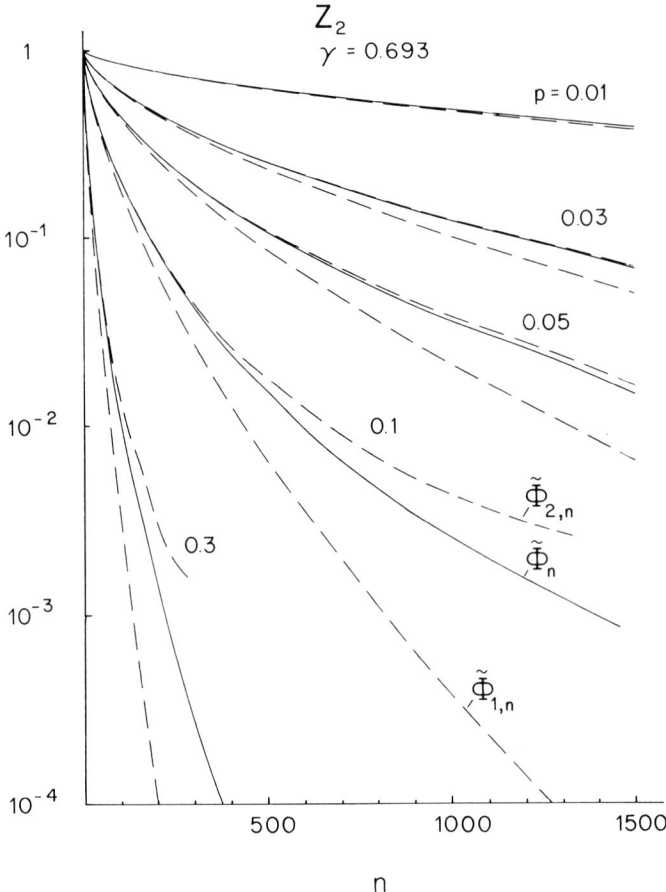

Fig. 25. Same as Figure 24, for $\gamma = 0.693$.

the bimolecular reactions considered here deviations from $1/t$ indicate loss of homogeneity.

7.1. BIMOLECULAR REACTIONS ON REGULAR LATTICES AND ON FRACTALS

Here we start with the $A + A \to 0$ reaction, since it will turn out to be *less* influenced by fluctuations than the strictly bimolecular $A + B \to 0$, $A_0 = B_0$, reaction. This reaction is of importance in energy transfer problems, where it describes exciton up-conversion and annihilation processes [28, 61, 62, 161, 162]. From the previous study of pseudo-unimolecular reactions, we found that in the short-time regime the kinetic exponential was modified by the appearance of $S(t)$. Thus one may expect that the $A + A \to 0$ decay will follow a $1/S(t)$ law at longer times [24, 28, 29] and thus, one may find as an approximation to the decay,

$$\Phi_n^{AA} \simeq (1 + 2pS_n)^{-1}. \qquad (7.1)$$

In [24] we have established through numerical simulations that Equation (7.1) correctly describes the decay behavior both on regular lattices and on fractals (Sierpinski gaskets). For p we used the value $p = 0.1$ and we started from configurations containing some 2×10^4 particles. We had thus to fix the extent of the spatial regions such as to have $N_T = 2 \times 10^5$. For the dynamical processes periodic boundary conditions were used, so that the systems appeared infinite. The particles were placed randomly on the lattice, following a binomial (yes–no) distribution. At each step the walkers moved one by one to neighboring sites. Any two particles that during this process happened to occupy the same site were immediately removed.

In Figure 26 we display the results for the decay following the $A + A \to 0$ reaction on the linear chain, the square, the simple cubic lattices and for the two-dimensional Sierpinski gasket. The full lines give the simulation results obtained by averaging over 100 distinct initial conditions and walks each, whereas the dashed lines are the approximation, Equation (7.1). We choose to plot $\ln \Phi_n^{AA}$ vs $\ln n$. In these scales, according to Equations (7.1), (4.10) to (4.12) and (4.24), at long times the decays should turn into straight lines. From the figure this behavior is clearly apparent. Moreover, it turns out that Equation (7.1) is almost quantitative in the range investigated [23, 24].

The same behavior obtains also for fractals, as we have exemplified in [23], by analyzing the deviations of the Smoluchowski-type approximation, Equation (7.1) from the simulated decays. An exact solution to the many-body problem involved in the $A + A \to 0$ reaction was found by Torney and McConnell for $d = 1$ in the continuum, diffusion-equation limit [162]. In the range of Figure 26 their expression is hardly distinguishable from the discrete, random-walk result. Hence, in all cases investigated, the long-time decay Φ_n^{AA} follows an algebraic from $\Phi_n^{AA} \sim n^{-\varepsilon}$ (with $\varepsilon = \tilde{d}/2$ for $\tilde{d} < 2$ and $\varepsilon = 1$ for $\tilde{d} > 2$). Heuristically one may view Equation (7.1), the solution of a many-body problem, as being related to the probability of encounter of two particles, which itself is expressible, via S_n, by the volume visited by each. This point will be used in the next subsection when we will address the question of $A + A$ decays under CTRW conditions.

One must realize that the relative simplicity of the result (7.1) does not carry over to the strictly bimolecular $A + B \to 0$, $A_0 = B_0$ reaction. For this reaction type the situation is different, due to spatial fluctuations, which get *enhanced* by the chemical reaction. The reason for this effect is that at longer times, due to the progress of the reaction, large regions containing only A or only B molecules appear. Then the diffusion no longer provides an efficient stirring, and the reactions proceed more slowly since mainly only molecules at the boundaries of the A and B regions are prone to react. The expected decay at longer times then

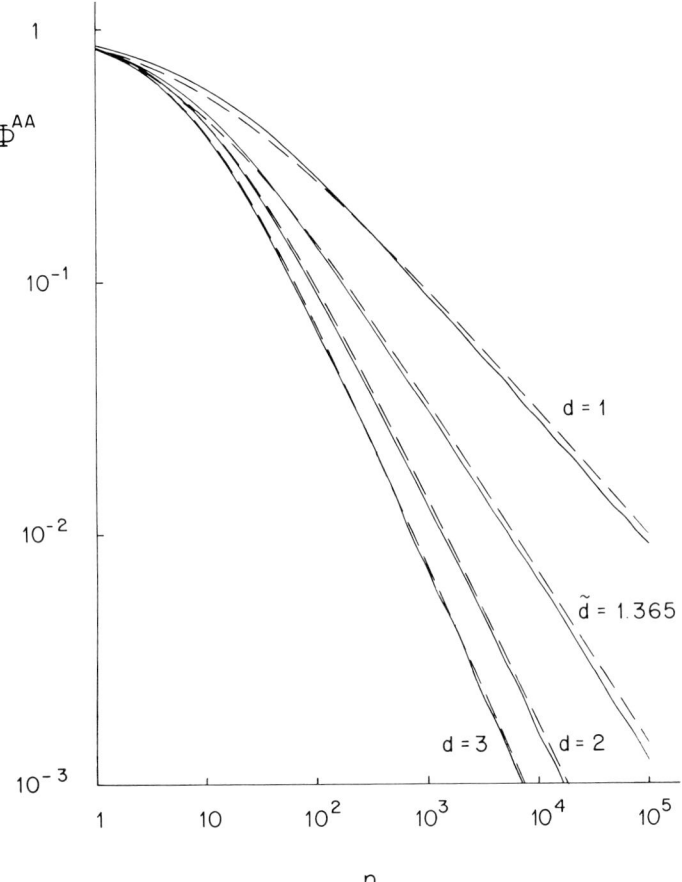

Fig. 26. Decays due to the bimolecular reaction $A + A \to 0$. The full lines are the simulation results for the linear chain ($d = 1$), for the square ($d = 2$), for the simple cubic lattice ($d = 3$) and for a two-dimensional Sierpinski gasket ($\tilde{d} = 1.365$). The dashed lines are Smoluchowski-type approximations (Equation (7.1)).

follows [23, 24, 30, 108, 163, 164]:

$$\Phi_n^{AB} \sim \begin{cases} n^{-\tilde{d}/4} & \text{for } \tilde{d} < 4 \\ n^{-1} & \text{for } \tilde{d} > 4. \end{cases} \quad (7.2)$$

The marginal dimension for the $A + B \to 0$, $A_0 = B_0$ reaction is thus four [30, 163].

In Figure 27 we present a snapshot of a simulation calculation on a Sierpinski gasket ($\tilde{d} = 1.36$), after 3×10^5 steps have elapsed. In the calculation we started with some 10^5 particles, placed such that the initial concentrations $p_A = p_B$ were 0.1. The separation of the A and B species into distinct regions is clearly evident.

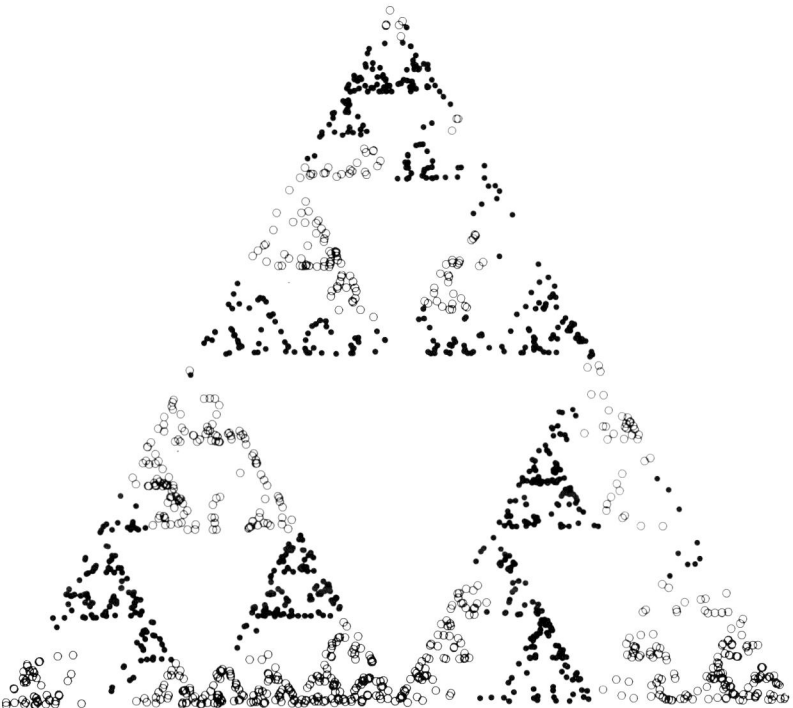

Fig. 27. Snapshot of the strict bimolecular reaction $A + B \to 0$, $A_0 = B_0$ on a two-dimensional Sierpinski gasket ($\tilde{d} = 1.365$) at the 12th iteration stage (some 10^6 lattice sites). The initial concentrations are $A_0 = B_0 = 0.05$ and correspond to some 10^5 particles. The picture shows the situation after 3×10^5 steps. The remaining A and B particles (around 700 each) are indicated by circles and dots. (Reproduced from Ref. 115)

In this regime the decay law Φ_n^{AB} has crossed over from the Smoluchowski-type pattern, Equation (7.1), valid for smaller n to the decay behavior of Equation (7.2). As we have demonstrated in [23] and [24], a similar behavior obtains also for other Sierpinski gaskets, imbedded in spaces of higher Euclidean dimensions; we have considered Sierpinski gaskets with spectral dimensions $\tilde{d} = 1.55$, 1.65 and 1.77. Similar findings were also reported for strictly bimolecular reactions on stochastic fractals [164]. We remark that with increasing $\tilde{d}$ the Smoluchowski-type region of Equation (7.1) increases. Hence, an increase in the spectral dimension pushes the crossover to the form given by Equation (7.2) to later times.

To conclude this subsection let us summarize our findings for bimolecular reactions $A + A \to 0$ and $A + B \to 0$ ($A_0 = B_0$) on regular lattices and on fractals. For the first we find agreement with the Smoluchowski-type form, Equation (7.1), for the whole decay range studied, whereas the second obeys such a form only in the initial time domain and crosses over to a lower decay $t^{-\tilde{d}/4}$ ($\tilde{d} < 4$) at longer times. Thus, bimolecular reactions display a richer behavior than that predicted by the standard kinetic approach, a behavior which we deem to be experimentally accessible.

Following the general scheme of this overview, we now extend the models to continuous-time and also investigate the role of the energetic disorder on the relaxation patterns.

7.2. BIMOLECULAR REACTIONS: CTRW AND ULTRAMETRIC SPACES

In the previous subsection we have shown that the $A + A \to 0$ decays are related to S_n, the mean number of distinct sites visited by a walker, through a relation of the form (7.1). This relation reads, by using Equations (4.10) to (4.12) and Equation (4.24):

$$\Phi_n^{AA} \sim \begin{cases} n^{-\tilde{d}/2} & \text{for } \tilde{d} < 2 \\ n^{-1} & \text{for } \tilde{d} > 2. \end{cases} \tag{7.3}$$

From our viewing S_n as a measure of explored volume it is therefore tempting to envisage that in the CTRW scheme S_n gets replaced in Equation (7.1) by $S(t)$, and therefore, for CTRW one should find:

$$\Phi^{AA}(t) \sim \begin{cases} t^{-\gamma \tilde{d}/2} & \text{for } \tilde{d} < 2, \gamma < 1 \\ t^{-\gamma} & \text{for } \tilde{d} > 2, \gamma < 1. \end{cases} \tag{7.4}$$

Equation (7.4) is then another example of subordination [120, 141, 153].

Our numerical simulations support Equation (7.4) well [151]. In Figure 28 we present the decay of the $A + A \to 0$ reaction under CTRW conditions. We start from walkers on a simple cubic lattice, $d = 3$, and use $\psi_2(t)$, as given by Equation (5.10), and thus $\gamma = 1/2$. In Figure 28 we present the corresponding decay law and contrast it with the simple RW results. Whereas at longer times the simple RW decay follows t^{-1}, for the CTRW with $\psi_2(t)$ we find at longer times a $t^{-1/2}$ dependence, as may be verified by inspection of Figure 28, in which these asymptotic slopes are also indicated. As a further example we have performed simulations on several Sierpinski gaskets and on the linear chain [151]. The long-time decay behavior indeed follows the form $t^{-\tilde{d}/4}$ for CTRW with $\psi_2(t)$ instead of $t^{-\tilde{d}/2}$ for the simple RW decay. All findings are consistent with $\Phi^{AA}(t) \sim [S(t)]^{-1}$, i.e. with Equation (7.4).

Turning now to the strictly bimolecular $A + B \to 0$ reaction we note that we have found, for a fixed stepping frequency, that the $n^{-\tilde{d}/4}$ decay is well-obeyed on Sierpinski gaskets embedded in spaces of different Euclidean dimensions d, see the discussion after Equation (7.2). Here the role of the CTRW remains to be assessed. As in previous cases, we expect the decay form to subordinate through the time variable to $\psi(t)$, and expect hence:

$$\Phi^{AB}(t) \sim t^{-\gamma \tilde{d}/4} \quad (\text{for } \tilde{d} < 4, \gamma < 1). \tag{7.5}$$

Simulation calculations [151] for particles moving on a linear chain under the influence of the waiting-time distribution $\psi_2(t)$, Equation (5.10), support the conjectured subordination displayed by Equation (7.5). Hence, all analyzed

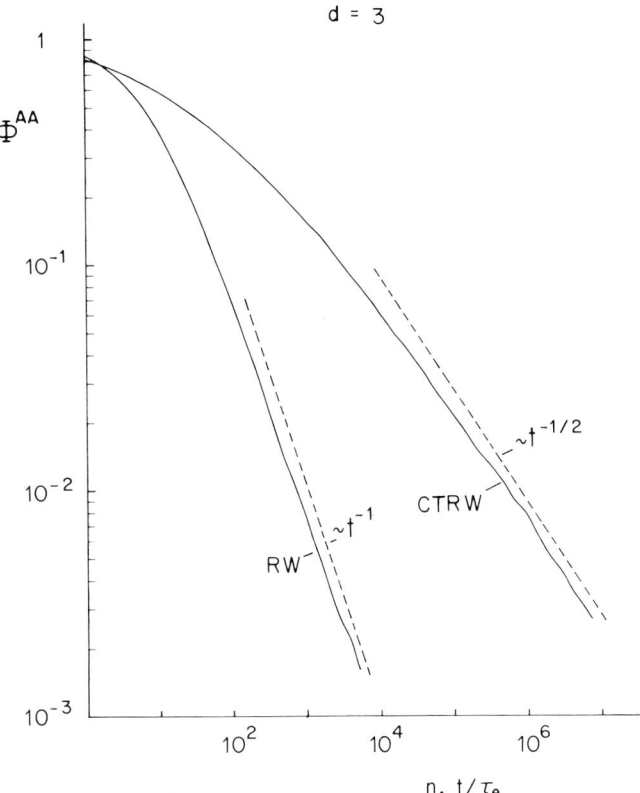

Fig. 28. Decay laws Φ^{AA} for the $A + A \to 0$ reaction on a simple cubic lattice, both for simple RW and for CTRW with $\psi_2(t)$. The initial particle density is $p = 0.1$. The full lines are the simulation results averaged over 100 runs. The theoretical long-time slopes are indicated by dashed lines.

reactions, apart from the very particular pseudo-unimolecular trapping problem display subordination under CTRW conditions, i.e. the cooperative random aspects lead to a multiplicative combination of the scaling coefficients in the final relaxation forms.

To conclude our study of bimolecular reactions we also consider the influence of energetic randomness, as displayed by ultrametric spaces (UMS). We begin with the $A + A \to 0$ reaction. At longer times, we expect, according to Equations (6.7), (6.8) and (7.1), the relaxation pattern

$$\Phi^{AA}(t) \sim \begin{cases} t^{-\gamma} & \text{for } \gamma = (kT/\Delta) \ln z < 1 \\ t^{-1} & \text{else.} \end{cases} \tag{7.6}$$

In the simulations we start by randomly distributing A-particles on the UMS $\mathbf{Z}_2$. We use a structure with 15 hierarchical levels, so that for a typical concentration of $p = 0.2$ some 10^4 A particles are involved. At fixed time intervals the molecules attempt a (temperature-activated) crossing of the hierarchical energy barriers in accordance with the imposed temperature parameter $\Delta\beta$, the individual stepping procedure following exactly the pattern described in Section 6 for the pseudo-unimolecular decays.

In Figure 29 we display the decay Φ_n^{AA}, when the temperature factor $\Delta\beta$ varies from 0.25 to 2. The full lines give the results of the simulation, averaged over 100 distinct initial conditions and walks. From the $\ln \Phi_n^{AA}$ vs $\ln n$ scales of Figure 29, the tendency towards the decay form of Equation (7.6) is well documented. In Figure 29 we have also indicated the asymptotic slopes which we expect from an analogy between $\tilde{d}$ and 2γ, i.e. we have plotted as dashed lines $t^{-\gamma}$ for $\gamma < 1$ and t^{-1} for $\gamma > 1$. These dashed lines are given as a guide to the eye. In the cases $\gamma < 0.6$ and $\gamma > 2$ the deviations from this expected asymptotic behavior are

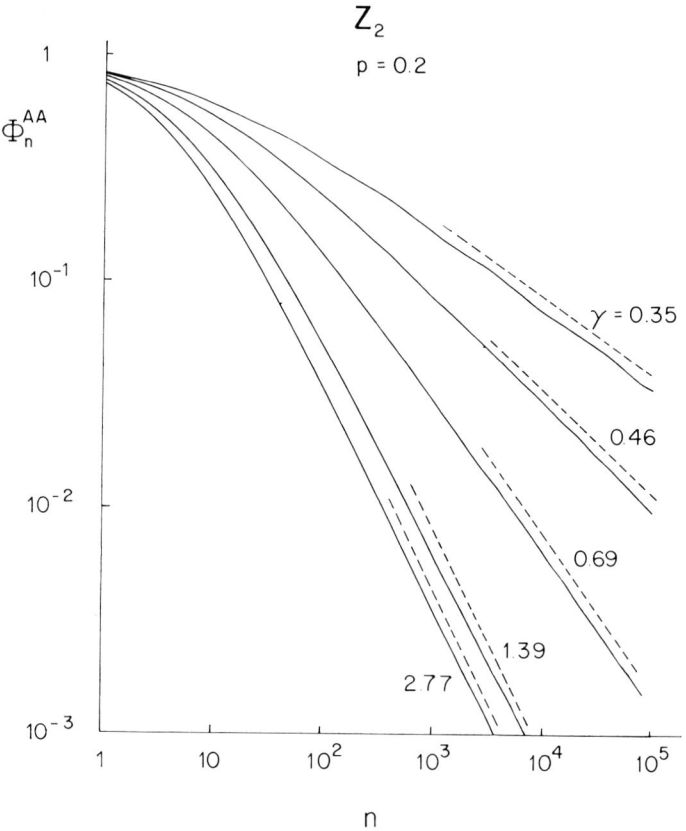

Fig. 29. Decay laws Φ_n^{AA} for the bimolecular reaction $A + A \to 0$ on the ultrametric space $\mathbf{Z}_2$. The initial density of the A particles is $p = 0.2$ and the temperature parameter $\Delta\beta$ equals 0.25, 0.5, 1, 1.5 and 2. The full lines are the simulation results, whereas the dashed lines indicate as a guide to the eye the expected asymptotic slopes.

very small. A somewhat larger deviation is found for $\gamma = 1.39$, in which case a three-orders-of-magnitude decay does not yet suffice to match the expected t^{-1} law. This fact again stresses the central role played by the value $\gamma = 1$, around which, as also found in the pseudo-unimolecular reactions, the deviations from simple decay patterns are most conspicuous. The same behavior is also apparent in simulations on the UMS $\mathbf{Z}_3$, in which analogous deviations exist around $\gamma = 1$. As a rule, however, in the γ-range displayed here we again find support for the parallelity between UMS and the related fractal spaces, $\tilde{d} = 2\gamma$. Thus the $A + A \rightarrow 0$ reaction displays, as a function of temperature, a continuous change between being 'poorly stirred' at low temperatures towards being 'well-stirred' at high temperatures.

These aspects connected to the question of the 'well-stirred reactor' also show up in the strictly bimolecular scheme $A + B \rightarrow 0$ ($A_0 = B_0$). One expects here, based on our previous knowledge of the $A + B$ reaction and on the relation $\tilde{d} = 2\gamma$;

$$\Phi_n^{AB} \sim \begin{cases} t^{-\gamma/2} & \text{for } \gamma < 1 \\ t^{-1} & \text{for } \gamma > 2 \end{cases} \tag{7.7}$$

with a crossover region for γ between 1 and 2. Simulation calculations [165] show, indeed, that Equation (7.7) offers a very reasonable description of the long-time behavior for $\gamma < 1$ and for $\gamma > 2$. Furthermore, the crossover character of γ values around and above unity is well marked; for $\gamma = 1.10$ after two orders of magnitude in the decay the Φ_n^{AB} pattern is intermediate between the two asymptotic slopes, being somewhat nearer, however, to the $t^{-\gamma/2}$ form. In the whole range of γ values investigated, the parallellity between $\tilde{d}$ and 2γ is conspicuous.

Summarizing our findings for bimolecular reactions on UMS we note that as in the case of random walks and reactions on fractals and on regular lattices, the $A + A \rightarrow 0$ reaction is less influenced by fluctuations. Here its decay behavior at longer times follows qualitatively the $t^{-\gamma}$ form ($\gamma < 1$), which may be explained through Smoluchowski-type rates. For the $A + B \rightarrow 0$ strictly bimolecular reaction, the situation is different because the course of the reaction enhances the fluctuations to such a point, that at longer times the 'well-stirred reactor' condition is no longer valid. The UMS is then composed of large-scale clusters containing only A or only B molecules, and the decay proceeds at a much slower pace, since then the molecules have to surmount high barriers in order to react. The situation is reminiscent of the large-scale regions which form in low-dimensional spaces as a result of the strictly bimolecular reaction $A + B \rightarrow 0$ (*vide supra*). Here, the long-time decay also follows an algebraic form, but with a different, non-Smoluchowski-type power law, $\Phi^{AB} \sim t^{-\gamma/2}$, for $\gamma < 1$, see Equation (7.7). As in the case of pseudo-unimolecular reactions on UMS, the bimolecular reactions tend towards a 'normal' Smoluchowski-type decay (here $\Phi \sim t^{-1}$), when the temperature increases.

To conclude this section devoted to bimolecular reactions on regular lattices, fractals and UMS, also under continuous-time conditions, we note that several microscopic models of bimolecular type lead to power-law forms, Equations (2.17) and (7.2) to (7.7). In general, such decays may be well distinguished from other relaxation behaviors, such as stretched exponentials, by a sufficiently large dynamical range of measurements. On the other hand, distinguishing between different power-law decays may not be easy. Thus, additional experimental information such as concentration dependence may be necessary in order to pinpoint the microscopic relaxation behavior of a specific material which displays a power-law form.

8. Conclusions

In this presentation we have aimed to stress the role of the disorder, such as found in glasses, on the course of dynamical processes, here exemplified through pseudo-unimolecular and bimolecular chemical reactions. The ordered state in chemical kinetic schemes is the one given by the 'well-stirred reactor' in which the spatial distribution of reactants is completely homogeneous. To such a distribution correspond clear-cut, simple relaxation patterns: the exponential decay for pseudo-unimolecular, the $1/t$ form for bimolecular reactions.

As we have shown in Section 2, deviations from such relaxation behaviors are widespread, and extend from stretched-exponential to algebraic decays. As we have demonstrated in the subsequent sections, one may readily obtain such decay forms from theoretical approaches which include randomness. Here the models may be either purely parallel (and involve only an ensemble average over exponentially decaying forms as in Section 3) or include sequential aspects (such as random walks, Sections 4 to 7).

Randomness as found in glasses, displays not only spatial but also temporal and energetic facets. We have found it advisable to use models tailored to disorder. Thus spatial randomness may be modeled through fractal structures, temporal randomness through waiting-time distributions in the framework of continuous-time random walks (CTRW) and energetic randomness through ultrametric spaces (UMS). The advantage of these models is that they are very flexible and mathematically tractable. In all cases considered we were able to display the decay laws for the pseudo-unimolecular and for the bimolecular reactions investigated, either in analytically closed form or as a result of computer simulations. Disorder helps to differentiate between several reactions schemes, whose decays are similar under 'well-stirred' conditions. Thus the target problem, in which the minority species is stationary, has a more rapid decay than the trapping problem, in which only the minority species moves. The $A + A \rightarrow 0$ reaction has a marginal dimension of 2, whereas the marginal dimension of the $A + B \rightarrow 0$, $A_0 = B_0$ strictly bimolecular reaction is 4.

Underlying the basic models for disorder are dilatational symmetries, which lead to unexpected connections. Thus the γ parameter which relates the temperature, the activation energies and the branching ratio of a UMS has connotations of an effective spectral dimension, $\gamma = \tilde{d}/2$ for a related fractal space. The same parameter γ may be envisaged as arising from a multiple trapping (MT) scheme, and is immediately reinterpretable as determining a waiting-time distribution in the CTRW picture. Furthermore, CTRW processes may be applied to fractals; apart from special cases such as trapping, the CTRW parameter γ and the spectral dimension $\tilde{d}$ combine multiplicatively in the relaxation patterns: the processes subordinate. It is certainly of great importance to follow these symmetries in order to possibly unravel universal aspects of disorder.

Dynamical processes in glasses represent a challenge, and it is of importance to see under which conditions experimentally reported decay patterns may be understood to arise from particular models. Since several models lead to similar relaxation patterns, one needs for a correct description both relaxation measurements in an extended decay range and also structural information on the microscopic behavior. In our opinion further progress in the field necessitates a close cooperation between theory and experiment.

Acknowledgments

The authors are much indebted to Professor K. Dressler for continuous encouragement and help. We thank Dr. R. Hilfer for discussions and P. Bocklet for technical assistance. The support of the Deutsche Forschungsgemeinschaft and of the Fonds der Chemischen Industrie and a grant of computer time from the ETH-Rechenzentrum are gratefully acknowledged. Permission to reproduce the figures as indicated is kindly acknowledged.

References

1. P. W. Anderson in *Ill-Condensed Matter* (eds R. Balian, R. Maynard and G. Toulouse), North Holland, Amsterdam (1979), p. 162.
2. T. S. Kuhn, *The Structure of Scientific Revolutions* (2nd edn), Univ. of Chicago Press, Chicago (1970).
3. B. B. Mandelbrot, *The Fractal Geometry of Nature*, Freeman, San Francisco (1982).
4. K. J. Falconer, *The Geometry of Fractal Sets*, Cambridge Univ. Press (1985).
5. H. Scher and M. Lax, *Phys. Rev.* **B7**, 4491 (1973).
6. H. Scher and M. Lax, *Phys. Rev.* **B7**, 4502 (1973).
7. H. Scher and E. W. Montroll, *Phys. Rev.* **B12**, 2455 (1975).
8. G. Pfister and H. Scher, *Adv. Phys.* **27**, 747 (1978).
9. E. W. Montroll and M. F. Schlesinger in *Nonequilibrium Phenomena II: From Stochastics to Hydrodynamics* (eds J. L. Lebowitz and E. W. Montroll), North Holland, Amsterdam (1984).
10. N. Bourbaki, *Eléments de mathématique, Topologie générale*, Chap. IX, CCLS, Paris (1974).
11. A. D. Gordon, *Classification*, Chapman and Hall, London (1981).
12. W. H. Schikhof, *Ultrametric Calculus*, Cambridge Univ. Press (1984).
13. S. Kirkpatrick, *Rev. Modern Phys.* **45**, 574 (1973).
14. J. W. Essam, *Rep. Progr. Phys.* **43**, 843 (1980); H. Kesten, *Percolation Theory for Mathematicians*, Birkhäuser, Boston (1982).

15. G. Deutscher, R. Zallen, and J. Adler (eds), *Percolation Structures and Processes*, Ann. Isr. Phys. Soc. Vol. 5, Hilger, Bristol (1983).
16. D. Stauffer, *Introduction to Percolation Theory*, Taylor and Francis, London (1985).
17. J. L. van Hemmen and I. Morgenstern, 'Heidelberg Colloquium on Spin Glasses', *Lect. Notes in Phys.* **192**, Springer, Berlin (1983).
18. J. Laidler, *Chemical Kinetics*, McGraw-Hill, New York (1950).
19. H. Eyring, S. H. Lin, and S. M. Lin, *Basic Chemical Kinetics*, Wiley, New York (1980).
20. F. Wilkinson, *Chemical Kinetics and Reaction Mechanisms*, Van Nostrand, New York (1980).
21. P. C. Jordan, *Chemical Kinetics and Transport*, Plenum, New York (1980).
22. J. W. Moore and R. G. Pearson, *Kinetics and Mechanism*, Wiley, New York (1981).
23. A. Blumen, G. Zumofen and J. Klafter in *Structure and Dynamics of Molecular Systems* (eds R. Daudel *et al.*), Reidel, Dordrecht (1985), p. 71.
24. G. Zumofen, A. Blumen and J. Klafter, *J. Chem. Phys.* **82**, 3198 (1985).
25. J. M. Smith, *Chemical Engineering Kinetics*, McGraw-Hill, Kogakusha, Tokyo (1981).
26. T. R. Waite, *Phys. Rev.* **107**, 463 (1957); **107**, 471 (1957).
27. J. Klafter and A. Blumen in *Random Walks and their Applications in the Physical and Biological Sciences* (eds M. F. Shlesinger and B. J. West), American Institute of Physics, New York (1984), p. 173.
28. P. W. Klymko and R. Kopelman, *J. Phys. Chem.* **87**, 4565 (1983).
29. B. Ya. Balagurov and V. G. Vaks, *Zh. Exp. Teor. Fiz.* **65**, 1939 (1973) [English transl.: *Sov. Phys. JETP* **38**, 968 (1974)].
30. D. Toussaint and F. Wilczek, *J. Chem. Phys.* **78**, 2642 (1983).
31. J. Friedrich and A. Blumen, *Phys. Rev.* **B32**, 1434 (1985).
32. R. Kohlrausch, *Ann. Phys. (Leipzig)* **12**, 393 (1847).
33. G. Williams and D. C. Watts, *Trans Faraday Soc.* **66**, 80 (1970); G. Williams, *Adv. Polymer Sci.* **33**, 59 (1979).
34. J. Jäckle, *Rep. Prog. Phys.* (in press).
35. A. K. Jonscher, *Nature* **267**, 673 (1977).
36. K. L. Ngai, *Comments Solid State Phys.* **9**, 127 (1979); **9**, 141 (1980).
37. A. A. Jones, J. F. O'Gara, P. T. Inglefield, J. T. Bendler, A. F. Yee, and K. L. Ngai, *Macromolec.* **16**, 658 (1983).
38. G. D. Patterson, *Adv. Polym. Sci.* **48**, 125 (1983).
39. G. Fytas, T. Dorfmüller, and C. H. Wang, *J. Phys. Chem.* **87**, 5041 (1983).
40. G. Fytas, A. Patkowski, G. Meier, and T. Dorfmüller, *J. Chem. Phys.* **80**, 2214 (1984).
41. R. Richert and H. Bässler, *Chem. Phys. Lett.* **118**, 235 (1985).
42. V. L. Vyazovkin, B. V. Bol'shakov, and V. A. Tolkatchev, *Chem. Phys.* **75**, 11 (1983).
43. T. Doba, K. U. Ingold, and W. Siebrand, *Chem. Phys. Lett.* **103**, 339 (1984).
44. A. Plonka, J. Kroh, W. Lefik, and W. Bogus, *J. Phys. Chem.* **83**, 1807 (1979).
45. T. Förster, *Z. Naturf.* **A4**, 321 (1949).
46. M. Inokuti and F. Hirayama, *J. Chem. Phys.* **43**, 1978 (1965).
47. D. Rehm and K. B. Eisenthal, *Chem. Phys. Lett.* **9**, 387 (1971).
48. F. K. Fong (ed), *Radiationless Processes in Molecules and Condensed Phases*, Top. Appl. Phys. **15**, Springer, Berlin (1976); W. M. Yen and P. M. Selzer (eds), *Laser Spectroscopy in Solids*, Springer, Berlin (1981); W. M. Yen, this volume.
49. A. I. Burshtein, *Usp. Fiz. Nauk* **143**, 553 (1984) [*Sov. Phys. Usp.* **27**, 579 (1984)]; *J. Lumin.* **34**, 167 (1985); **34**, 201 (1985).
50. K. Kemnitz, T. Murao, I. Yamazaki, N. Nakashima, and K. Yoshihara, *Chem. Phys. Lett.* **101**, 337 (1983).
51. F. Willig, A. Blumen, and G. Zumofen, *Chem. Phys. Lett.* **108**, 222 (1984).
52. U. Even, K. Rademann, J. Jortner, N. Manor, and R. Reisfeld, *Phys. Rev. Lett.* **52**, 2164 (1984).
53. J. R. Miller, *Chem. Phys. Lett.* **22**, 180 (1973).
54. J. V. Beitz and J. R. Miller, *J. Chem. Phys.* **71**, 4579 (1979).
55. H. A. Stoddart, M. Pollak and J. Tauc in *Proc. 17th Int. Conf. Phys. Semicond.* (eds J. D. Chadi and W. A. Harrison), Springer, New York (1985).

56. A. Blumen, *Nuovo Cimento* **B63**, 50 (1981).
57. J. R. Morgan and M. A. El-Sayed, *J. Phys. Chem.* **87**, 2178 (1983).
58. J. Friedrich and D. Haarer, *Angew. Chemie Int. Ed. Engl.* **23**, 113 (1984); see also this volume.
59. W. Breinl, J. Friedrich, and D. Haarer, *J. Chem. Phys.* **81**, 3915 (1984).
60. J. Tauc, *Semicond. and Semimetals* **21B**, 299 (1984).
61. P. Evesque, *J. Physique* (*Paris*) **44**, 1227 (1983).
62. P. Argyrakis and R. Kopelman, *J. Chem. Phys.* **83**, 3099 (1985) and references therein.
63. G. H. Weiss, *Separation Sci. Techn.* **17**, 1609 (1982–83).
64. R. G. Palmer, D. L. Stein, E. Abrahams, and P. W. Anderson, *Phys. Rev. Lett.* **53**, 958 (1984).
65. D. L. Dexter, *J. Chem. Phys.* **21**, 836 (1953).
66. D. L. Huber in *Laser Spectroscopy in Solids* (eds W. M. Yen and P. M. Selzer), Springer, Berlin (1981), p. 85.
67. S. I. Golubov and Yu. V. Konobeev, *Fiz. Tverd. Tela* **13**, 3185 (1971) [Engl. trans. *Sov. Phys. Solid State* **13**, 2679 (1972)].
68. V. P. Sakun, *Fiz. Tverd. Tela* **14**, 2199 (1972) [Engl. trnasl. *Sov. Phys. Solid State* **14**, 1906 (1973)].
69. D. L. Huber, D. S. Hamilton, and B. Barnett, *Phys. Rev.* **B16**, 4642 (1977).
70. A. Blumen and J. Manz, *J. Chem. Phys.* **71**, 4694 (1979).
71. A. Blumen, *J. Chem. Phys.* **74**, 6926 (1981).
72. H. Kellerer and A. Blumen, *Biophys. J.* **46**, 1 (1984).
73. I. M. Rozman, *Izv. Akad. Nauk SSSR Ser. Fiz.* **36**, 922 (1972) [Engl. transl. *Bull. Acad. Sci. USSR Phys. Ser.* **36**, 833 (1972)].
74. J. Klafter and A. Blumen, *Chem. Phys. Lett.* **119**, 377 (1985).
75. A. Blumen, *J. Chem. Phys.* **72**, 2632 (1980).
76. A. Blumen, J. Klafter, and G. Zumofen, *J. Chem. Phys.* **84**, 6679 (1986).
77. C. L. Yang, P. Evesque, and M. A. El-Sayed, *J. Phys. Chem.* **89**, 3442 (1985).
78. B. B. Mandelbrot, *Les objets fractals: forme, hasard et dimension*, Flammarion, Paris (1975).
79. W. Sierpinski, *Compt. Rend.* (*Paris*) **160**, 302 (1915); **162**, 629 (1916); *Oeuvres Choisies* (S. Hartman *et al.*), Editions Scientifiques, Warsaw (1974).
80. P. Urysohn, *Veh. Koning. Akad. Wetensch.* (*Amsterdam*) 1. Sect. **13**, 4 (1927).
81. S. Alexander and R. Orbach, *J. Phys. Lett.* **43**, L625 (1982).
82. P. Pfeuty and G. Toulouse, *Introduction to the Renormalization Group and to Critical Phenomena*, Wiley, London (1977).
83. D. Dhar, *J. Math. Phys.* **18**, 577 (1977).
84. D. Dhar, *J. Math. Phys.* **19**, 5 (1978).
85. R. Rammal and G. Toulouse, *J. Phys. Lett.* **44**, L13 (1983).
86. S. Havlin and D. Ben-Avraham, *J. Phys.* **A15**, L311 (1982).
87. M. E. Cates, *Phys. Rev. Lett.* **53**, 926 (1984).
88. M. Muthukumar, *J. Chem. Phys.* **83**, 3161 (1985).
89. P. Pfeifer and D. Avnir, *J. Chem. Phys.* **79**, 3558 (1983).
90. J. M. Drake and J. Klafter, *J. Lumin.* **31–32**, 642 (1984).
91. W. D. Dozier, J. M. Drake, J. Klafter, and P. M. Chaikin (to be published).
92. S. Alexander, C. Laermans, R. Orbach, and H. M. Rosenberg, *Phys. Rev.* **B28**, 4615 (1983).
93. T. A. Witten and L. M. Sander, *Phys. Rev. Lett.* **47**, 1400 (1981).
94. T. A. Witten and L. M. Sander, *Phys. Rev.* **B27**, 5686 (1983).
95. F. Family and D. P. Landau (eds), *Kinetics of Aggregation and Gelation*, North Holland, Amsterdam (1984).
96. J. Klafter and A. Blumen, *J. Chem. Phys.* **80**, 875 (1984).
97. H. Hilfer and A. Blumen, *J. Phys.* **A17**, L537; L783 (1984).
98. R. Hilfer and A. Blumen, in *Fractals in Physics* (eds. L. Pietronero and E. Tossatti), North Holland, Amsterdam (1986) p. 33.
99. H. E. Stanley, *J. Stat. Phys.* **36**, 843 (1984).
100. J. A. Given and B. B. Mandelbrot, *J. Phys.* **A16**, L565 (1983).

101. A. Maritan and A. L. Stella, in *Fractals in Physics* (eds. L. Pietronero and E. Tossatti), North Holland, Amsterdam (1986) p. 107.
102. M. N. Barber and B. W. Ninham, *Random and Restricted Walks*, Gordon and Breach, New York (1970).
103. G. H. Weiss and R. J. Rubin, 'Random Walks: Theory and Selected Applications', *Adv. Chem. Phys.* **52**, 363 (1983).
104. Papers presented at the Symposium on Random Walks, *J. Stat. Phys.* **30**, No. 2 (1983).
105. M. F. Shlesinger and B. J. West (eds), *Random Walks and their Applications in the Physical and Biological Sciences*, Amer. Inst. Phys., New York (1984).
106. G. Zumofen and A. Blumen, *Chem. Phys. Lett.* **88**, 63 (1982).
107. H. E. Stanley, K. Kang, S. Redner, and R. L. Blumberg, *Phys. Rev. Lett.* **51**, 1223 (1983).
108. K. Kang and S. Redner, *Phys. Rev. Lett.* **52**, 955 (1984).
109. R. Kopelman, P. W. Klymko, J. S. Newhouse, and L. W. Anacker, *Phys. Rev.* **B29**, 3747 (1984).
110. J. Klafter, A. Blumen and G. Zumofen, *J. Phys. Chem.* **87**, 191 (1983).
111. E. W. Montroll and J. T. Bendler, *J. Stat. Phys.* **34**, 129 (1984).
112. M. F. Shlesinger and E. W. Montroll, *Proc. Natl. Acad. Sci. USA* **81**, 1280 (1984).
113. S. Redner and K. Kang, *J. Phys.* **A17**, L451 (1984).
114. A. Blumen, G. Zumofen, and J. Klafter, *Phys. Rev.* **B30**, 5379 (1984).
115. A. Blumen, G. Zumofen, and J. Klafter, *J. Physique* Colloque, Tome 46, C7—3 (1985).
116. A. Blumen and G. Zumofen, *J. Chem. Phys.* **75**, 892 (1981).
117. G. Zumofen and A. Blumen, *J. Chem. Phys.* **76**, 3173 (1982).
118. H. B. Rosenstock, *Phys. Rev.* **187**, 1166 (1969); *SIAM J. Appl. Math.* **27**, 457 (1974).
119. G. H. Weiss, *Proc. Natl. Acad. Sci. USA* **77**, 4391 (1980).
120. P. G. de Gennes, *C. R. Acad (Paris)* Ser II, **296**, 881 (1983).
121. B. Movaghar, G. Sauer, D. Würtz, and D. L. Huber, *Solid State Commun.* **39**, 1179 (1981) and *J. Stat. Phys.* **27**, 473 (1982).
122. S. Redner and K. Kang, *Phys. Rev. Lett.* **51**, 1729 (1983).
123. J. K. Anlauf, *Phys. Rev. Lett.* **52**, 1845 (1984).
124. M. D. Donsker and S. R. S. Varadhan, *Comm. Pure Appl. Math.* **28**, 525 (1975) and **32**, 721 (1979).
125. P. Grassberger and I. Procaccia, *J. Chem. Phys.* **77**, 628 (1982).
126. R. F. Kayser and J. B. Hubbard, *Phys. Rev. Lett.* **51**, 6281 (1982).
127. J. Klafter, G. Zumofen, and A. Blumen, *J. Physique Lett.* **45**, L49 (1984).
128. G. Zumofen, A. Blumen, and J. Klafter, *J. Phys.* **A17**, L479 (1984).
129. S. Havlin, M. Dishon, J. E. Kiefer, and G. H. Weiss, *Phys. Rev. Lett.* **53**, 407 (1984).
130. G. Zumofen, A. Blumen, and J. Klafter, 'Random Walks on Fractals' in *Structure and Dynamics of Molecular Systems* (eds R. Daudel, J. P. Korb, J. P. Lemaistre, and J. Maruani), Reidel, Dordrecht (1985), p. 87.
131. E. W. Montroll and G. H. Weiss, *J. Math. Phys.* **6**, 167 (1965).
132. S. H. Glarum, *J. Chem. Phys.* **33**, 639 (1960).
133. P. Bordewijk, *Chem. Phys. Lett.* **32**, 592 (1975).
134. J. E. Shore and R. Zwanzig, *J. Chem. Phys.* **63**, 5445 (1975).
135. J. L. Skinner, *J. Chem. Phys.* **79**, 1955 (1983).
136. S. Alexander, *Ann. Isr. Phys. Soc.* **5**, 149 (1983) (ref. 15).
137. R. Orbach, *J. Stat. Phys.* **36**, 735 (1984).
138. S. Alexander, O. Entin-Wohlman, and R. Orbach, *J. Physique Lett.* **46**, L549 (1985).
139. S. Alexander, O. Entin-Wohlman, and R. Orbach, *J. Physique Lett.* **46**, L555 (1985).
140. A. Blumen, J. Klafter, and G. Zumofen, *Phys. Rev.* **B28**, 6112 (1983).
141. J. Klafter, A. Blumen, and G. Zumofen, *J. Stat. Phys.* **36**, 561 (1984).
142. I. M. Lifshitz, *Adv. Phys.* **13**, 483 (1964).
143. D. Bedeaux, K. Lakatos-Lindenberg, and K. E. Shuler, *J. Math. Phys.* **12**, 2116 (1971).
144. V. Kenkre, E. Montroll, and M. Shlesinger, *J. Stat. Phys.* **9**, 45 (1973).
145. V. Kenkre and R. Knox, *Phys. Rev.* **B9**, 5279 (1974).

146. J. Klafter and R. Silbey, *J. Chem. Phys.* **72**, 843 (1980).
147. J. Klafter and R. Silbey, *Phys. Rev. Lett.* **44**, 55 (1980).
148. A. Blumen and G. Zumofen, *J. Chem. Phys.* **77**, 5127 (1982).
149. K. L. Ngai and F. S. Liu, *Phys. Rev.* **B24**, 1049 (1981).
150. M. F. Shlesinger, *J. Stat. Phys.* **36**, 639 (1984).
151. A. Blumen, J. Klafter, and G. Zumofen, in *Fractals in Physics* (eds. L. Pietronero and E. Tossatti), North Holland, Amsterdam (1986) p. 399.
152. M. F. Shlesinger, *J. Stat. Phys.* **10**, 421 (1974).
153. A. Blumen, J. Klafter, B. S. White, and G. Zumofen, *Phys. Rev. Lett.* **53**, 1301 (1984).
154. G. Zumofen, J. Klafter, and A. Blumen, *J. Chem. Phys.* **79**, 5131 (1983).
155. E. W. Montroll, *J. Phys. Soc. Japan Suppl.* **26**, 6 (1969).
156. H. Scher, *J. Physique Coll.* **42**, C4—547 (1981).
157. A. Blumen, J. Klafter, and G. Zumofen, *J. Phys.* **A19**, L77 (1986).
158. S. Grossmann, F. Wegner, and K. H. Hoffmann, *J. Physique Lett.* **46**, L575 (1985).
159. A. T. Ogielski and D. L. Stein, *Phys. Rev. Lett.* **55**, 1634 (1985).
160. B. A. Huberman and M. Kerszberg, *J. Phys.* **A18**, L331 (1985).
161. D. C. Torney and H. M. McConnell, *Proc. Roy. Soc. London* **A387**, 147 (1983).
162. D. C. Torney and H. M. McConnell, *J. Phys. Chem.* **87**, 1941 (1983).
163. A. A. Ovchinnikov and Ya. B. Zel'dovich, *Chem. Phys.* **28**, 215 (1978).
164. P. Meakin and H. E. Stanley, *J. Phys.* **A17**, L173 (1984).
165. G. Zumofen, A. Blumen, and J. Klafter, *Ber. Buns. Gesell.* (in press).

INDEX

absorption 16, 19
activation energy 246
adiabatic 160
aggregation 214
AHV-P model 65
alcohol glass 174
algebraic time-dependence 203
amorphous materials, optical spectra 66
amorphous metals 69
amorphous potential 163
amorphous state 25
anomalous TLSs 66
area of the hole 164
as functions of Δ 104
asymmetry parameter 71
asymmetry parameter distribution 73
average velocity of sound 99
averaged linewidth, analytical approximations 118
averaging of linewidths 68, 114
averaging of the line shape 114, 125
averaging over λ, Δ 114
averaging over λ, Δ and ΔV 118
averaging procedures 68
avoided-crossing patterns 243

back-transfer 207
bath 79
BeF_2 51
BeF_2 glass 49
Bethe lattice 246
Bi^{3+} 43, 47
bimolecular reactions 202, 250
bimolecular reactions: continuous-time random walk (CTRW) 256
binomial 253
Born approximation 92
borosilicate 31
branching ratio 261
broadening of optical lines 81

calcium phosphate 31
case 1 138
case 2 139
Ce^{3+} 43
characteristic frequency 75
characteristic frequency ω 99
chemical reactor 200
classification 246
coherent-potential approximations 199

coloring of glasses 29
compact exploration 247, 248
compact visitation 234
comparison with experiment 137
components of the total system 79
computer simulation of structure of glasses 49
concentration of guest 77
concentration quenching 46
conjugate index 94
continuous random network 69
continuous-time random walks 201, 235
correlation function, evolution matrix 131
correlation functions 87, 88
correlation functions for the optical line shape 87
correlation functions, equations of motion 91, 95
coupled electrostatically 68
coupling of the guest molecule via orbitlattice interaction 68
Cr^{3+} 39, 45, 53, 54
Cr^{3+} doped glasses 40
critical ΔV 107, 116
crossover 106, 137, 140
crystal field theory 33
cumulant expansion 242
Coulomb interaction 33
cumulants 233
cumulants expansion 224

Debye density of states 99
Debye frequency 99
Debye model 97, 135
Debye velocity of sound 99
Debye-Waller factor 151
decoupling approximations 14
deformation potentials 4, 74, 144
deformation tensor 73
degree of polarization 185
dendogram 246
density of TLS states 157, 158
density of states of the TLS 156
dependence of the linewidth on ΔV 106
dephasing 104
dephasing processes 160
dephasing rate 7, 9, 10
deterministic fractals 217
deuterated quinizarin 181
deuteration effect 179
deuterium isotope effect 177
diagonal coupling 5
diagonal modulation 5, 7, 8, 16, 19

diagonalize the system Hamiltonian 83
dielectric relaxation 202
dielectric susceptibility 87
dielectrit measurements 66
diffusion of energy 47
diffusion-limited aggregate (DLA) 216, 217
1,4-dihydroxyanthraquinone 174
dila(ta)tion 213
dilatation 215
dimension 231
dipolar 11
dipolar coupling 118, 139
dipole moment correlation function 130
dipole moment operator 93
dipole moment operator, many particly representation 129
dipole-quadrupole coupling 118
direct interaction between the guest and the phonons 78
direct transfer 206
disclination lines 70
discontinuities in the linewidth 104
disentanglement theorem 91
distribution 253
distribution function of 164
distribution of ΔV 137
distribution of ΔV, case 1 137
distribution of ΔV, case 2 137
distribution of TLS states 159
distribution of parameters λ, Δ 113
distributions for the TLSs 144
Donsker-Varadhan form 234
doped amorphous solids 152
doped glasses 149
double index 88
double index notation 93
double-well potential 71
Dy^{2+} 43
dynamic 160
dynamic scattering processes 161

e_g orbitals 34
effective deformation potential 142-143
effective spectral dimension 261
effective-medium 199
eigenvalues R_k of **N** 97
eigenvalues dependence on system parameters 104
eigenvalues of **N** 101
eigenvalues, analytical approximations 107
eigenvalues, approximate calculation 132
eigenvectors $\{\mathbf{x}^k\}$ 97
elastic coupling 140
electric multipolar 2
electron scavenging 200
electron-hole recombination 201
electrostatic coupling 140

energetic disorder 246
energetic randomness 257
energy barriers 246
energy level scheme 81
energy transfer 58
equations of motion 14, 15
Er^{3+} 41
error function 238
estimate of the temperature dependence of the linewidth 143
Eu^{2+} 43, 47, 49, 50, 51, 56
exchange interactions 208
exciton up-conversion 252
exponential-logarithmic form 210
exponential-logarithmic relaxation 205

fast 17
fast modulation 6
field effects and spectral diffusion phenomena 188
first-passage time 224, 227
FLN 49, 50, 54, 55, 56, 57
fluorescence lifetimes 42
fluorescence line narrowing (FLN) 48, 67, 149
fluorophosphate 31
formation of a glassy state 25
Förster decay 210
Förster expression 204
Förster-decay laws 222
Förster-type decays 208
four level system 84, 127, 128
fractals 56, 200, 213, 230
fraction 231
fraction scattering processes 160
fracton dimension 215
fracton model 143
fractons 232
free defects 202
frequency domain experiments 114
frequency matrix 90
frozen-in disorder 204

Garum model 229, 235, 242
gelation 214
generalized Langevin equations 88
generalized master equation 235
generating-function formalisms 235
glass 43
glass transition 155
glass transition temperature 25
glass, continuum approximation 70
glass, vibrations 70
glass: non-equilibrium nature of 155
glassy state 25
Green's function 14
guest 142
guest molecule, interaction with TLS 78

guest molecules 66, 68

Hamiltonian 68
Hamiltonian of the vibrations 69
Hamiltonian, single TLS 72
Hausdorff dimension 215
Heisenberg-limit 67
heat bath 142
heat capacity 65
heat capacity, time dependence 65
Hilber-Schmidt scalar product 89
hole-burning 1, 52, 56, 149, 193, 205
hole-burning spectroscopy 154
hole-burning spectrum 151
hole-burning, mechanisms 66
hole-burning, non-photochemical 66
hole-burning, photochemical 66
hole-burning, transient 67
holewidth, logarithmic time dependence 68
holewidth, pressure dependence 68
homogeneous broadening 36
homogeneous line 66
homogeneous line, measurement 126
homogeneous linewidths 2, 159
homogeneous zero-phonon absorption line 151

impurity, coupled to several TLSs 128
impurity, coupled to a single TLS 93
inhomogeneous broadening 35
inhomogeneous line 66
inhomogeneous linewidths 2
inorganic glasses 1
inorganic materials 115
integro-differential equations 91
interaction between TLSs 66, 76
interaction between TLSs and vibrations 73
interaction of the guest molecule with a single TLS 80
interaction of the guest molecule with several TLSs 83
interaction sphere 138
*inter*chain interactions 205
internal rearrangements 199
*intra*chain 205
interaction Hamiltonian between the guest and the TLSs 78
inversion symmetry 190
irrelevant subspace 89
irreversibility of hole width 161
irreversible reactions 202
isotope effect 184

J-multiplets 33, 34
Judd-Ofelt parameters 34, 43

kinetic approach 201, 255
kinetic scheme 202

Laplace transform 238
laser spectroscopy of ions in glasses 48
ligand field theory 33
line shape 114
line shape average 114
line shape formula 134
line shape, dominant contribution 111
line shape, prefactor 134
line shape, vector and matrix notation 133
line-broadening 5
linear Stark effect 194
linear response theory 87
lineshape 9
linewidth 1, 5, 6, 137
linewidth at extremely low temperatures 122
linewidth of ions in glasses 55
linewidth of transition R 104
linewidth, athermal contribution 135
linewidth, thermal contribution 135
linewidths in inorganic and organic material 116
Liouvillian of the total system 90
liquid viscosity, η 25
lithium aluminum borate 31
lithium borate 31
lithium lime silicate 31
local host-guest configuration 149
localization 199
logarithmic decay law 166, 171
long-range interaction 118
long-time behavior 12
long-time tails 206, 237
low-temperature properties of glasses 65
LS terms 33
Lycurgus Cup 30
Lyo-Orbach result 108

many particle treatment 73
many particle Hamiltonian 79
many-particle representation of H_s 128
many-particle treatment 142
marginal dimension 247
Markov approximation 91
master equation 235
measurements of the homogeneous optical linewidths 125
memory function 90, 236
merocyanine 204
metric space 246
microenvironments 199
Mn^{2+} 39
Mn^{3+} silicate glass 38
model 142
model Hamiltonian H, basis system of H_s 85
model Hamiltonian 68
modification of the density of the vibrational degrees 143
modified distribution of TLSs 143

modified distributions 125
modulation 2
molecules with inversion symmetry 189
molecules without inversion symmetry 192
Monte Carlo 49
Mori's formalism 87, 88, 130
motional narrowing 6, 105, 109
moving targets 223
multiple trapping (MT) 201, 247
multipolar 10
multipolar interactions 208

N 97
narrowing of lines 82
Nd^{3+} 41, 42, 46, 54, 57
Nd^{3+}, Er^{3+} 40
network 25
network former 26
network modifying cations 26
Ni^{2++} 39
non-radiative relaxation 44
numerical line shapes 141

off-diagonal coupling 5
off-diagonal modulations 8, 18, 19
OH^- molecular vibrations 45
one-phonon emission 16, 19
one-phonon-absorption 7
operator, electric dipole moment 87
optical absorption 87
optical hole-burning 66
optical line shape 133
optical transition to TLS 167
optical transitions 81, 82
Orbach processes 1
organic glasses 1, 69
overlap parameter 72

parallel processes 236
parallel relaxation 206
parameter values 137
Pb^{2+} 43
percolation 199
percolation clusters 214
perdeuterated glass 177
persistent spectral holes 163
phenomenological exponent $\varkappa$ 143
phenomenological parameter K 75
phonon coupling constant 99
phonon echo 66
phonon saturation 56
phonon sideband 151
phonon-assisted tunneling 2, 6
phonon-emission 7, 8
phonon-exchange interactions 2
phonons, induced transitions 82
phonons, spontaneous emission 82, 104

photo-induced TLS systems 169
photoactive TLS 188
photon echoes 56, 67
phthalocyanine 151
phthalocyanine in a PMMA matrix 195
Poisson distributions 227
Poisson process 236
polarization diffusion 184
polarization of holes 186, 187
polymer 204
potential wall 72
power broadening 67
Pr^{3+} 56
pressure-induced line-broadening of holes 195
pressure-induced, irreversible barrier crossing 196
principal term 136
principal value 97, 101
principal value terms 103, 132
probability distribution 137
product states 128
projection 89
projection operator 89
proton transfer reaction 174
proton tunneling processes 177
protonated alcohol glasses 175
protonated quinizarin 180
pseudo-unimolecular 203

quadrupolar 12
quadrupole-dipole coupling 139
quasi-exponential 203
quinizarin 175, 177, 178, 182, 185
quinizarin system 186

Racah parameters 35
Raman processes 1
Raman spectrum of 43
random close-packed structure 69
random network 69
random number generator 118
random walks 223
rapid modulation 17
rare earth ions 30
rare earth series 33
rare earth spectra 40
reaction dynamcis 199
real parts of the eigenvalues 104
relaxation absorption 66
relaxation channels 114
relaxation of photochemical holes 173
relaxation rate 114
relaxation rates of TLSs 111
relaxational transitions 81
relevant subspace 89
renormalization 214
residual optical width 163

resonant one-phonon processes 66
restricted geometries 209
Rosenstock approximation 229
Rosenstock form 225
ruby glasses 30
Russell-Saunders approximation 33
Russell-Saunders state 34

saturation 65
scalar product 89
scale-invariance 213
scaling 214
self-energy 236
self-similar 213
semi-invariants 250
sensitization 46
sequential schemes 223
shape term 135
short-time behavior 9
Sierpinski carpet 219
Sierpinski gasket 214, 220, 255
Sm^{2+} 43
Smoluchowski-type approximation 253
Smoluchowski-type decay 259
Smoluchowski-type form 255
ST terms 34, 35
$S(t)$ 97
Stark broadening 193
Stark effect 189, 195
Stark shifts 190
Stokes shift parameter 151
Sugano-Tanabe diagrams 35
site distribution function 152-153
site-density function 207
site-memory function 152, 153, 154
site-selective photochemistry 151
slow modulation 6, 17
soda lime silicate 31
solvent cage 149, 187
spectral diffusion 66, 68, 144
spectral function 9, 14, 17, 18
spectral intensity 114
spectral shift 6, 9, 17
spin doublet 16, 18, 19
spin-lattice relaxation rate 10, 16
spin-lattice relaxation time 5
spin-orbit interaction 35
spiropyran 204
spontaneous structural relaxation processes 174
statistical fractals 215
stochastic fractals 215
strain 4, 14
stretched-exponential 232
structural relaxation processes 163
subordination 240, 256
substitutional disorder 236
supercooled liquids 25

survival probability 225
system 79
system Hamiltonian 68
system Hamiltonian, diagonalized form 84
system operators 88
system part 80

t_{2g} orbitals 34
target 248
target annihilation 226, 248
techniques 49
tellurite 31
temperature dependence of averaged linewidths 115
temperature dependence of spectral holes 161
temporal *scaling* 237
temporal disorder 242, 244
ternary glasses 27
tetracene 174, 175
tetracene system 187
thermal conductivity 65
thermal expansion data 66
thermal prefactors 135
Ti^{3+} 36
time domain experiments 114
time evolution of the optical width 166
Tm^{3+} 41
TLS 1, 2, 3, 68, 71, 156,
TLS and spin 1/2 system 73
TLS barriers 161
TLS lattice 163
TLS system 164, 167
TLS, couplines to phonons 73
TLS, coupling to acoustic phonons 66
TLS, interaction with guest molecule 78
TLS – TLS interaction 77
TLSs 142
TLSs, intrinsic to glasses 66
Toblerone 219, 221
total Hamiltonian 78, 86
transfer of excitation 46
transition energies 82
transition metal ions 30, 34
transition metals 33
transition operator 94
transition rates 103
trapping 223, 234, 248
TRFLN 58, 59
triangle inequality 246
triangular generators 218
trigonometric prefactor 135, 136
triplet-triplet annihilation 202-203
tunnel states 177
tunneling matrix element 71
tunneling model 65
tunneling parameter distribution 73
tunneling relaxation rates 163

tunneling systems 71
two-level system 1, 156
two-level systems (TLSs) 65

ultrametric spaces 246
ultrasonic absorption 65
under external pressure 193
unimolecular reaction 202

variations of lifetimes 54
velocity of sound, temperature dependence 65, 66

vibrational lines 151
vibrations of the glass 142
Vo^{3+} 54
Vogel–Fulcher law 155

waiting-time distribution 235, 237
well-stirred reactor 203, 259, 260
Williams-Watts form 229, 234

Yb^{3+} 58, 59

zero-phonon line 153